Optical Wireless Communication Theory and Technology

Series Editor

Xizheng Ke, School of Automation and Information Engineering, Xi'an University of Technology, Xi'an, Shaanxi, China

The book series Optical Wireless Communication Theory and Technology aims to introduce the key technologies and applications adopted in optical wireless communication to researchers of communication engineering, optical engineering and other related majors. The individual book volumes in the series are thematic. The goal of each volume is to give readers a comprehensive overview of how the theory and technology in a certain optical wireless communication area can be known. As a collection, the series provides valuable resources to a wide audience in academia, the communication engineering research community and anyone else who are looking to expand their knowledge of optical communication.

Xizheng Ke

Space-Time Coding Theory in Optical Wireless MIMO Systems

Science Press
Beijing

Springer

Xizheng Ke
School of Automation and Information Engineering
Xi'an University of Technology
Xi'an, Shaanxi, China

ISSN 2731-5967 ISSN 2731-5975 (electronic)
Optical Wireless Communication Theory and Technology
ISBN 978-981-97-6448-8 ISBN 978-981-97-6449-5 (eBook)
https://doi.org/10.1007/978-981-97-6449-5

Jointly published with Science Press.
The print edition is not for sale in China (Mainland). Customers from China (Mainland) please order the print book from: Science Press.

Translation from the Chinese Simplified language edition: "无线光MIMO系统中的空时编码理论" by Xizheng Ke, © Science Press 2014. Published by Science Press. All Rights Reserved.

This Springer imprint is published by the registered company Springer Nature Singapore Pte Ltd.
The registered company address is: 152 Beach Road, #21-01/04 Gateway East, Singapore 189721, Singapore

Contents

Chapter 1
Introduction

As cellular mobile technology, the Internet, and multimedia services continue to develop, the demand for wireless communication-channel capacity continues to increase. Owing to the increasing scarcity of spectrum resources, high data-transmission rates can only be accessed using efficient signal-processing technologies.

The new generation of mobile-communication technology can provide data-transmission rates of 100 Mbit/s or higher, supporting services from voice to multimedia services (including real-time streaming services), and the data-transmission rate can be adjusted according to the different needs of these services; thus, improving spectrum utilization has become an urgent task [1–4].

In wireless-communication systems, the main methods for improving spectrum utilization are orthogonal frequency-division multiplexing (OFDM) technology, multiple-input multiple-output (MIMO) technology, smart antennas, and adaptive-coding technology. MIMO, sometimes referred to as space diversity, can counteract the fading by taking advantage of the fact that signals received at different locations fade independently. The use of MIMO technology makes space a resource that can be used to improve communication performance and significantly increase communication capacity. The use of space–time coding is a viable and effective method for achieving or approaching the capacity of MIMO wireless channels.

The core idea is to use multipath capability to achieve higher spectrum utilization and performance gains. Space–time coding combines multiple antenna arrays with spatial and temporal diversity, allowing the channel capacity to grow linearly with the minimum number of transmitting and receiving antennas, if the channel status information can be accurately estimated at the receiver end. Space–time coding can be used to provide transmit diversity and power gain without sacrificing bandwidth.

X. Ke, *Space-Time Coding Theory in Optical Wireless MIMO Systems*,
Optical Wireless Communication Theory and Technology,
https://doi.org/10.1007/978-981-97-6449-5_1

1.1 Wireless Laser Communication

Wireless laser communication is a wireless-communication method that uses a laser beam as the carrier for information transmission and can achieve high-speed data transmissions. It has the advantages of no electromagnetic interference, flexible networking, easy installation and maintenance, high communication reliability, good confidentiality, and an excellent performance-to-price ratio. It can transmit data, voice, images, etc., at a variety of rates, and has become popular in today's information-technology field. It can reduce the initial construction investment of a communication network, so that the project cost is significantly reduced. As a backup support for optical-fiber communication and microwave communication, it is considered to be the best solution to the "last mile" problem.

In recent years, with the development of semiconductor laser and photodetector technology, wireless laser communication has developed considerably [5–10]. However, laser communication faces many serious challenges, particularly in terms of the following:

(1) The scattering and absorption effects of the atmosphere on the laser signal cause the optical signal to have a large intensity attenuation in transmission (even in a clear atmosphere). Atmospheric attenuation can vary widely from 0.1 to 300 dB/km; therefore, the stability of a wireless optical communication system is closely related to the system power margin [11].
(2) Atmospheric turbulence causes random spot drift and light-intensity flickering. Turbulent motion produces random spatial and temporal distributions in the atmospheric refractive index. Random changes in the refractive index lead to random interference effects on the light waves, which cause rapid fluctuations in the intensity of the received optical signal. The light-intensity flicker effect causes the signal intensity to fluctuate rapidly. When the scintillation is large, the amplitude can fade over 30 dB. Particularly in strong turbulence situations, the optical signal can be seriously disturbed or even off-target, resulting in a large bit-error rate and short-term communication interruptions.
(3) Although the frequencies of optical carriers are high and the signal bandwidth of optical communication systems can exceed 1 THz, in practice, the transmission rate is often limited to 10 Gbit/s or less, owing to optical dispersion or the speed of the electronics.
(4) High-precision acquisition, tracking, and pointing (ATP) techniques are difficult to achieve; particularly when using narrow beams.
(5) The optical power provided by the transmitting antenna is limited, considering its impact on human-eye safety.

Various measures have been used to solve these problems in wireless laser communications. The main measures used to overcome the effects of atmospheric turbulence include increasing the optical transmitting power, large-aperture reception, adaptive optics, channel coding, partially coherent optical transmissions, and optical MIMO technology.

Although increasing the laser transmitting power can provide a sufficient power margin, atmospheric laser-communication systems generally use semiconductor lasers as the light source, and it is difficult to make the power of current high-speed modulated lasers sufficiently large. For human-eye safety, the laser power of atmospheric laser-communication systems should meet the provisions of the international safety standards for the visual eye, and the power should be limited to a certain extent.

Large-aperture reception uses aperture smoothing to suppress the received light-intensity fluctuation. However, large-aperture receivers are difficult to process and manufacture, and the devices are larger and more expensive, which makes them unsuitable for laser-communication systems with restricted terminal sizes.

Adaptive-optics methods can effectively compensate for the light-intensity attenuation caused by phase undulation; however, the technology is complex, expensive, and difficult to use in practical communication systems.

In conventional atmospheric laser communications, cascaded channel coding (e.g., Reed–Solomon (RS) error-correction codes, turbo codes, and low-density parity-check (LDPC) codes) [11–14] has been used to improve the communication performance in weakly turbulent channels. Although conventional channel-coding techniques can suppress turbulence effects, they are inadequate for suppressing the effects of strong turbulence and dense fog, and the system channel capacity is limited.

Partially coherent light is effective at suppressing strong turbulence; however, the degree of coherence between beams affects the orbital angular-momentum spectrum, which becomes more diffuse as the coherence deteriorates [15].

The optical MIMO technique transmits information through multiple incoherent laser beams along different paths to multiple receivers and uses incoherent superposition. Compared with the above-mentioned methods, optical MIMO technology has the advantages of simple implementation, low cost, and easy engineering application; thus, it has become a research hotspot in the atmospheric laser-communication field in recent years. However, the theory and models of optical MIMO still need to be improved, especially under strong turbulence, and further research and exploration of optical MIMO technology are needed. This book deals mainly with space–time coding and its related techniques in optical MIMO systems.

The main techniques used to increase the transmission rate are wavelength-division multiplexing (WDM) [16], polarization multiplexing (PDM) [17], imaging communication [18], multiple-input multiple-output (MIMO) [19, 20], high bit-rate modulation [21], high-order modulation techniques [22, 23], and channel-coding methods [5, 12, 13, 19, 24–26] (e.g., LDPC codes, RS codes, and turbo codes).

Although several other techniques can increase the system transmission rate, they cannot suppress the turbulence effect as effectively as MIMO techniques. In addition, using higher-order modulation techniques and better channel-coding methods to improve the system channel capacity can only be approximated by the existing single-input single-output (SISO) systems. MIMO increases the channel capacity of a communication system by obtaining higher spectrum utilization by simultaneously transmitting and receiving diversity gains.

In wireless laser communication, high-precision acquisition, tracking, and pointing (ATP) technology is one of the key technologies for improving the performance of communication systems. Research in this field has focused on adaptive-optical systems, the structure and beam collimation of optical antennas, and high-precision alignment [27]. The proposed optical MIMO technology not only reduces the requirements for tracking accuracy, but also reduces the complexity of the system hardware.

In wireless laser communications, MIMO technology exponentially increases the channel capacity of the original system without adding additional system complexity and effectively suppresses atmospheric-turbulence effects. With the current level of semiconductor-device technology and precision optical components, the channel capacity of MIMO systems using spatial-diversity technology is much higher than that of SISO systems for the same transmit power and transmission bandwidth. This increased channel capacity can be used to increase the information transmission rate, improve the reliability of the communication system by adding redundancy, or make a reasonable compromise between the two.

The space–time coding method combines channel coding, modulation, transmission, and transmit and receive diversity and combines them effectively to improve the transmission performance of MIMO systems. Space–time coding is based on MIMO techniques, which introduce time- and space-domain correlations between the transmit signals from different antennas. Introducing these correlation-processing techniques provides the basis for signal recovery.

Space–time coding combines the advantages of space and time diversity, providing both diversity and coding gains, and can be considered an advanced diversity technique. The idea of space–time coding originates from radio-frequency (RF) communications, which can increase the channel capacity of communication systems and suppress channel fading. With this in mind, the concept of space–time coding was introduced to atmospheric laser communications to address the challenges faced by wireless laser communications.

Although wireless laser and RF communications have some similarities, they exhibit significant differences. Their main differences are as follows:

(1) When propagating through the atmosphere, lasers are more complexly affected by the atmospheric environment than RF communications.
(2) The noise mechanisms that affect the two modes of communication are not identical.
(3) The modulation methods used to transmit the signals are different.
(4) The form, meaning, and representation of the transmitted signals are not the same.

1.2 Research Status of Optical MIMO Technology

1.2.1 Status of Foreign Development

Since 1960, extensive research has been conducted on wireless laser communications and considerable progress has been made [12, 28–31]. International organizations engaged in wireless laser communication research mainly include the European Space Agency (ESA) in Europe, the Communications Research Office of the Postal Department (CTL) and the Space Development Corporation (NASDA) in Japan, the United States Aerospace Administration (NASA) and the United States Air Force. In 1980, three countries successfully conducted experiments on atmospheric laser-communication transmission. In the same year, Lincoln Laboratory conducted an end-to-end demonstration satellite-communication experiment with a high code rate using the outlier method [31], which had a transmission rate of 220 Mbit/s. In 1985, the US Air Force developed a half-duplex optical-link experiment between a ground station and the LiDAR In-space Technology Experiment (LITE) system [29]. Astro Terra also developed a full-duplex wireless laser-communication system with an uplink bitrate and 1.24 Gbps downlink bitrate of 155 Mbps and 1.24 Gbps, respectively. In the late 1980s, the European Space Agency established a laser-communication link between two satellites to validate inter-satellite laser-communication unit technology [28, 30].

In the 1990s, with the development of technologies, such as wavelength-division multiplexing (WDMA), adaptive optics, and erbium-doped fiber amplifiers (EDFA), wireless laser communication improved in terms of transmission capacity and distance, accelerating its commercialization. In 1995, Japan successfully conducted a star–ground link optical-communication experiment between a ground station and the ETS-VI engineering test satellite. In the same year, the US Strategic Missile Defense Organization developed an all-weather tracking and scanning test system with a 1 Gbit/s information rate and a bit-error rate (BER) of 10^{-6}.

With funding from the US Department of Defense, Thermo Trex conducted communications between two islands with a range of 150 km [32]. In 1996, the company conducted experiments on optical communication between ground stations and aircraft. Currently, foreign manufacturers working on atmospheric laser communications include Air Fiber, Light Pointe, Terabeam, and Canon.

In 1982, Nakai studied coding and modulation techniques in optical channels, which laid the foundation for the development of optical communications [33]. Later, RS codes [34], trellis-coded modulation (TCM) codes [35], Bose–Chaudhuri–Hocquenghem (BCH) codes [36], turbo codes [37, 38], LDPC codes [39], as well as interleaving techniques [40] and cascaded codes [41, 42], were applied to wireless laser communications. Although the problems existing in wireless optical communication have been solved to a certain extent, it is still difficult to meet the requirements of high capacity, high-speed long-distance communication.

The theory of multi-beam transmission was first studied by Lesh and Wilson at the California Institute of Technology (Caltech) in 1997 [43]. Multiple incoherent laser beams are emitted at the transmitting end from a number of transmitting mirrors, where the distance between apertures is greater than the atmospheric coherence length. They are superimposed incoherently at the far-field receiving end to suppress turbulence effects in the atmospheric transmission of the laser and to effectively compensate for the atmospheric channel.

An incoherent laser beam is created by splitting a single laser beam into multiple beams and ensuring that the difference in range between the beams is greater than the coherence length of the laser. The essence of multi-beam emission is a non-coherent superposition technique. Early research on multi-beam technology focused on the detection of atmospheric optical parameters, theoretical simulations, and experimental studies of multi-beam intense laser transmissions in the atmosphere. Research on multi-beam emission and reception techniques for wireless laser communications was conducted after the concept of multi-beam emission was introduced [43].

Since the concept of multi-beam emission was proposed, the Lincoln Laboratory at MIT and the Jet Propulsion Laboratory (JPL; USA) have conducted atmospheric transmission experiments with multiple beams and obtained a large number of experimental results [44–47]. From October 1995 to May 1996, Kim et al. developed the TerralinkTM 8-155 atmospheric-scintillation measurement experiment [44]. The experimental results showed that when the receiving antenna had a certain aperture, as the number of emitted beams increased, the fluctuation of the received light intensity caused by atmospheric turbulence sequentially decreased. When a single beam was used, the intensity fluctuation was large and the received intensity-distribution curve in the far field was closer to a negative exponential distribution. As the number of transmitted beams increased, the received intensity-distribution curve approached a log-normal distribution.

Theoretical analytical models for beam transmission in turbulent atmospheres have been presented in the literature [48, 49]; however, many problems in the practical application of the model still need to be investigated.

In 1996, Ibrahim [50] investigated the performance of wireless laser communication using receive diversity, based on the assumption that the input-signal fading of each small-aperture receiver was independent of the others. In 2000, Bell Labs successfully developed a prototype system for atmospheric laser communication using multi-beam transmission techniques [51].

In 2002, the Jet Propulsion Laboratory (JPL) in the USA used multiple beams to conduct optical-transmission experiments over a distance of 46.8 km. The results showed that as the number of beams increased, the light-intensity fluctuation at the receiving end decreased [52]. The JPL conducted a study on merging the output signals after using spatial-diversity reception, and conducted field experiments [53–55]. The experimental results showed that diversity reception had good tracking and capture capabilities and was robust and easily scalable.

In 2001, the Communications Engineering Research Laboratory at the JPL and the Laser Communications Research Group at Stanford University, USA, conducted intensive research on diversity reception and optical-signal detection and achieved good results [56]. In the same year, Haas et al. [57] from MIT, USA, investigated the space–time coding problem in atmospheric laser communication from an information-theoretic perspective for a log-normal-distributed atmospheric channel (weakly undulating turbulence) model. They proposed a space–time coding criterion for atmospheric laser communication, and derived its BER performance limits. However, no specific space–time coding method applicable to turbulent channels was given, and the theory is only suitable for the outlier approach.

Outlier detection requires that the frequencies of the receiver's local oscillating laser and the transmitting laser must be very stable, and the frequency stability must be at least on the order of 10^{-11}, affecting its practicality [58]; therefore, direct detection is still an effective photoelectric detection method. Space–time coding guidelines and BER performance limits are also given in the literature [59]; however, difficulties still exist when using this method for atmospheric laser-communication systems with direct detection.

In 2002, Zhu et al. [60, 61] used a combination of theoretical analysis and numerical simulations to apply time-domain detection techniques, likelihood-ratio detection theory, and diversity-reception techniques from RF communications to atmospheric laser communications and investigated the effectiveness of these theories for suppressing turbulence effects. However, these studies mainly focused on weakly turbulent channels and did not consider the problem of strong turbulence leading to communication-link disruptions or the improvement in system performance due to coding gains.

In 2003, Simon and Vilnrotter et al. improved the conventional Alamouti code [62], providing a method for implementing space–time coding in atmospheric laser communications using an intensity modulation/direct detection (IM/DD) style; however, the modified Alamouti code could only be used for weakly undulating turbulent channels. During the same year, Alqudah and Kavehrad et al. proposed an orthogonal space–time coding for indoor wireless optical channels [61].

In 2004, Feng et al. [63] analyzed the optical-intensity undulation characteristics of a multi-beam transmission link between a star and the earth and showed that multi-beam transmissions could effectively suppress the optical-intensity undulation in the entire region from weak to strong turbulence. In the same year, for an IM/DD atmospheric-turbulence channel, Wilson and Brandt-Pearce et al. derived and simulated the BER performance of MIMO optical communications using quaternary pulse-position modulation (QPPM) [20]. However, this method used repetitive code with low code rates. Under the same conditions, Brandt-Pearce and Wilson et al. [64] designed a coding scheme with equal Hamming weights using the pulse-position modulation (PPM) technique, and analyzed the BER performance limits. However, signal detection using this system is difficult, and the BER is relatively large.

In 2005, Simon and Vilnrotter et al. proposed another space–time coding scheme suitable for IM/DD [65]; however, the scheme was only applicable to on–off keying (OOK) modulation and 2-PPM modulation. In 2006, Angulta et al. [66] analyzed a communication system using multi-beam transmissions and proposed two modes of space–time coding. Both modes achieved a higher signal-to-noise gain than single-beam transmissions.

In 2007, Angulta et al. [67] investigated the factors affecting light-intensity undulation in multi-beam transmissions. The ability to reduce the light-intensity undulation is related to the spatial correlation of each beam, transmission distance, turbulence conditions, emitter spacing, and receiver aperture. The smaller the spatial correlation, the greater the ability of the multi-emission beam link to suppress light-intensity flicker.

Navidpour et al. [68] of Pennsylvania State University, USA, derived a formula for calculating the BER when simultaneous diversity was used at the transceiver end. The results showed that the performance of spatial diversity was severely affected by the spatial correlation of each subchannel. In the same year, Cao et al. [69] proposed a MIMO system suitable for atmospheric laser communication using pre-amplification for PPM modulation, but without using space–time coding and with a weak ability to suppress turbulence effects.

Additionally in 2007, García-Zambrana analyzed the BER performance of Simon's proposed space–time block code in a strongly turbulent channel [70]. Although Simon's proposed space–time block code was able to suppress atmospheric turbulence well, the improved performance reduced the multiplexing gain and failed to maximize the system transmission rate.

In 2008, Neda et al. [71] evaluated a terrestrial optical-communication link for multi-beam transmit–multi-APD (avalanche photodiode) detection reception, deriving performance bounds for the link and discussing the choice of gain size for the merging of multi-APD output signals. In the same year, for IM/DD optical communications, the interruption capacity of MIMO in a Poisson fading channel with scattering noise and limited transmit power was analyzed [72]; however, the effect of thermal noise on the system performance was not considered.

In the same year, Letzepis et al. investigated the performance of MIMO systems in free-space optical communication in the Gaussian regime using PPM modulation [73]. Although the system performance under different scenarios was investigated [72, 73], the studies did not address how the optimal transmission mechanism (including the transmit-power allocation, coding method, and transmit-aperture arrangement structure) could be used to achieve the maximum channel capacity for MIMO transmissions.

In 2008, Safari et al. [74] investigated the performance of repetition codes and space–time block coding using OOK modulation. The simulation results showed that the BER performances of the two were very close in the case of strong turbulence, and the former outperformed the latter in the case of weak turbulence. In the same year, Raza et al. [75] combined medium-access control (MAC)-layer and physical-layer coding (orthogonal space–time coding) to propose a cross-coding scheme that used training sequences to estimate the channel parameters.

In 2009, Ntogari et al. [76] investigated the performance of space–time coding in an indoor environment by combining IM/DD techniques, showing that space–time block codes could improve the channel capacity, expand the reception area, and reduce the transmission power. In 2010, Bayaki et al. [77] presented a closed form for the probability of pairwise errors in free-space optical communication space–time codes with two lasers and arbitrary detector reception. They also proposed a simple criterion for the design of space–time codes in optical communication.

In the same year, under the same conditions, Antonio et al. [78] proposed a space–time lattice code for strongly turbulent channels and analyzed its error performance. Therefore, further research on the construction, design, performance and applicability of space–time codes in IM/DD-style atmospheric laser communications is needed.

1.2.2 Status of Domestic Development

Research related to atmospheric laser-communication technology began in China as early as the 1960s. Major research units included the Harbin Institute of Technology, Changchun University of Technology, Wuhan University, University of Electronic Science and Technology, Institute of Optoelectronics of the Chinese Academy of Sciences, Peking University, and National University of Defense Technology. These research institutes have conducted numerous theoretical and experimental studies on the key technologies and theories of atmospheric laser communication. However, compared with the United States, Japan, and Europe, China's research on wireless laser-communication technology is lacking, especially in the areas of coding, atmospheric-channel compensation, and tracking and capturing technologies.

In 1999, the Northwest Institute of Nuclear Technology conducted a study on the transmission of multi-beam lasers in the atmosphere. The results showed that the received light intensity increased by a factor of two when multiple beams were used, compared to single-beam transmissions [79]. In 2003–2004, the University of Electronic Science and Technology studied the theory of multiple-beam transmissions and showed that multiple beams could effectively suppress atmospheric-turbulence effects [80, 81].

In 2004, the University of Defense Technology [82–84] studied the channel model, space–time coding, and temperature field of multi-beam light sources for multi-beam transmission. The Shanghai Institute of Optics and Precision Machinery, Chinese Academy of Sciences, implemented a 155/622 Mbit/s multi-emitter laser communication system [85]. The system used transmit diversity, but not receive diversity, and did not employ space–time coding techniques.

In 2005, the Huazhong Institute of Optoelectronics Technology [86] and Yanshan University [87] studied the diversity of technology in atmospheric laser communications and analyzed its system performance. This was also reported recently by Anhui University [88] and Beijing University of Aeronautics and Astronautics [89]. In 2006, Wuhan University used quadrature amplitude modulation (QAM) to apply space–time coding in RF communications to wireless optical communications, and verified the performance of space–time block coding through simulation experiments [90].

Since 2007, the Xi'an University of Technology has conducted research on space–time coding technology in atmospheric laser communications and conducted experiments on optical MIMO communications. Later, Huazhong University of Science and Technology [91, 92] and Changchun University of Science and Technology [93, 94] also conducted research on spatial diversity, temporal diversity, optical MIMO technology, and space–time coding in wireless optical communications.

1.3 Optical MIMO Systems

MIMO technology has its origins in RF communications and is a major breakthrough in smart-antenna technology for wireless mobile communications. Because RF communication and wireless laser communication differ, the theory and model of the original MIMO technology are very different from those of wireless laser-communication systems. Optical MIMO technology can not only effectively suppress atmospheric-turbulence effects but also exponentially increase the system channel capacity without an additional increase in the transmit power and bandwidth.

1.3.1 Components of an Optical MIMO System

In a wireless optical MIMO system, the original MIMO technology ideas are applied to atmospheric laser communications, considering channel coding, modulation, transmit and receive diversity, and their organic integration. A wireless optical MIMO system is mainly composed of space–time coding, modulation and drive circuits, a laser array, transmit aperture, receive aperture, detector array, demodulation, and signal processing, as shown in Fig. 1.1.

The basic principle of an optical MIMO system is that the binary bits of data sent by a source are first grouped according to certain rules, and then the grouped data are mapped into a constellation, which is then mapped to obtain multiple modulation symbols. These modulated symbols are fed into a space–time encoder for space–time coding, resulting in a space–time coding matrix. In effect, the original

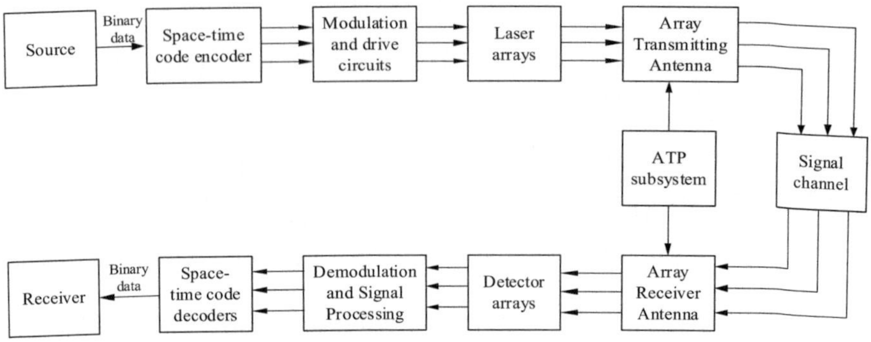

Fig. 1.1 Components of an optical MIMO communication system

data are decomposed into multiple sub-streams after the space–time encoding. The multiple sub-streams are then converted into multiple optical signals that are sent simultaneously by multiple optical-transmitting antennas.

At the receiving end, multiple optical-receiving antennas are used to receive and convert the optical signals into electrical signals, which are then decoded and converted into multiple data sub-streams, which are combined to obtain a copy of the transmitted signal. Wireless optical MIMO systems use multiple optical antennas to simultaneously transmit and receive multiple parallel independent data sub-streams to increase the channel capacity and fade resistance of the system [95].

1.3.2 *Classification of Wireless Optical MIMO Technologies*

According to the information-theory principle, if copies of the original transmitted signal are provided to the receiver with different degrees of attenuation, this will help the received signal make the correct decision. This method of improving the correct-decision rate of the received signal by providing multiple copies of the transmitted signal is known as diversity.

Diversity techniques do not require training codes, thus, reducing overhead. If a signal on one radio-propagation path experiences deep fading, another relatively independent path may still contain a strong signal. Then, two or more signals can be selected from the multipath signal. The benefit of diversity is that it has an improved instantaneous signal-to-noise ratio (SNR) and average signal-to-noise ratio for the receiver.

1.3.2.1 Diversity Techniques

Diversity is used to reduce the depth of fading that can occur during transmission by combining different copies of the same transmitted signal for reception. It is a fundamental technique for suppressing fading in wireless communication systems, overcoming channel fading, and reducing system BERs to a certain extent. In wireless mobile communications, diversity is more often used to reduce fading caused by multipaths and to improve the reliability of the system without increasing the transmit power or channel bandwidth [96].

The basic principle of diversity is that multiple signal sequences carrying information are received from several differing channels (in frequency, space, or time), and the multiple signals usually fade independently because of the comparable strength of the different signal copies. The signals fade to different depths during transmission. By combining them appropriately at the receiver end, a signal copy of sufficient strength is guaranteed to be transmitted to the receiver, thereby improving signal-transmission reliability. If diversity is not used in the presence of noise interference, the transmitter must have a sufficient transmit-power margin to ensure a proper link connection in the event of deteriorating channel conditions.

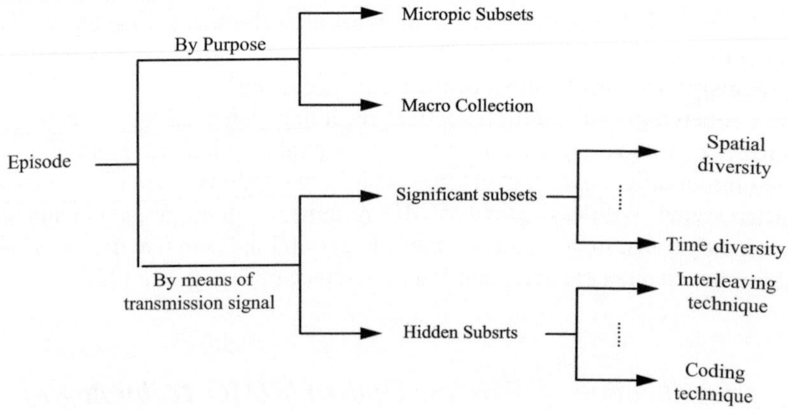

Fig. 1.2 Classification of diversity technologies

Diversity is a decentralized transmission technique that enables the receiver to obtain multiple statistically independent fading signals carrying the same information. The combination of multiple statistically independent fading signals received by the receiver reduces the effects of fading. Therefore, the most important condition for obtaining the diversity effect is that the signals should not be correlated with each other.

As can be seen from Fig. 1.2, diversity can be divided into implicit and explicit diversity, according to the signal-transmission method. Depending on the purpose, the diversity can be divided into micro and macro diversity; the amount of fading to be overcome is the main difference between them.

Implicit diversity refers to the incorporation of some coding methods or the use of interleaving and other signal-processing techniques, to achieve diversity. The commonly used techniques are mainly interleaving technology and coding and decoding technology, where coding and decoding technology refers to the use of a variety of coding and decoding methods. Interleaving technology refers to the method of spreading out the successive bits in a message.

The apparent diversity technique is a method of adding redundancy to the signal by transmitting multiple identical signals at the transmitter end to reduce the transmit power. This includes spatial and temporal diversity. They are implemented by combining multiple devices receiving signals at different spatial, frequency, and polarization directions; hence, the term manifold diversity.

Space diversity is achieved using multiple antennas at the receiver and transmitter ends, and includes both transmit and receive diversity. Space diversity does not result in any loss of bandwidth utilization, which is superior to time and frequency diversity. Receive diversity refers to the use of multiple antennas at the receiving end to receive signals. It has been found that receive diversity has a limited anti-interference effect and supplementary anti-interference methods are necessary [97].

The diversity effect of spatial diversity is significantly influenced by the number of diversity weights at the transmitter and receiver ends, and the system has a better diversity effect when the diversity M at both ends is greater. When M is greater than a certain value, the rate of increase in diversity gain also decreases, while the difficulty of diversity increases. Spatial diversity does not introduce redundancy in the time or frequency domain, and the diversity technique can reduce the signal-transmission power; however, the use of space–time codes with transmit diversity can largely improve the frequency utilization and increase the channel capacity.

Time-diversity techniques generally use temporal interleaving in conjunction with channel coding and are therefore of limited use for slow fading, but effective for fast fading. In wireless communications, time diversity is achieved by combining coding with interleaving techniques with error control, where the signal at the receiver contains a large number of redundant signals, owing to error-correction coding [97].

In angular diversity, the multipath channels arriving at the receiver may arrive from different directions, owing to different topographies, etc. In this case, directional antennas can be used at the receiver to point to different signal-arrival directions, which ensures that the multipath signals received by each directional antenna are uncorrelated, thus achieving angular diversity. This method is more commonly used in the RF field.

Frequency diversity is RF communication using two or more microwave frequencies with a certain frequency interval to send and receive the same information simultaneously. Synthesis or selection involves using signals located in different frequency bands after the fading channel considering the statistical uncorrelated characteristics, that is, the difference in the fading statistics of different frequency bands, to achieve anti-frequency selective fading. This can be achieved by modulating the information to be transmitted on separate uncorrelated frequency carriers [98].

In practice, two or more diversity schemes are often combined to achieve multidimensional diversity [96]; for example, in space–time coding techniques, where spatial and temporal diversity are combined to achieve the required system performance.

1.3.2.2 Subsetting and Merging Techniques

The rational introduction of diversity into wireless optical communications can effectively suppress fading caused by atmospheric turbulence [99]. At a given point in time, the beam path of one wavelength may experience deep fading and flickering, whereas the beam path of another wavelength may allow a stronger received signal. Diversity techniques allow multiple simultaneous transmissions of the same signal, which effectively circumvents interruptions in the communication link and prevents all paths from being severely affected by atmospheric fading simultaneously, which is the basic idea of spatial diversity [100].

Diversity technology is the study of making full use of the multipath signal energy in transmission to improve the transmission reliability. It investigates how the fundamental parameters of a signal can be spread out and collected in the time, frequency, and space domains [101–103]. After obtaining several independent branched signals at the receiving end, the diversity gain can be obtained by combining the following techniques.

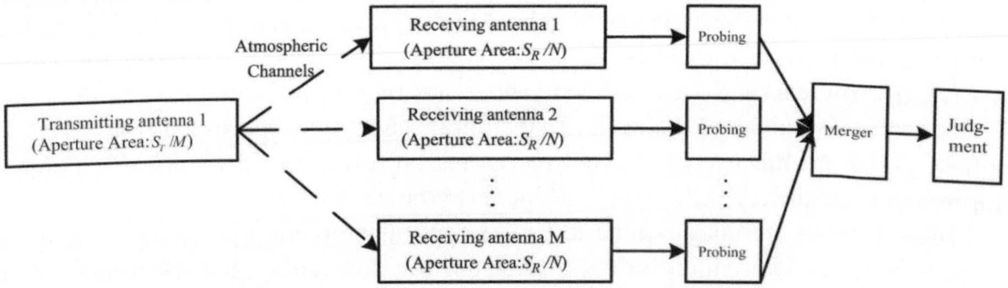

Fig. 1.3 Block diagram of equal-gain-combined receive diversity [105]

(1) Equal-gain-combined receive diversity

Equal-gain combining (EGC), also known as phase equalization, corrects for the phase shift of the channel, but not the amplitude. It is optimal in terms of maximizing the signal-to-noise ratio, assuming that each signal has the same signal-to-noise ratio [104–106]. It outputs the result as a superposition of signal amplitudes in each direction. EGC is used when it is inconvenient to measure the amplitude of the received signal in some systems, as shown in Fig. 1.3.

Assuming that there are $N \geq 2$ receiving antennas, the total receiving-aperture area of the N receiving antennas is S_R, the aperture area of each receiving antenna is S_R/N, and they are combined for equal gain. It is assumed that the receiving-aperture area is equal to that of a non-diversity single-ended–transmitting single-ended–receiving system, which makes it easier to compare the system performance under the same conditions.

(2) Transmit and receive diversity, equal-gain combined

As shown in Fig. 1.4, we assume that the number of transmit antennas $M \geq 2$, the number of receive antennas $N \geq 2$, the aperture area of each transmit mirror is A_T/M, the transmit power is $1/M$ of the transmit power of a non-diversity single-ended–transmit single-ended–receiver system, and the area of each receive antenna is A_R/N.

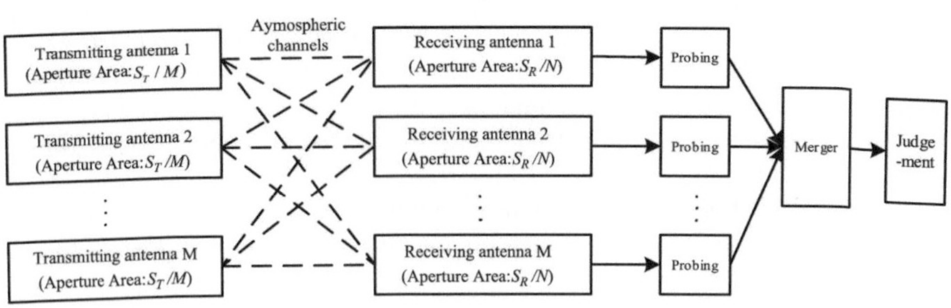

Fig. 1.4 Block diagram of equal-gain-combined transmit and receive diversity [105]

It has been demonstrated [107] that the single transmit–receive diversity in Fig. 1.3 is equivalent in effect and has the same power gain as the spatial-diversity structure using $M \times N$ receive antennas shown in Fig. 1.4, which fully demonstrates the superiority of MIMO systems.

(3) Maximum-ratio combined

At the receiving end, a number of diversity branches, phase-corrected and weighted by an appropriate gain factor, are summed in phase and fed into the detector for detection. At the receiver end, each uncorrelated diversity branch is phase-corrected and re-summed by an appropriate variable-gain weighting, and then fed to the detector for coherent detection. The variable-gain weighting factor for the ith branch can be set as the ratio of the signal amplitude to the noise power for the diversity path.

The maximum-ratio combining (MRC) scheme requires only linear processing of the received signal at the receiver end and then uses maximum-likelihood detection to restore the original information. The decoding process is simple and easy to implement. The combined gain is proportional to the number of diversity branches N.

(4) Selection-combining

When using selection-combining (SC) technology, the output signals of the N receivers are first fed into the selection logic, which selects the baseband signal with the highest baseband signal-to-noise ratio as the output from the N received signals. Each additional diversity branch contributes only an inverse multiple of the total number of diversity branches to the signal-to-noise ratio of the selective diversity output.

(5) Switching and merging

The receiver scans all diversity branches and selects a branch where the signal-to-noise ratio is above a specific preset threshold. The signal is selected as the output signal until the signal-to-noise ratio of the signal drops below the set threshold. At that point, the receiver starts rescanning and switches to another branch; this is also known as scan-merge.

Because switch-merging is not a continuous selection of the best instantaneous signal, it may be inferior to selective merging. However, because switching-combining does not require the continuous simultaneous monitoring of all diversity branches, it is a much simpler method. For both selective and switching combinations, the output signal is equal to only one of the diversity tributaries, and there is no need to know the channel-state information.

(6) Threshold merging

The threshold-merging strategy searches for the first component with a signal-to-noise ratio greater than the preset threshold, outputs it until the channel component is below the threshold value, and then searches again.

1.3.2.3 Spatial-Multiplexing Techniques for MIMO Systems

Spatial multiplexing is primarily used to increase the transmission rate of a system by transmitting mutually independent signals at the transmitter end and decoding them at the receiver end using interference-suppression techniques. The spacing between the transmitting and receiving antennas must be greater than the correlation distance to achieve spatial multiplexing, such that the individual subchannels at the transmitting and receiving ends are independently fading uncorrelated channels. If the path fadings between the transmitting and receiving antennas are independent of each other, multiple parallel spatial subchannels can be constructed. By transmitting different information streams between these parallel subchannels, the data-transmission rate can be improved. This method is called spatial multiplexing.

Spatial multiplexing uses multiple antennas at both the transmitter and receiver ends to make full use of the multipath component of spatial propagation, so that the transmitted signals of multiple data channels occupy only one frequency band. Thus, the channel capacity increases linearly with an increasing number of antennas. This type of channel increases the channel capacity without consuming additional transmit power or taking up additional bandwidth; it is, therefore, an effective solution for improving the channel capacity and system performance.

The implementation of spatial multiplexing is shown in Fig. 1.5, where the signal to be transmitted is first converted into several disjoint parallel sub-signal streams by serial and parallel conversion and sent out simultaneously on the same frequency at each antenna. Each pair of transmitting antennas generates a different spatial signal in the channel for the receiver, which uses the difference in these signals to differentiate the data streams at each transmitter for decoding. Spatial multiplexing requires the spacing between the transmitting and receiving antennas to be greater than the correlation distance, to ensure that each subchannel is an independent uncorrelated channel.

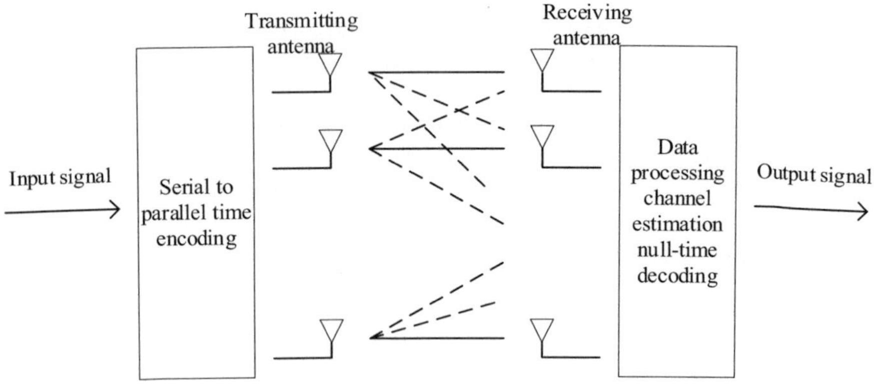

Fig. 1.5 Schematic diagram of spatial multiplexing

The main receiver-side decoding algorithms for spatial multiplexing are the forced-zero algorithm, minimum mean squared error algorithm, and maximum-likelihood decoding algorithm [108]. The minimum mean squared error algorithm is also a linear decoding algorithm, which can minimize the error caused by the interference between the noise and the co-channel signal. It has good noise immunity, although the quality of the separated signal is not very good, and the best performance; however, it is difficult to apply in practice, owing to its high computational complexity.

Spatially multiplexed data are split into multiple data sub-streams, each sent from a different antenna, similar to diversity transmissions. The difference is that spatial multiplexing achieves full ergodic capacity over independent co-distributed Rayleigh channels [109]; however, it does not provide the same diversity gain as diversity. Spatial multiplexing sends independent information streams for higher transmission rates, whereas transmit diversity sends redundant information streams for higher system reliability. Real systems require a reasonable compromise between transmission rate and reliability.

The system is capable of spatial-diversity gain and spatial-multiplexing gain, and the trade-off between system diversity and multiplexing is essentially a trade-off between the system error probability and system data-transmission rate. As the system is limited by its own degrees of freedom, increasing the spatial-diversity gain reduces the spatial-multiplexing gain. Therefore, we need to consider both gains and make a reasonable trade-off in our design, considering the actual environment.

The vertical Bell-Laboratories layered space–time (V-BLAST) wireless communication architecture encodes, maps, and interleaves M bit streams and then transmits them independently at M antennas to maximize diversity. Diagonal BLAST (D-BLAST) is essentially the same as V-BLAST, in terms of coding and modulation, except that after processing, the signal is divided into different blocks of signals, which are simultaneously assigned to different antennas. Thus, for less–than-ideal transmission on one line, the overall signal transmission is not affected too badly; however, the system is correspondingly more complex.

BLAST can make full use of scarce spectrum resources, but is less feasible for downlinks because the number of transmitting antennas is less than or equal to the number of receiving antennas. Furthermore, because all antennas transmit different signals, the information in each link is unique, and signal integrity is not guaranteed when a link is broken.

1.3.3 Classification of Space–Time Coding in Wireless-Optical MIMO

Space–time signal processing uses M antennas at the transmitter and receiver ends to compensate effectively in time, and space–time codes are the main techniques for space–time signal processing. Spatiotemporal codes are coding techniques used for multiple transmit antennas. The main idea is to integrate coding techniques into spatial and temporal subsets such that the information flow is coded simultaneously in space and time, thus improving the performance of the communication system. This spatial–temporal correlation enables the receiver to suppress MIMO channel fading and reduce the BER.

Compared with spatially uncoded systems, spatial–temporal coding can achieve transmit diversity and a certain coding gain without sacrificing bandwidth. There are various methods of space–time coding according to the coding structure, such as space–time block coding (STBC), space–time trellis coding (STTC), layered space–time codes (LSTC), and differential space–time block codes (DSTBC) [110].

According to their different channel environments, space–time coding schemes can be divided into two categories. One category requires the receiver to accurately estimate the channel characteristics, and includes schemes such as layered space–time codes (LSTC), space–time trellis codes (STTC), and space–time block codes (STBC); the other category does not require channel estimation, and includes schemes such as unitary space–time codes (USTC) and differential space–time codes (DSTC).

1.3.3.1 Space–Time Grouping Codes

The first orthogonal packet codes proposed by Alamouti were for two transmitting and two receiving antennas [111]. Their full diversity scheme was implemented by a very simple maximum-likelihood decoding algorithm, using the orthogonal property between two sequences generated by two transmitting antennas, and was called space–time block coding (STBC) [112]. Later, Tarokh proposed an orthogonal packet code with arbitrary transmit antennas [113, 114], and this technique was introduced in both wideband code-division multiple access (W-CDMA) and CDMA2000. The decoding of space–time block codes is simple and easy to implement, as shown in Fig. 1.6, which displays a conceptual diagram of a two-transmit, two-receive orthogonal space–time block code scheme.

Assuming an STBC system with two transmitting and one receiving antennas, $x = (x_1, x_2)$ is a set of modulated symbols that are transmitted simultaneously from each of the two transmitting antennas at a given time, and arrive at the receiving antenna via an atmospheric channel with a fading channel factor of $h = (h_1, h_2)$.

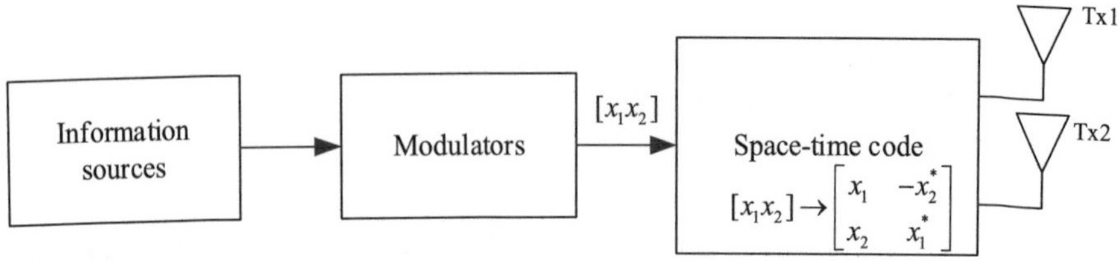

Fig. 1.6 STBC transmitter with diversity technology

Assuming constant symbol spacing for successive transmissions, the received signal is

$$r_1 = h_1x_1 + h_2x_2 + n_1 \tag{1.1}$$

$$r_2 = h_1x_2^* + h_2x_1^* + n_2, \tag{1.2}$$

where n_1 and n_2 are additive Gaussian white noise with a power spectral density of $N_0/2$ and a zero mean, the source of which can be considered to be the noise generated by the photodetector and its attached electronic circuitry.

For a maximum-likelihood decoding method at the receiver,

$$\tilde{x} = \arg\min \|\{r_1, r_2\} - a\tilde{x}\|^2 \quad \tilde{x} \in C, \tag{1.3}$$

where r_1, r_2 denotes the received signals at moments t and $t + 1$, C is the set of all possible sets of modulation symbols, $(\tilde{x}_1, \tilde{x}_2)$, and $a = |h_1|^2 + |h_2|^2$. The system shown in Fig. 1.7 can be extended to have N receiving antennas, in which a maximum of $2N$ diversity orders can be obtained. We have

$$r_n = H_n \cdot x + n_n. \tag{1.4}$$

At this point, the maximum-likelihood decoding at the receiver side can be expressed as

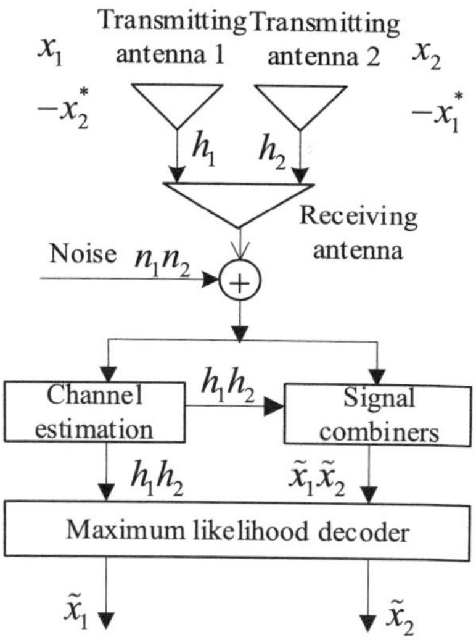

Fig. 1.7 Two-antenna space–time coded receiver

$$\tilde{x} = \arg\min \sum_{n-1}^{N} \|\{r_1, r_2 \cdots r_n\} - H_n\tilde{x}\|^2 \quad \tilde{x} \in C. \tag{1.5}$$

STBC simplifies decoding by mapping the input symbol groups to the time and space domains and by constructing an orthogonal structure. It has the advantage of being simple to compile, using a maximum-likelihood decoding algorithm, and can achieve the same diversity gain as maximum-ratio combining (MRC) reception. The impact on the decoding complexity is minimal, whether increasing the data rate or the number of transmitting antennas.

1.3.3.2 Hierarchical Space–Time Codes

Forschini first proposed the concept of hierarchical space–time coding [115, 116], a coding method that allows for greater channel capacity. This technique combines spatial-dimensional transmission with a one-dimensional coding method so that the decoding complexity at the receiver side only increases linearly with the number of antennas. Layered space–time coding (LST) transforms a high-speed information stream into a lower-speed data stream by means of a series–parallel variation, which is encoded by a common parallel-channel coder and then layered in parallel, with the coded signal modulated and transmitted independently at multiple antennas.

LST is divided into several differentiated types of layered space–time structures: error-free coded LST, such as the vertical-layered space–time (VLST) technique, or the vertical Bell-Labs layered space–time (V-BLAST) technique, with error-controlled coding, such as threaded-layered space–time (TLST), diagonal-layered space–time (DLST), and horizontal-layered space–time (HLST).

In Fig. 1.8, the input sequence is separated into different streams by a series–parallel converter, each of which is then modulated independently and transmitted from a transmitting antenna. The received signal is separated into N signals by a spatial separator, each of which is demodulated and then arranged into a single signal by a multiplexer and output. The vertical structure consists of a matrix column only in the space–time domain.

To improve the system performance, it is possible to combine one-dimensional convolutional coding, which can also be multidimensional in time. At the receiver side, a linear judgment-feedback equalizer implements hierarchical judgment-feedback interference cancellation, followed by hierarchical space–time decoding.

In Fig. 1.9, the signal is first encoded by the encoder and then passed through the serial–parallel converter; at this point, the signal is converted into N_t independent signals and is assigned to each antenna. After independent modulation, interleaving from the corresponding transmitting antenna is used to transmit. Interleaving occurs in the process of reading and writing data, in accordance with the form of "line reading listed" or "column reading line out."

At the receiving end, each antenna receives the signal after the detector detects the useful signal; these signals are interleaved and then combined by the parallel–serial

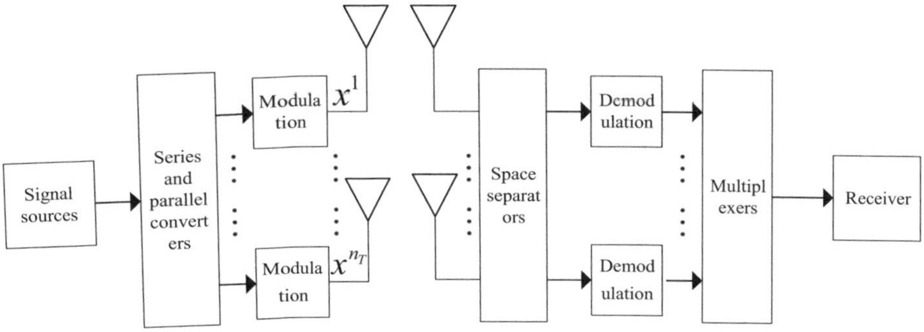

Fig. 1.8 Block diagram of the vertical layered space–time structure

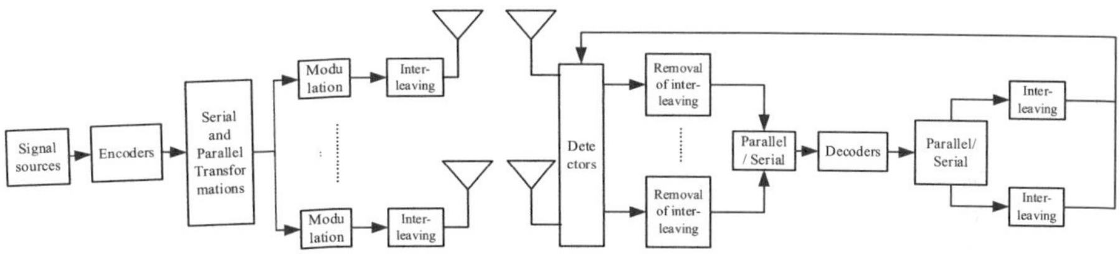

Fig. 1.9 Horizontal hierarchical space–time structure of a single code

converter into one way and decoded output. The decoded signal is used to update the a priori probability of the transmitted signal, and is fed back to the detector to calculate the symbol valuation.

In Fig. 1.10, the information stream is first converted into *N* different sub-streams by a serial–parallel conversion. Then, each sub-stream is encoded independently, modulated, interleaved, and finally transmitted by the corresponding transmitting antenna. At the receiving end, the receiving antenna and detector detect noisy signals that are then de-interleaved and decoded. Simultaneously, the decoded signal is inter-leaved again to update the prior probability of the transmitted signal and is fed back to the detector to calculate the symbol valuation. Unlike in Fig. 1.9, each information stream can be coded differently.

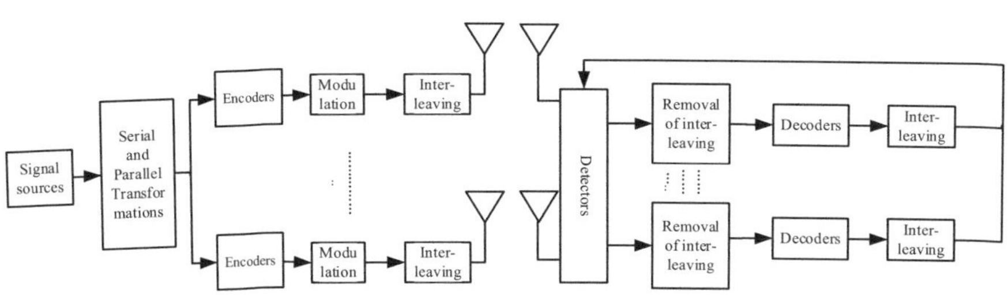

Fig. 1.10 Horizontal hierarchical space–time structure with each layer independently coded

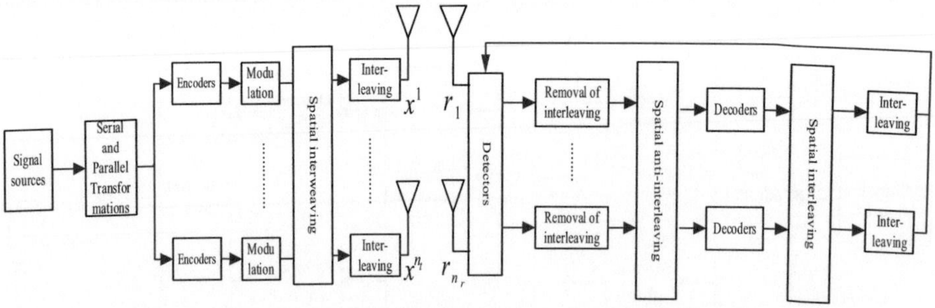

Fig. 1.11 Diagonally layered space–time code and spiral layered space–time code structures

In Fig. 1.11, the information is first transformed into N different sub-streams by a series–parallel converter, each of which is coded and modulated differently. The information is then combined and transmitted in a diagonal or spiral pattern in a spatial interleaver, and the interleaved signals are interleaved in time, so that they are transmitted at different times by different antennas.

At the receiving end, the receiving antenna and detector detect the noisy signal, which is then de-interleaved twice (temporal and spatial de-interleaving) and decoded for output. The decoded signal is interleaved again in space and time, fed back to the receiver to update the prior probability of the transmitted signal, and fed back to the detector to calculate the symbol valuation.

Diagonally layered space–time (DLST) and threaded layered space–time (TLST) codes are structurally similar, in that both are based on horizontally layered space–time with the introduction of a spatial interleaver before the temporal interleaver. This introduces spatial diversity and allows for better performance than horizontally layered codes.

Diagonally layered space–time codes suffer from spectral-utilization loss because a part of their transmission matrix is zero-element complementary. In all three receivers with error-control coding, the detector provides a joint judgment valuation of the N transmitted symbol sequences. In Figs. 1.10 and 1.11, it can be seen that the transmitter structure in the horizontal hierarchical space–time structure can be divided into two types, according to the coding order.

Because hierarchical space–time codes can generally be decoded, using detection algorithms based on the forced-zero and minimum mean squared deviation criterion with high complexity, the study of new decoding algorithms for hierarchical space–time codes, especially for diagonal hierarchical space–time codes, is still a current research direction [117].

1.3.3.3 Differential Space–Time Grouping Codes

Although space–time block codes are less difficult to compile and implement than most space–time coding techniques, they all require the channel-state information

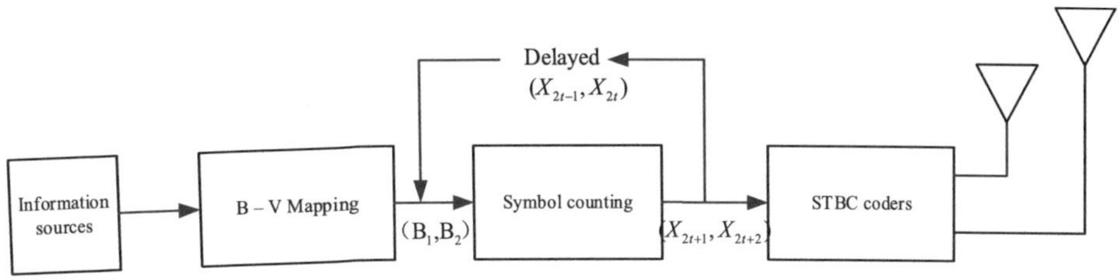

Fig. 1.12 Differential STBC encoder

(CSI) to be explicitly known at the receiver end. In practice, it may not be possible to estimate the channel-state information accurately in all cases, or it may be very difficult. Therefore, there is an urgent need to develop a space–time code that can accurately compile the code without requiring channel-state information at the receiver and transmitter ends.

A new type of space–time code, differential space–time block codes (DSTBC), is proposed based on the theory of orthogonal space–time block codes, which can compile codes without channel estimation, thus greatly reducing the complexity of the receiving equipment. This significantly reduces the complexity of the receiving equipment. Figure 1.12 shows the encoder block diagram of a differential space–time block code.

In Fig. 1.12, if the signal emitted at time $2t - 1$ is (X_{2t-1}, X_{2t}), the signal emitted at time $2t$ is $(-X_{2t}^*, X_{2t-1}^*)$, according to the emission principle of the space–time block code. Currently, $2t + 1$ $2m$ bits reach the encoder, where the B-V mapping is achieved by linearly combining the vectors (R_1, R_2), which can then be calculated symbolically to give

$$(X_{2t+1}, X_{2t+2}) = R_1(X_{2t-1}, X_{2t}) + R_2(-X_{2t}^*, X_{2t-1}^*). \tag{1.6}$$

Thus, at time 2t + 1, the transmitted signal is (X_{2t+1}, X_{2t+2}), and this process is repeated until the end of the transmitted signal bits.

1.3.3.4 Space–Time Trellis Codes

Space–time trellis codes (STTC) obtain large diversity gains when using relatively simple decoding algorithms; however, non-full-rate STTCs may introduce bandwidth expansion, whereas space–time block codes do not provide coding gains, so new space–time codes are being actively investigated. The space–time trellis code is a new space–time code proposed by Tarokh and Seshadri, based on the delayed diversity technique proposed by Wittneben [106]. Because the space–time trellis code combines coding, modulation, and transmitting and receiving diversity into one, its appearance enables existing space–time codes to obtain greater diversity gain along with coding gain.

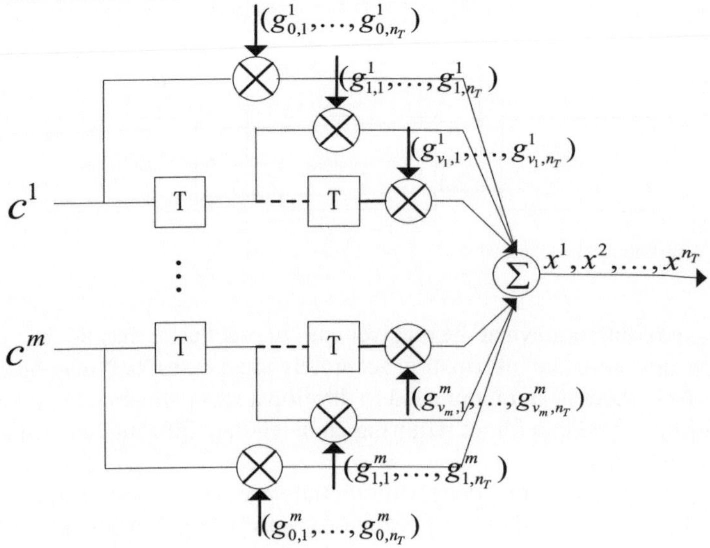

Fig. 1.13 Space–time trellis-code encoder

In a space–time trellis code, the encoder maps the binary data into different modulation symbols according to different state points in the trellis diagram. An STTC encoder, as shown in Fig. 1.13, has a transmit antenna number of N_t.

The input message stream is represented as

$$c = (c_0, c_1, \ldots c_t, \ldots). \tag{1.7}$$

c_t is a set of m bits of information at time t, as shown in the following equation:

$$c_t = (c_t^1, c_t^2, \ldots c_t^m). \tag{1.8}$$

The STTC encoder shown in Fig. 1.13 maps the input sequence to a sequence of modulated signal emissions:

$$x_t = (x_t^1, x_t^2, \ldots x_t^{n_t})^T. \tag{1.9}$$

In the encoder shown in Fig. 1.13, a sequence of m bits is fed into an encoder consisting of m feedforward shift registers T. Then, the sequence entered at the kth moment is

$$c^k = (c_0^k, c_1^k, \ldots c_t^k). \tag{1.10}$$

The kth shift register is fed and multiplied by a null-time trellis-code encoder coefficient, $g_{j,i}^k$. The multiplier output of all shift registers is summed modulo M to obtain the encoder output of

$$x = (x^1, x^2, \ldots x^{n_t}). \tag{1.11}$$

x_t^i represents the output of the encoder to transmitting antenna i at time t. The following equation can be obtained:

$$x_t^i = \sum_{k=1}^{m}\sum_{j=0}^{v_k} g_{j,i}^k c_{t-j}^k \bmod M \quad (\mathrm{i} = 1, 2, \ldots n_t). \tag{1.12}$$

$g_{j,i}^k (k = 1, 2 \cdots m; j = 1, 2 \cdots, v_k; i = 1, 2, \cdots, n_t)$ is an element of the M–phase-shift keying (PSK) signal constellation atlas and v_k is the memory length of the kth shift register.

The output signal x_t^i in Eq. (1.11) is the modulated emission signal, which forms the space–time symbol for the emission at time t.

$$x_t = (x_t^1, x_t^2, \ldots x_t^{n_t})^T \tag{1.13}$$

The sum of the full memory length of the encoder is v, that is,

$$v = \sum_{k=1}^{m} v_k, \tag{1.14}$$

where v_k is the memory length of the kth encoder branch:

$$v_k = \left| \frac{v + k - 1}{\log_2^M} \right|. \tag{1.15}$$

The total number of encoder states is 2^v. The sequence of m multiplicative coefficient sets is called the generating sequence.

A 2×2 quadrature phase-shift keying (QPSK)-STTC system was used as an example, as shown in Fig. 1.14. Using the QPSK modulation constellation (1), each binary bit can be mapped into a different phase on the constellation diagram (0 maps to 0, 1 maps to $\pi/2$, 2 maps to $3\pi/2$, and 3 maps to π), and the transmitted digital information is represented by the phase change of the carrier. In four-state trellis-coded modulation, there are four states (states 0, 1, 2, and 3) at the initial moment.

Suppose that starting from state 0 when the input is 10, the output is 02; the next time, the initial state is shifted to state 3 corresponding to 10; if 01 is input, the output is 21, and so forth. Assuming that information is input at a certain moment, the input at time k is represented by $b_k a_k$; then, the input bits at time $k-1$ can be represented as $b_{k-1}a_{k-1}$, and both bits are simultaneously stored in the register (x_1^k, x_2^k). Represents the output state of the momentary lattice code, which corresponds to the number of pairs on the right side of the lattice diagram in (2). It can be seen that, for an initial value of 0 and input $b_k a_k$ of 00, 01, 10, and 11, the corresponding coded outputs

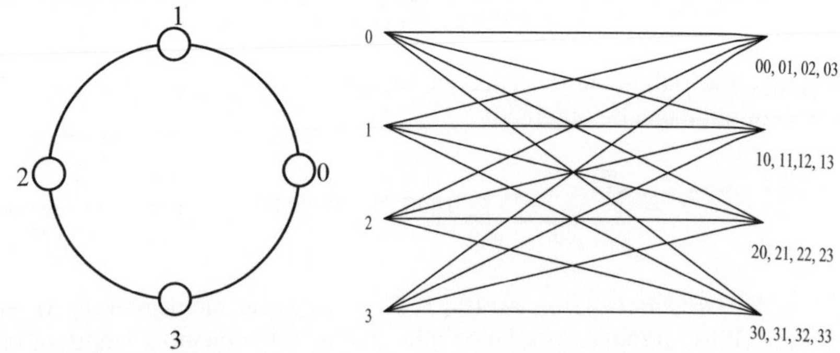

(a) 4-PSK constellation diagram (b) Trellis diagram

Fig. 1.14 Coding trellis diagram

(x_1^k, x_2^k) are $(0, 0)(0, 1)(1, 0)(1, 1)$ to obtain the outputs at states 1, 2, and 3. From this, the output expression can be expressed as

$$(x_1^k, x_2^k) = b_{k-1}(2, 0) + a_{k-1}(1, 0) + b_k(0, 2) + a_k(0, 1). \tag{1.16}$$

Outputs x_1^k and x_2^k correspond to the QPSK constellation points and are sent simultaneously by both transmitting antennas. The encoding trellis diagrams for eight and 16 states using 4-PSK are shown in Fig. 1.14, where two pairs of antennas are used to send the data symbols. The numbers in each state indicate the output of the encoder when moving from that state to another state. The first number indicates the sign of the constellation point sent on the first antenna, and the second number indicates the sign of the constellation point sent out on the second antenna. Zero to three represent the 4-PSK constellation point, whereas zero to seven or zero to M are used if 8-PSK or M-PSK are used (Fig. 1.15).

Although spatiotemporal trellis codes can achieve greater coding and diversity gains, they suffer from transmission-rate limitations. This is because the complexity of the decoder increases exponentially as the number of bits contained in a single symbol increases; this is a bottleneck that limits space–time trellis codes (Table 1.1).

1.3.3.5 Unitary Space–Time Codes

Unitary space–time codes (USTC) are a type of space–time code constructed based on previous research [118]. They address the problem that it is difficult or even impossible to estimate the channel matrix at the receiving end, when the channel parameters are changing rapidly or the number of transmitting and receiving antennas is large.

Unlike other space–time codes, which are based on the assumption that the receiver can accurately estimate the channel characteristics, unitary space–time codes

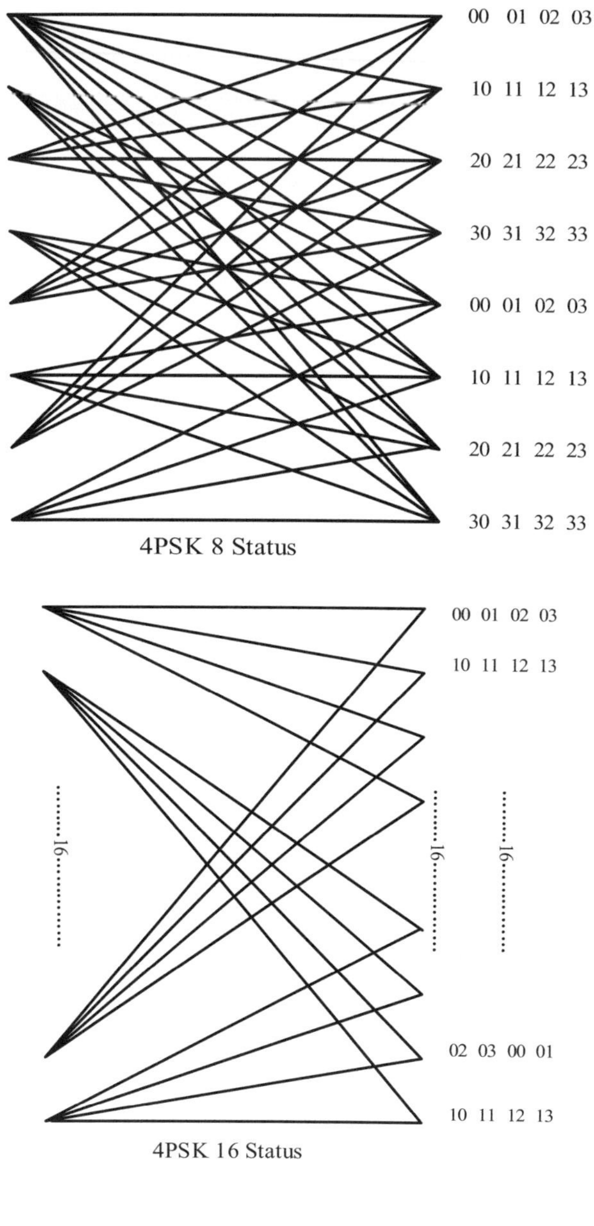

Fig. 1.15 4-PSK eight-state and 16-state trellis diagrams

do not require channel estimation. The unitary space–time code is similar in form to the space–time block code, but requires the transmitting code matrix to be a unitary matrix, which no longer optimizes the Euclidean distance, but rather optimizes the matrix parameterization of the correlation matrix; the smaller the parameterization value, the better the performance.

Table 1.1 Comparison of performance of space–time codes [118]

	STBC	STTC	LSTC
Number of transmitting and receiving antennas	Fixed (2 rounds and 2 receipts)	Variable	Equal number of transmitting and receiving antennas
Difficulty of coding the generation matrix	Simple	Complex	Easier
Diversity gain	Large	Large	General
Difficulty of decoding	Simple	Complex	Simple
Difficulty in achieving	Low	High	Low
Performance	Compares well	Very good	General
Development prospects	Simple and practical but difficult to improve	Difficult practical application and high compilation-code complexity	Codec methods could be improved

Although the unitary space–time code has some practical significance as a space–time code solution for fast time-varying channels, it is difficult to construct and store a large number of unitary matrices; therefore, it is difficult to construct a unitary space–time code with good performance for practical applications.

1.3.4 Evolution from SISO to MIMO Channels

The SISO wireless channel model is typically used in communication systems and is the basis for all derived channels. The key point that distinguishes a MIMO channel model from a SISO is the spatial characteristics of the wireless channel [119]. It is not possible to rely on a SISO channel model without considering the space–time characteristics of the MIMO system. The transmission environment of a MIMO channel must be studied in detail and in depth. Based on its spatial characteristics and the SISO-channel vector model, a vector model of the MIMO system was derived [120].

The SISO wireless channel system, that is, the single-transmitter–single-receiver antenna-communication system, has a channel impulse response of

$$h(t,\tau)=\sum_{i=1}^{L}\alpha(t,\tau)e^{j\varphi_i(t,\tau)}\delta[\tau-\tau_i(t)], \tag{1.17}$$

where $\varphi(t,\tau)$ denotes the sum of the phase shifts of the multipath propagation channels.

The channel model for a system with one transmitter antenna and m receiver antennas at the receiver is called SIMO. According to [121], the SIMO channel can be treated as an m merged SISO scalar channel with an impulse response of

$$h(t,\tau) = [h_1(t,\tau, h_2(t,\tau))], \ldots, h_M(t,\tau)]^T, \tag{1.18}$$

where $h_i(t,\tau)$ is the impulse response of the ith channel of the SISO. Then, the impulse response of the SIMO channel is given by

$$h(t,\tau) = \sum_{i=0}^{l-1} \alpha(\theta_i,\phi_i)\beta_i^{j(2\pi f_i \tau_i + \phi_i)}\delta[\tau - \tau_i(t)], \tag{1.19}$$

where β_i denotes the path gain of the ith subchannel, τ_i denotes the path delay of the ith subchannel, $\alpha(\theta_i,\phi_i)$ denotes the direction of arrival of the ith subchannel, f_i denotes the Doppler shift, ϕ_i denotes the additional phase shift, and $\alpha(\theta,\varphi)$ denotes the array manipulation vector, with the following expression:

$$\alpha(\theta,\phi) = [1, \alpha_1(\theta,\phi), \ldots, \alpha_{M-1}(\theta,\phi)]^T, \tag{1.20}$$

where the ith component is

$$\alpha_i(\theta,\phi) = e^{j2\pi(x_i \sin\theta\cos\phi + y_i \sin\theta\sin\phi + z_i\cos\theta)}.$$

Comparing the SIMO channel with the SISO channel, the scalar channel extends to the vector channel, thereby showing the spatial-channel characteristics of the system.

The MISO channel is similar to the idea of a SIMO channel formation. It uses n transmit antennas and a single receive antenna, so the channel can be regarded as a vector channel formed by combining n SISO scalar channels [122]. The MISO channel impact response can be obtained as

$$h(t,\tau) = [h_1(t,\tau), h_2(t,\tau), \ldots, h_N(t,\tau)]^T, \tag{1.21}$$

where $h_i(t,\tau)$ is the impulse response of the ith subchannel. Subsequently, the impulse response of the MISO channel is

$$h(t,\tau) = \sum_{i=0}^{l-1} \alpha(\theta_i,\phi_i)\beta_i^{j(2\pi f_i \tau_i + \phi_i)}\delta[\tau - \tau_i(t)], \tag{1.22}$$

where $\alpha(\theta_i,\phi_i)$ denotes the direction of the going wave. Based on the SIMO and MISO formations, a MIMO channel with n and m transceiver antennas can be considered a matrix channel with $n \times m$ SISO scalar channels.

$$h(t,\tau) = \begin{bmatrix} h_{11}(t,\tau) & h_{12}(t,\tau) & \cdots & h_{1n}(t,\tau) \\ h_{21}(t,\tau) & h_{22}(t,\tau) & \cdots & h_{2n}(t,\tau) \\ \vdots & \vdots & \ddots & \vdots \\ h_{m1}(t,\tau) & h_{m2}(t,\tau) & \cdots & h_{mn}(t,\tau) \end{bmatrix}, \tag{1.23}$$

where $h_{ij}(t,\tau)$ is the impulse response of the subchannel (SISO channel scalar) between the ith transmitter antenna and the jth receiver antenna.

1.4 Key Technologies in Wireless-Optical MIMO Systems

As can be seen from its composition, the optical MIMO system is based on the classical system (SISO). It has both differences and links with the classical system. The key technologies are described in detail below, based on the composition of the optical MIMO system.

1.4.1 Empty-Time Coding

Space–time coding is a joint coding technique that combines transmit diversity in the spatial domain with channel coding in the temporal domain [123]. Space–time coding can be divided into space–time trellis codes (STTC), space–time block codes (STBC), layered space–time codes (LSTC), unitary space–time codes (USTC), and differential space–time block codes (DSTBC–time trellis codes is a joint encoding method that combines emission diversity with mesh encoding modulation. This coding scheme achieves full diversity gain and high coding gain without sacrificing the system bandwidth, thus improving transmission quality. However, due to its high decoding complexity, it hinders its practical use.

Space–time block codes use the orthogonal design principle to distribute the transmit signal format at each transmitting antenna and are, in fact, a joint spatial- and temporal-domain orthogonal block-coding method. The space–time block code enables full diversity gain to be obtained after decoding at the receiver, and ensures that the decoding operation is only a simple linear combination with low decoding complexity. Hierarchical space–time coding combines spatial-multiplexing and receiver-diversity techniques and has received considerable attention, owing to its simple and practical structure and extremely high spectrum utilization.

1.4.2 Laser Arrays

Lasers are a key component of laser communications. They are used to generate the laser signals that form the beam transmitted through an optical antenna. The characteristics of the laser and the placement of the array have a direct impact on the quality and distance of the communication; therefore, their selection is very important.

To ensure that the signal from the laser in the array is correctly distinguished by the detector at the receiving end, the distance between two lasers in the array must not be too small. Moreover, the optical power received at the receiving end from the transmitter must meet the minimum requirements for processing by the signal detector.

The currently used light sources are semiconductor lasers, CO_2 lasers, He–Ne lasers, Nd:YAG lasers, and excimer lasers. Semiconductor lasers are most commonly used, owing to their simple structure, vibration resistance, small size, light weight, high efficiency, long life, and easy modulation. In terms of wavelength selection, visible light lacks confidentiality and infrared light is too long and lacks a corresponding high-speed detector. However, efficient, high-speed, and relatively inexpensive detectors are available in the near-infrared band, and a high atmospheric-transmission window exists in the 790-nm band in the near-infrared spectrum [117].

1.4.3 Detector Arrays

Spatial-diversity reception can be used to achieve high transmission rates, that is, multiple detectors can be used to receive the signal. The placement of the detector array should be consistent with the structure of the laser array at the transmitter, and the distance between detectors should be identical to the distance between lasers; otherwise, the intensity of the light signal reaching the receiving side of the detector is too weak, resulting in the signal not being received correctly at the receiving end.

To ensure that the signal received at the receiving end meets the non-correlation condition, the distance between the photodetectors D_r should meet $\sqrt{\lambda Z} < D_r < \theta Z$, where θ is the divergence angle of the beam, λ is the wavelength of the light, and Z is the distance travelled. The typical correlation distance is 50 cm.

Theoretically, to improve reception sensitivity, an outlier-detection method should be used [118]. However, outlier detection requires that the frequencies of the receiver's locally oscillating laser and the transmitting laser must be very stable, and the frequency stability must be at least 10^{-11}, which affects its practicality [57].

Direct detection remains a good method for photoelectric detection. For direct detection, the core is the photodetector, which converts the optical/electrical signal. Currently, the most commonly used photodetectors are Si photodiodes, PIN photodiodes, and avalanche photodiodes (APD). They are all photodetectors, which play a very important role in the detection of weak and fast light signals. Si photodiodes have high quantum efficiency, low noise, fast response, low power consumption, small size, wide operating-wavelength range, the highest response rate in the 0.8–0.9 μm band, and other characteristics.

The PIN photodiode has a very short carrier-crossing time, owing to the presence of an intrinsic layer, which results in a very fast response, and can detect light signals with a high modulation frequency (10^9-Hz class). It uses the avalanche effect of photogenerated carriers in the high-electric-field region to obtain the photocurrent gain. Currently, the materials used to fabricate APDs are mainly silicon and germanium, and practical devices have extremely short response times; that is, response frequencies of several gigabits, gains of up to 10^2–10^3, and very small dark currents.

1.4.4 Optical-Transceiver Antennas

In RF communication, many types of antenna arrays exist, based on the arrangement of the antenna units, such as linear arrays, circular arrays, and two-dimensional plane arrays. The linear array has better directionality and can form a narrower main flap in a certain direction. However, it is unable to maintain consistent performance over a range of 360° and suffers from phase blurring over a large range [114].

A uniform circular-array antenna is composed of antenna units uniformly distributed on a circumference or multiple concentric circumferences; its horizontal structure is shown in Fig. 1.2. Relative to a line array, a circular array can estimate the pitch angle and achieve 360° azimuthal coverage. A circular array forms a beam assignment that is not as good as a two-dimensional planar array; however, it requires fewer antenna units. Correspondingly, the amount of data is lower, and the processing method is simpler. A uniform circular array has circular symmetry, which makes it easy to install.

In a circular array, the central array element has a greater effect on the performance improvement of the array directional map, particularly when the number of elements is odd [1]. Sketches of circular arrays with three, four, and eight array elements are shown in Fig. 1.16. Owing to its inherent circular rotational symmetry, the array allows the beam to scan uniformly in a plane by simply rotating the weighting vectors of each cell. It can also overcome some other inherent disadvantages of linear arrays, such as the unbalanced mutual-coupling effect between antenna units in RF communication, difficulty in achieving wide-angle scan matching, and large size.

Uniform circular-array antennas have superior structural characteristics and good direction-finding performance; thus, they have attracted attention in military and civil use. However, the manifold of a circular array is more complicated than that of a uniform line array.

An optical antenna system is an important part of wireless laser communication and consists of an array of transmitting and receiving antennas. The main function of the transmitting antenna is to compress the beam-divergence angle, which collimates

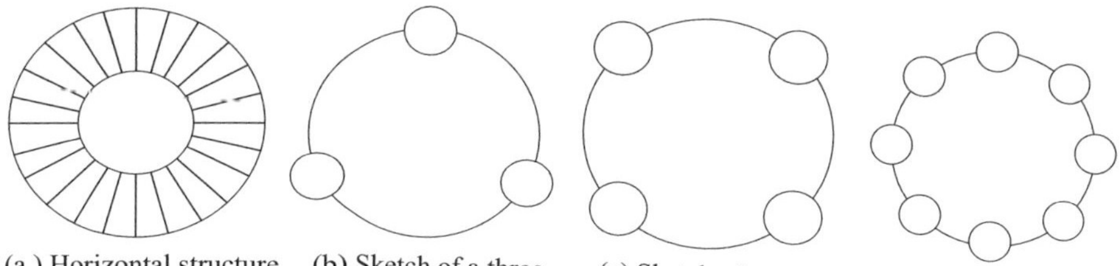

(a) Horizontal structure of a circular-array antenna
(b) Sketch of a three-element circular array
(c) Sketch of a four-element circular array
(d) Sketch of an eight-element circular array

Fig. 1.16 Sketches of circular arrays

and expands the beam. The role of the array-receiving antenna is to receive a weak light signal and gather it on the detector surface, thereby increasing the effective receiving area of the detector. The design of an optical array antenna is also a question that needs further exploration.

1.4.5 Modulation Techniques

It is essential to choose appropriate modulation, coding, and demodulation methods, as this can have a significant impact on system performance. A modulator is an electrical/optical converter. Two main types of modulation techniques are suitable for laser communication: one is suitable for correlation detection, and includes frequency-shift keying (FSK), phase-shift keying (PSK), and quadrature amplitude modulation (QAM). The other is suitable for direct detection, and includes on–off keying (OOK), phase-amplitude modulation (PAM), pulse-position modulation (PPM), pulse-interval modulation (PIM), and pulse-position keying (PPK).

OOK is one of the easiest modulation methods to implement in IM/DD systems. However, this modulation method is too inefficient and cannot meet the high-speed transmission requirements when the background light is strong or high-speed communication is performed. PPM modulation is orthogonal and has higher optical-power utilization. It can achieve high-speed communication and is less affected by external noise and multipath dispersion than OOK technology.

1.4.6 ATP Subsystems

Acquisition, tracking, and pointing (ATP) is one of the key technologies of wireless laser-communication systems, and usually consists of three parts: capture, tracking, and aiming. Spatial capture involves pointing the antenna in the direction of the arrival field and adjusting the normal vector over the surface area of the antenna to match the beam-arrival angle. The capture range can be $\pm$ 1° to $\pm$ 20° or higher. This is often achieved using an array of charge-coupled devices (CCDs), using servo-actuators with pass filters and real-time signal processing to capture coarse-tracking targets.

The beam is tracked and aimed by the fine-tracking system, whose main function is to aim and track the target in real time after the target has been captured. This is typically achieved using a four-quadrant quantum-dot (QD) infrared detector or quantum-dot APD (Q-APD) high-sensitivity position sensor with a corresponding servo-control system for electronics [117].

Targeting requires a field angle of a few hundred μ rad, tracking accuracy of μ rads, and tracking sensitivity of a few milliwatts. Commonly used ATPs include target-imaging location-tracking systems, warp-tracking platforms, etc.

1.4.7 Spatial Optical-Fiber Coupling Technologies

Space-optical coupling technology is an important part of space-optical communication technology. The light signal received by a free-space optical-communication system is often very weak, and the spot drift, phase fluctuation, and light-intensity fluctuation caused by atmospheric turbulence seriously affect the quality of the light beam coupled into the fiber. They limit the communication performance of the laser-communication system and the reliability of the laser link.

Atmospheric turbulence causes beam drift and amplitude fluctuations in the light field, resulting in unstable light intensity being coupled into the fiber; angle-of-arrival fluctuations increase the amount of light coupled at an angle greater than the critical angle on the receiving surface, resulting in a significant reduction in the power coupled into the fiber; and wavefront phase fluctuations degrade the spatial coherence of the light field. These effects result in reduced efficiency and undulation of the spatial light-to-fiber coupling, leading to an increased BER to the point of communication disruption.

In 1990, the Lincoln Laboratory of the Massachusetts Institute of Technology, USA, proposed an active coupling scheme, based on fiber-optic chapter motion [124]. The fiber-position error is evaluated in terms of the variation of the coupled optical power while receiving the optical signal, and the fiber position is corrected by feedback to a position-control motor, reducing the coupling loss to only 2 dB. In 1998, Ruilier [125] theoretically analyzed the average coupling efficiency and coupling-efficiency undulation variance of spatial light-to-single-mode fiber under turbulence, but did not consider the intensity flicker caused by turbulence or the aperture-averaging effect of the receiving antenna.

In 2001, Sayano et al. [126] optically coupled an optical signal into a single-mode fiber, using a phase-sensitive sensor to detect changes in the incident-light angle and driving a piezoelectric-ceramic tilting mirror to compensate, with a coupling loss of approximately 3 dB. In 2005, Dikmelik et al. from Johns Hopkins University, USA, analyzed the effect of turbulence-induced random phase perturbations on the average coupling efficiency of free-space optical single-mode fibers under weak turbulence, derived an expression for the fiber-coupling efficiency, based on atmospheric turbulence parameters, and quantified the link distance at the point where atmospheric effects on the coupling efficiency became unbearable [127].

In 2006, Lazzaroni [128] investigated the fading probability of coupled signals in the context of spatial optical communication, assuming that the allowed random angular jitter at a fading level of 3 dB and a fading probability of 10^{-2} should be less than 0.164 times the mode-field radius.

In 2006, Xiang et al. [129] from the University of Electronic Science and Technology established a model for the average coupling efficiency and coupling-power undulation of spatial optical-to-single-mode fiber based on weak turbulence theory. They considered the effects of turbulence undulation, turbulent wavefront distortion, and tracking error on the coupling system, and analyzed the average coupling efficiency and coupling-power undulation upstream and downstream. In 2007, Deng Ke et al. [130] analyzed the factors affecting the coupling efficiency of a spatial optical-to-single-mode fiber based on mode-field matching theory.

In 2007, Ma Jing et al. of the Harbin Institute of Technology [130] analyzed the factors affecting the coupling efficiency of spatial light-to-single-mode fiber. By optimizing the relative aperture of the coupling lens, efficient coupling was achieved. Theoretical calculations yielded a maximum coupling efficiency of 82.54% for 1310-nm light and 82.69% for 1550 nm at relative apertures of 0.203 and 0.211, respectively, and the actual measured maximum coupling efficiency was 61%. In 2007, Ma Jing et al. analyzed the effect of the local wavefront phase difference on the coupling efficiency of single-mode fibers in free-space optical communication [131].

In 2009, Ma Jing et al. studied the effect of random angle jitter on the coupling efficiency of spatial light to single-mode fiber. The results showed that the fiber alignment error introduced additional coupling losses, and the average BER and power cost increased significantly when the standard deviation of the random lateral offset was greater than 0.2 times the fiber-mode field radius [132].

In 2010, Ai Yong of Wuhan University proposed a finite-state machine (FSM) fine-tracking control system to suppress the arrival-angle fluctuations caused by atmospheric turbulence when transmitting lasers in turbulent atmospheres, which reduces the efficiency of spatial optical-single-mode fiber coupling. The experiments on a 16-km link showed that the spot center-of-mass coordinates were basically normally distributed, the maximum undulation amplitude was suppressed by 78%, and the mean squared deviation of the undulation was reduced by four tenths. The fiber-coupling efficiency was improved by six times, and the relative fluctuation of the optical power was significantly reduced [133].

Spatial-optical coupling into single-mode fibers, owing to the small core of single-mode fibers (only a few microns), greatly reduces the spatial-optical coupling efficiency. To improve the communication performance of spatial-optical links, methods such as adaptive optics, lens arrays, and tapered fibers have been developed to improve the efficiency of spatial-optical coupling.

In 2002, Weyrauch [134] conducted experiments on adaptive-optics spatial laser-to-single-mode fiber coupling using a VLSI-controlled micromechanical variational mirror to correct the wavefront. He obtained spatial laser-to-single-mode fiber coupling efficiencies of approximately 60% and multimode-fiber coupling efficiencies of approximately 75%. In 2007, Kim et al. [135] investigated methods to correct for wavefront phase distortion due to atmospheric turbulence.

In 2009, Puryear [136] showed that losses due to turbulence in the atmospheric channel could be suppressed by spatial-diversity reception, wavefront pre-correction, and coherent detection. In the same year, Buske [137] investigated wavefront distortion caused by atmospheric turbulence, leading to image-speckle expansion and optical-resolution loss, and used an adaptive-optics system to compensate for the phase distortion caused by atmospheric turbulence.

In 2010, Sinefeld et al. [138] used an iterative feedback algorithm to compensate the phase by estimating 15th-order Zernike polynomial coefficients. In addition, Sarmadi et al. [139] proposed a blind search technique for wavefront-phase distortion correction using a genetic algorithm to control deformation mirrors to correct the wavefront and enhance the output signal. In 2010, Podanchuk introduced a holographic Shack-Hartmann wavefront sensor that compensated for phase aberrations using a two-channel array of holographic lenses [140].

In 2005, Huan et al. [141], from Changchun University of Technology, studied an adaptive-compensation method for atmospheric turbulence in space laser communication systems. In 2008, Yang Huizhen et al. [142] from the Institute of Optoelectronics Technology, Chinese Academy of Sciences, studied an adaptive-optics technique without a wavefront sensor, which could avoid the effect of light-intensity flicker caused by atmospheric turbulence on wavefront correction in atmospheric optical communication.

In 2010, Bo [143] proposed an adaptive-optics technique based on a stochastic parallel gradient-descent algorithm to achieve closed-loop compensation of wavefront-phase distortion without a wavefront sensor. However, the slow convergence speed of the algorithm restricts its application in systems with high real-time requirements. In 2010, Xia et al. [144], of the University of Electronic Science and Technology, reconstructed a distorted wavefront and then used a reflective deformation mirror to correct it.

In 2010, Qiao [145], from the University of Electronic Science and Technology of Guilin, analyzed the working principle and performance index of the deformation mirror in adaptive optics and pointed out that the wavefront-compensation effect of an adaptive-optical system consisting of a detector, controller, and deformation mirror could reach 90%. Also in 2010, Li [146], from the Wuhan Institute of Technology, studied a genetic algorithm–based closed-loop adaptive-control optical system to compensate for the aberrations of the laser beam.

In 2011, Han [147], from the Harbin Institute of Technology, analyzed the relationship between the coupling efficiency and turbulence intensity, transmission distance, and zenith angle. A stochastic parallel gradient optimization algorithm was used to control the distortion mirror in real time to optimize the single-mode fiber-coupling efficiency.

In 2009, Heinemann et al. [148], from the Fraunhofer Center for Laser Technology Research, USA, investigated the coupling technique of fiber arrays. In 2010, Daniel [149] used microlens arrays to improve fiber-coupling efficiency. In 2011, Belmonte [150] analyzed coherent optics in atmospheric-turbulence communication systems with adaptive-field conjugate receivers and used the maximum-ratio combining (MRC) method to achieve diversity reception. This compensated for the wave-front distortion caused by atmospheric turbulence and improved the fiber-coupling efficiency.

In 2008, Yuan [151], of the Huazhong University of Science and Technology, designed a new type of optical antenna using a 7 × 1 fiber coupler to couple space light into multiple multimode fibers. The computational results showed that the multi-aperture optical antenna had a higher coupling efficiency than the single-aperture optical antenna under a turbulent atmosphere. In 2010, Wu et al. [152], from the Institute of Mechanical Research and Institute of Optoelectronics, Chinese Academy of Sciences, analyzed the coupling efficiency of Zernike-mode adaptive-optical correction. The effect of multiple coherent fiber arrays on the coupling-efficiency improvement was also compared, and a hybrid technique combining the two was proposed to further improve the coupling efficiency.

In 2011, Leitgeb [153] studied the use of optical cones in free-space optical (FSO) receivers to improve the spatial-optical coupling efficiency. He observed through theoretical simulations that the coupling efficiency was higher using optical cones than lenses. However, the reflection angle of the optical cones was found to be the key to high coupling efficiency in experiments. Moreover, coupling into multimode fibers was much more efficient than coupling into single-mode fibers. The experimental results showed that a solid light cone could effectively receive a focused light signal after a beam shift; however, further research is needed to optimize the cone shape and improve the coupling efficiency.

References

1. Ma L (2002) Intelligent antennas in wireless communication-IS95 and third generation CDMA applications. Machinery Industry Press, Beijing, pp 20–31
2. Huang HX, Wu ZL (2001) DOA and Doppler frequency estimation algorithms for signal sources based on uniform circular array. J Electron 29(5):619–620
3. Khalighi MA, Brossier JM, Jourdain GV, Raoof K (2001) Water filling capacity of Rayleigh MIMO channels. In: 12th IEEE International symposium on personal, indoor and mobile radio communications, 2001, vol 10, 1, pp 156–158
4. Gao J (2008) The application of MIMO technology in ultra-wideband systems. Xi'an University of Electronic Science and Technology, Xi'an, pp 10–11
5. Djordjevic BI, Vasic B, Neifeld AM (2007) LDPC coded orthogonal frequency division multiplex over the atmospheric turbulence channel. Opt Express 15(10):6332–6346
6. Paterson C (2005) Atmospheric turbulence and orbital angular momentum of single photons for optical communication. Phys Rev Lett 94(15):153901-1–153901-4
7. Polishuk A, Arnon S (2004) Optimization of a laser satellite communication system with an optical preamplifier.Opt Soc Am 21(7):1307–1315

8. Ricklin JC, Davidson FM (2003) Atmospheric optical communication with a Gaussian Schell beam. Opt Soc Am 20(5):856–866
9. Tang Y, Shieh W, Yi X (2007) Optimum design for RF-to-optical up-converter in coherent optical OFDM systems. IEEE Photonics Technol Lett 19(7):483–485
10. Wang HQ, Ke XZ (2008) Free space optical communication based on vertically hierarchical space-time codes. China Laser 35(6):874–878
11. Chen J, Huang XD, Yuan XH (2006) Characterization of atmospheric laser communication systems based on dual-mode turbo codes. China Laser 33(11):1532–1536
12. Stéphanie S, Anne JV, Jean-Pierre C (2009) Soft decision LDPC decoding over chi-square based optical channels. J Lightwave Technol 27(16):3540–3545
13. Stephanie S, Damien F, Anne JV, Georges R, Jean-Pirre C, Philippe G (2009) LDPC code design and performance analysis on OOK chi-square-based optical channels. IEEE Photonics Technol Lett 21(17):1190–1192
14. He XM, Li XF, Zhang DY, Che YL (2008) Performance analysis of efficient error correction coding techniques in wireless optical communication. J Photonics 37(12):2427–2429
15. Liu YD, Gao CQ, Li F, Gao MW (2007) Analytical study of the orbital angular momentum of partially coherent light and its spectrum. Appl Opt 28(04):462–467
16. Xiang P, Wang R (2009) Dynamic wavelength routing and time slot allocation algorithms in WDM-TDM optical networks. J Electron Inf 31(3):679–681
17. Zhou GT, Zhang XG, Xi LX, Yu L, Shen Y (2005) Experimental study on adaptive compensation of two-level polarization mode dispersion in 40Gb/s optical time division multiplexing system. J Opt 25(1):15–20
18. Pan WQ, Liu LR, Liu XM (2006) Imaging optical communication system and coding and decoding scheme. China Laser 33(2):213–220
19. Djordjevic IB (2007) LDPC-coded MIMO optical communication over the atmospheric turbulence channel using Q-ary pulse-position modulation. Opt Express 15(16):10026–10032
20. Wilson SG, Brandt-Pearce M, Cao Q (2004) Optical MIMO transmission with Q-ary PPM for atmosphere channel. IEEE Trans Commun 53(8):1090–1094
21. Awan MS, Brandl P, Leitgeb E, Nadeem F, et al (2009) Transmission of high data rate optical signals in fog and snow conditions. In: IEEE, Wireless VITAE'09, 2009, pp 702–706
22. Feng B, Guo F, Yu M (2009) Dynamic probabilistic packet marking based on PPM. In: Proceedings of the 2009 2nd Pacific-Asia conference on web mining and web-based application, WMWA 2009, Wuhan, China, 2009, pp 289–292
23. Tomaz J, Igor J, Muhammad SS, Gorazd K (2009) Simplified soft value extraction for M-PPM-modulated signals in FSO systems. Int J Electron Commun 63(7):595–599
24. Huang J-F, Yang C-C, Huang C-M (2009) On analyzing quasi-cyclic LDPC codes over modified welch-costas-coded optical CDMA system. J Lightwave Technol 27(12):2150–2158
25. Kamran K (1998) Turbo-coded optical PPM communication systems. J Lightwave Technol 16(1):18–26
26. Young KJ, Vincent PH (2001) Turbo-coded optical direct-detection CDMA system with PPM modulation. J Lightwave Technol 19(3):312–323
27. Danping L (2005) Research on high precision alignment information acquisition method in spatial optical communication. University of Electronic Science and Technology, Chengdu
28. Laurent B, Planche G (1997) SILEX overview after flight terminals campaigh. SPIE, 1997, 2990:10–22
29. Fenner WR (1866) Future trends in crosslink communications. SPIE 1993:1–8
30. Gotthard O (1997) SILEX program status-a major milestone is reached. SPIE 2990:2–9
31. Bondurant RS (1995) Overview of lasercom program at Lincoln laboratory. SPIE 23(81):2–3
32. Szajowski PF, Nykolak G, Auborn JJ (1998) 2·4km Free-space optical communication 1550 nm transmission link operating at 2.5 Gb/s-experimental results. Proc SPIE 3532:29–38
33. Divsalar D, Gagliardi RM, Yuen JH (1984) PPM performance for Rees-solomon decoding over an optical-Rf relay link. IEEE Trans Commun 32(3):302–305
34. Koh YTDF (1989) Interleaved concatenated coding for the turbulent atmospheric direct detection optical communication channel. IEEE Trans Commun 37(6):648–651

35. Park H, Barry JR (2004) Trellis-coded multiple-pulse-position modulation for wireless infrared communications. IEEE Trans Commun 52(4):643–651
36. Scki K, Mikami K, Katayama A, Suzuki S, Shinohara N, Nakabayashi M (2003) Single-chip FEC code using a concatenated BCH code for 10 Gb/s long-haul optical communication systems. In: IEEE 2003 Custom in Integrated Circuits Conference, 2003, pp 279–280
37. Cai Y Morris JM, Adalı T et al (2003) On Turbo code decoder performance in optical-fiber communication systems with dominating ASE noise. J Light Wave Technol 21(3):727–734
38. Chitra K, Ravichandran VC (2007) Turbo-coding of coherence multiplexed optical PPM CDMA system with balanced detection. Am J Appl Sci 4(5):264–268
39. Sahuguede S, Fafchamps D, Julien-Vergonjanne A, Rodriguez G, Cances J-P, Gallion P (2009) LDPC code design and performance analysis on OOK Chi-Square-Based optical channels. IEEE Photonics Technol Lett 21(17):1190–1192
40. Davidson FM, Koh YT (1988) Interleaved convolution coding for the turbulent atmospheric optical communication channel. IEEE Trans Commun 36:993–1003
41. Maurizio M, Essiambre R-J, Basch BE et al (2010) Concatenated coded modulation for optical communications systems. IEEE Photonics Technol Lett 22(16):1244–1246
42. Djordjevic IB, Xu L, Wang T (2010) Reverse concatenated coded modulation for high-speed optical communication. IEEE Photonics J 2(6):1034–1039
43. Wilson KE (1997) Results of the compensated Earth-Moon-Earth retro-reflector laser link (CEMERLL) experiment. TDA Progress Report 42-131, 1997:1–13
44. Kim II (1997) Scintillation reduction using multiple transmitters. SPIE, 1997:102–113
45. Kim II et al (1998) Scintillation measurements performed during the limited-visibility lasercom experiment. SPIE, 1998, 209–220
46. Kim II et al (1999) Measurement of scintillation for free space laser communication at 785 and 1554 nm. SPIE, 1999, pp 49–62
47. Kim II et al (1999) Measurement of scintillation and link margin for the terralink laser communication system. SPIE, pp 100–118
48. Strohbehn JW (1978) Laser beam propagation in the atmosphere. Spring Berlin Heidelberg, New York
49. Tatarskii VI (1961) Wave propagation in a turbulent medium. McGraw, New York
50. Ibrahim MM, Ibrahim AM (1996) Performance analysis of optical receivers with space diversity reception. IEEE Proc Commun 143(6):369–372
51. Nykolak G et al (2000) A 40Gbps DWDM free space optical transmission link over 4.4 km. SPIE 16–20
52. Biswas A, Wright MW (2002) Mountain-top-to-mountain-top optical link demonstration. Part I. IPN Progress Report, 2002, pp 42–149
53. Vilnrotter V, Lau CW, Andrews K et al (2006) Real-time combining of optical array signals. SPIE, 2006, 6105:610508-610501–610508-610511
54. Vilnrotter V, Lau CW, Andrews K et al (2005) Two-element optical array receiver concept demonstration. SPIE, pp 225–239
55. Vilnrotter V, Lau CW, Srinivasan M et al (2005) Optical array receiver for communication through atmospheric turbulence. IEEE J Lightwave Technol 23(4):1664–1675
56. Vilnrotter V et al (2001) Optical communications through atmospheric turbulence using photo-detector arrays. SPIE, pp 282–292
57. Haas SM et al (2001) Capacity of wireless optical communication. IEEE J Sel Areas Commun 21(8):1346–1357
58. Ma D, Wei Q, Zhuang Z (2003) Space laser communication and its applications. Semicond Optoelectron 24(2):139–144
59. Haas SM et al (2001) Space-time codes for wireless optical channels. IEEE Trans Inf Theory, 244
60. Zhu X, et al (2003) Mitigation of turbulence-induced scintillation noise in free-space optical links using temporal-domain detection techniques. IEEE Photonics Technol Lett 15(4):623–625

61. Yazan A, Alqudaha, Kavehrad M (2003) Orthogonal spatial coding in indoor wireless optical link reducing power and bandwidth requirements. SPIE, 237–245
62. Simon MK et al (2003) Alamouti-type space-time coding for free-space optical communication with direct detection. IPN Progress Report 42-156; 2003, 1–9
63. Feng P, Jing M, Liying T et al (2004) Scintillation characterization of multiple transmitters for ground-to-satellite laser communication. SPIE, pp 448–454
64. Brandt-Pearce M, Wilson S, Cao Q et al (2004) Code design for optical MIMO systems over fading channels. In: Proceedings of the 38th Asilomar conference on signals, systems & computers, Monterey, CA, pp 871–875
65. Simon MK, Vilnrotter V (2005) Alamouti-type space-time coding for free-space optical communication with direct detection. IEEE Trans Wireless Commun 4(1):35–39
66. Anguita JA, Neifeld MA, Vasic BV (2006) Multi-beam space-time coded systems for optical atmospheric channels. SPIE, 63041B-63041–63041B-63049
67. AnguitaJA, Neifeld MA, Vasic BV (2007) Spatial correlation and irradiance statistics in a multiple-beam terrestrial free-space optical communication link. Appl Opt 46(26):6561–6571
68. Mohammad Navidpour S, Uysal M, Kavehrad M (2007) BER performance of free-space optical transmission with spatial diversity. IEEE Trans Wireless Commun 6(8):2813–2819
69. Cao Q, Brandt-Pearce M, Wilson SG (2007) Free space optical MIMO system using PPM modulation and a single optical amplifier. Commun Network China 508–512
70. García-Zambrana A (2007) Error rate performance for STBC in free space optical communications through strong atmospheric turbulence. IEEE Commun Lett 11(5):390–392
71. Cvijetic N, Wilson SG, Brandt-Pearce M (2008) Performance bounds for free-space optical MIMO systems with APD receivers in atmospheric turbulence. IEEE J Sel Areas Commun 26(3):3–11
72. Chakraborty K (2008) Outage capacity of MIMO Poisson Fading Channel. IEEE Trans Inf Theory 54(11):4887–4907
73. Letzepis N, Holland I, Cowley W (2008) The Gaussian free space optical MIMO channel with Q-ary pluse position modulation. IEEE Trans Wireless Commun 7(5):1744–1753
74. Safari M, Uysal M (2008) Do we really need OSTBCs for free-space optical communication with direct detection. IEEE Trans Wireless Commun 7(11):4445–4448
75. Zaidi SAR, Hafeez M (2008) Cross layer design for orthogonal space time block coded optical MIMO systems. In: WOCN '08. Wireless and optical communications networks, 2008, pp 1–5
76. Ntogari G, Kamalakis T, Sphicopoulos T (2009) Performance analysis of space time block coding techniques for indoor optical wireless systems. IEEE J Sel Areas Commun 27(9):1545–1552
77. Ehsan Bayaki RS (2010) On space–time coding for free–space optical systems. IEEE Trans Wireless Commun 58(1):58–62
78. García-Zambrana A, Castillo-Vázquez C, Castillo-Vázquez B (2010) Space-time trellis coding with transmit laser selection for FSO links over strong atmospheric turbulence channels. Opt Express 18(6):5356–5366
79. Qiang XW, Wang TL, Wu NQ (1999) Multi-beam laser atmospheric transmission. Photonics 19(3):167–172
80. Liu W, Wu J (2003) Effect of atmospheric channel spacing and beam number on light intensity fluctuations. Opt Optoelectron Technol 1(4):15–17
81. Liu W, Wu J (2004) Effect of multiple partially coherent beams of light on light intensity scintillation through strong turbulence. Opt Eng 31(4):5–8
82. Ma D (2004) Research on multi-beam emission and reception technology in atmospheric laser communication. National University of Defense Technology, Chang Sha
83. Ma D, Hua W, Wei Q, Temperature field analysis of laser light source in multi-beam atmospheric laser communication.
84. Ma D-T, Wei Q-B, Zhuang Z-W (2004) Analysis of multi-beam transmission performance and channel modeling in atmospheric laser communication. J Opt 24(8):1020–1024

85. Chen G, Dong Z, Geng J et al (2004) A 155/622 Mb/s multi-emitter laser communication system. China Laser 31(5):583–587
86. Daidzhu D, Huafu G (2005) Application of multi-transmitters in atmospheric laser communication. Opt Optoelectron Technol 3(4):13–15
87. Wu C, Pan J, Yu R (2005) Optimal selection of diversity in atmospheric optical communication. Infrared Laser Eng 34(1):114–126
88. He J, Hu YJ (2007) Performance analysis of wireless optical MIMO system based on SVD algorithm. J Anhui Univ 31(3):34–37
89. Zheng Z, Li N (2008) Diversity detection technology in laser communication. J Beijing Univ Aeronaut Astronaut 34(3):280–284
90. Tian Z, Huaibei Z (2006) Structure and performance of space-time block code. Wireless Commun. Network Mob Comput, pp 1–4
91. Chen J (2007) Channel and performance study of laser atmospheric transmission. Huazhong University of Science and Technology, Wuhan
92. Deng T (2007) Key technologies for free-space optical communication systems. Huazhong University of Science and Technology, Wuhan
93. Zhang H (2008) Research on space-time coding based on emission diversity in atmospheric laser communication. Changchun University of Technology, Changchun
94. Chen C (2009) Research on the mechanism and suppression technology of atmospheric influence in wireless optical communication. Changchun University of Science and Technology, Changchun
95. Wang H, Ke X (2008) MIMO atmospheric laser communication and its key technologies. Laser J 29(3):52–54
96. Simon MK, Alouini MS (2000) Digital communication over fading channels: a unified approach to performance analysis. Wiley, pp 31–97
97. Naguib AF, Calderbank R (2000) Space-time coding and signal processing for high data rate Wireless communications. IEEE Signal Process Mag, 17(3):76–92
98. Sun W, He S, Zhao H (1999) Suppression of wideband millimeter wave radar target angular scintillation by frequency diversity technique. Mod Radar 02:14–17
99. Alamouti SM (1998) A simple transmit diversity technique for wireless communications. IEEE J Sel Areas Comm 16(8):1451–1458
100. Tarokh V, Jafarkhani H, Calderbank AR (1999) Space-time block codes from orthogonal designs. IEEE Trans Inform Theory 45(5):1456–1467
101. Ma D (2004) Research on multi-beam emission and reception technology in atmospheric laser communication. Ph.D. Thesis, National University of Defense Technology, 2004, pp 30–69
102. Ren L, Song M, Xi S et al (2004) MIMO technology in mobile communication. Mod Telecommun Technol 1:43–46
103. Zhang X et al (2000) Communication signal processing. National Defense Industry Press, Beijing, pp 134–145
104. Gesbert D, Akhtar J (2002) Breaking the barriers of Shannon's capacity: an overview of MIMO wireless system. Telektronikk Telenor J (2002)
105. Telatar E (1995) Capacity of multi-antenna Gaussian channels. ATT-Bell Labs Murray Hill, Tech Memo
106. Vucetie B, Yuan J, Wang X et al (2004) Space-time coding techniques. Machinery Industry Press, Beijing, vol 1, pp 80–129
107. Tarkh V, Jafarkhani H, Calderbank AR (1999) Space-time block codes from orthogonal designs. IEEE Trans Inform Theory 45(5):1456–1467
108. Tarkh V, Jafarkhani H, Calderbank AR (1999) Space-time block coding for wireless communications: performance result. IEEE J Sel Areas Comm 17(3):451–460
109. Foschini GJ (1999) Layered space-time architecture for wireless communication in fading environment when using multi-element antennas. Bell Labs Tech 1996, 1:41–59
110. Foschini GJ, Gans MJ (2003) Analysis and performance of some basic space-time architectures. IEEE J Select Areas Comm 21(3):303–320
111. Xi Y (2008) A comparative study of MIMO space-time coding. Inf Technol 6:104–105

112. Tarokh V, Seshadri N, Calderbank AR (1998) Space-time codes for high data rate wireless communication: performance criterion and code construction. IEEE Trans Inf Theory 44(2):744–765
113. Gang C, Yanmin C, Gaoting C et al (2000) Discussion on some issues of space laser communication technology. Infrared Laser Eng 29(3):44–48
114. Mecherle GS et al (1991) Homodyne PSK receivers with laser diode sources. SPIE, 99–108
115. Foschini G (1996) Layered space-time architecture for wireless communication in a fading environment when using multi-element antennas. Bell Labs Tech J, 1996:41–59
116. Foschini G, Gans M (1998) On limits of wireless communication in a fading environment when using multiple antennas. Wireless Pers Commun 6(3):311–335
117. Huang H-X, Wu Q-L (2001) DOA and Doppler frequency estimation algorithms for signal sources based on uniform circular array. J Electron 29(5):619–621
118. Li Z (2006) Research on DPPM modulation and demodulation for atmospheric laser communication. Xi'an University of Technology. Xi'an
119. An Y, Li W (2005) Analysis of wireless optical communication technology and application. Power Syst Commun 26(150):4–6
120. Chen S, Han Y (2001) Review of wireless optical communication technology, Guangxi Commun Technol (01):39–41
121. Xu L (2009) Research on MIMO channel modeling. Nanjing University of Technology, Nanjing
122. Liu X (2006) Space-time coding techniques in MIMO. Shanghai Jiaotong University, Shanghai
123. Wolniansky PW, Foschini GJ, Golden GD, Valenzuela RA, V-BLAST: an architecture for realizing very high date rates over the rich-scattering wireless channel. In: Proc, ISSSE-98
124. Swanson EA, Bondurant RS (1989) Fiber-based receiver for free-space coherent optical communication system. OFC, 6–9
125. Ruilier C (1998) A study of degraded light coupling into single mode fibers. Proc SPIE 3350:319–329
126. Sayano K, Nguyen I, Chan JK (2001) Demonstration of multi-channel optical CDMA for free space communications. Proc SPIE 4272:38–49
127. Dikmelik Y, Davidson FM (2005) Fiber-coupling efficiency for free-space optical communication through atmospheric turbulence. Appl Opt 44(23):4946–4952
128. Lazzaroni M, Zocchi FE (2004) Optical coupling from plane wave to step-index single-mode fiber. Opt Comm 237(1–7):37–43
129. Xiang J, Chen Y, Hu Y (2006) Effect of atmospheric turbulence on spatial optical coupling to single-mode fibers. Intense Lasers Part Beams 18(3)377–380
130. Deng K, Wang BZ, Wang X (2007) Analysis of spatial optical-single-mode fiber coupling efficiency factors. J Univ Electron Sci Technol 36(5):889–891
131. Jing M, Chong G, Li-Ying T (2007) Angle-of-arrival fluctuations in moderateto strong turbulence. Chin Phys 16(5):1327–1333
132. Ma J, Zhao F, Tan L et al (2009) Plane wave coupling into single-mode fiber in the presence of random angular Jitter. Appl Opt 48(27):5184–5189
133. Li M, Ai Y, Huang Y et al (2010) Research on angle-of-arrival undulation suppression in spatial optical communication. China Commun 1:138–142
134. Weyrauch T, Vorontsov MA, Gowens J W et al (2002) Fiber coupling with adaptive optics for free-space optical communication. Proc SPIE 4489:177–184 (S0277-786X)
135. Kim I, Goldfarb G, Li G (2007) Electronic wavefront correction for PSK free-space optical communications. Electron Lett 43:1108–1109
136. Puryear A, Chan VWS (2009) Coherent optical communication over the turbulent atmosphere with spatial diversity and wavefront predistortion. In: Global telecommunications conference, pp 1–8
137. Buske I, Riede W (2009) A mobile adaptive optics system for compensation of atmospheric turbulence-induced phase distortions. In: CLEO Europe—EQEC 2009. European 2009:1

138. Sinefeld D, Ella R, Zaharan O et al (2010) Wavefront aberration correction in a free-space optical communication link using only the fiber-coupled optical power as a feedback mechanism Electrical and Electronics Engineers in Israel (IEEEI), pp 890–891
139. Avanaki MRN, Hojjatolcslami SA, Ebrahimpour R et al (2010) Genetic algorithm for optimization of optical systems. In: Electrical engineering (ICEE), pp 172–176.
140. Podanchuk DV, Dan'ko VP, Kotov MM (2010) Holographic Shack-Hartmann wavefront sensor. In: Advanced optoelectronics and lasers (CAOL), pp 208–210
141. Li H, Zhang H, Yin F (2008) Adaptive compensation method for atmospheric turbulence in space laser communication systems. J Changchun Univ Sci Technol (Nat Sci Ed) 2:1–3
142. Yang, Huizhen, Cai, Dongmei, Chen, Bo et al (2008) Wavefront-free sensing adaptive optics and its application in atmospheric optical communication. China Laser 5:681–684
143. Chen B, Li XY, Jiang WH (2005) Optimization of an adaptive optical stochastic parallel gradient descent algorithm for atmospheric turbulence. China Laser 3:61–84
144. Xia LJ, Li XF (2010) Atmospheric optical communication wavefront correction experiments based on adaptive optics. Inf Electron Eng 03:331–335
145. Qiao RJ, Chen M (2010) Study of deformation mirrors in adaptive optics. Electron Des Eng 03:74–78
146. Li J, Xiao J (2010) Genetic algorithm-based closed-loop adaptive optics system. J Wuhan Inst Technol 01:47–49
147. Han LQ, Wang Q, Xin TKK et al (2010) Performance study of free-space optical communication systems based on adaptive optical compensation. Appl Opt 2:301–303
148. Heinemann S (2009) High brightness, fiber coupled single emitter arrays. Proc SPIE 7198(71980Q):1–5
149. Belmonte A, Kahn JM (2011) Field conjugation adaptive arrays in free-space coherent laser communications, 2011
150. Daniel VH, David MB, Nathan WR et al (2010) Fiber optic bundle array wide field-of-view optical receiver for free space optical communications. Opt Lett 35:3559–3561 (S1539-4794)
151. Tao F (2008) Research on a new optical antenna coupling system. Huazhong University of Science and Technology
152. Wu H, Yan H, Li X (2010) Modal correction for fiber-coupling efficiency in free-space optical communication systems through atmospheric turbulence. Optik 121:1789–1793
153. Leitgeb E, Stögner S, Löschnigg M (2011) Using tapers for efficient coupling of received FSO-signals into fibers. In: ICTON2011, 1–6

Chapter 2
Multi-aperture Transmitter/ Multi-aperture Receiver Technology in a Turbulent Atmosphere

Atmospheric turbulence reduces the reliability of atmospheric laser-communication systems, and light-intensity flicker increases the system bit-error rate. Although the characteristics of an atmospheric channel cannot be changed, methods such as large-aperture reception, automatic-gain control (AGC), adaptive optics, channel coding, multi-aperture transmission technology, and multi-aperture reception technology are effective methods to suppress the influence of atmospheric turbulence [1–7].

To address the influence of atmospheric turbulence on laser communications, this chapter discusses two methods for suppressing the effects of atmospheric turbulence: multibeam emission and multi-aperture reception. Theoretical analyses and experimental results show that multibeam transmissions can effectively suppress the light-intensity fluctuation at the receiving end and reduce the system bit-error rate; multi-aperture reception can produce an aperture-smoothing effect on the fluctuations of optical signals, and can also reduce the communication bit-error rate. Both are effective for suppressing the influence of atmospheric turbulence.

2.1 Performance Analysis of a Multibeam-Emission System

Atmospheric turbulence can cause light-intensity scintillation, increase the bit-error rate of atmospheric laser-communication systems, and degrade the laser-communication performance. Theoretical analyses and experimental results show [8, 9] that multi-aperture emission technology is an effective method to suppress the influence of atmospheric turbulence, and has been increasingly used in atmospheric laser-communication systems. Diversity technology suppresses signal attenuation in a turbulent flow by processing two or more uncorrelated signals, and multibeam-emission technology exhibits space diversity. Multibeam-emission technology can effectively suppress the light-intensity fluctuation at the receiving end and improve the system bit-error rate performance.

X. Ke, *Space-Time Coding Theory in Optical Wireless MIMO Systems*,
Optical Wireless Communication Theory and Technology,
https://doi.org/10.1007/978-981-97-6449-5_2

2.1.1 Intensity Fluctuation of Multibeam Emissions

Assuming that a laser beam adopts a single-mode Gaussian beam, the electric-field component u can be expressed as [10]

$$u(\vec{\rho}, z) = u_0 \frac{\omega_0}{\omega(z)} \exp[-\rho^2/\omega^2(z)] \cdot \exp\{-j[k\rho^2/(2R(z)) - kz]\}. \tag{2.1}$$

$\omega(z)$ is the spot radius of the Gaussian beam:

$$\omega(z) = \omega_0\sqrt{1 + (z/Z_0)^2}. \tag{2.2}$$

$R(z)$ is the iso-phase curvature radius of the Gaussian beam:

$$R(z) = z[1 + (Z_0/z)^2]. \tag{2.3}$$

Figure 2.1 shows a schematic diagram of multi-beam emissions. The transmitter uses multiple independent laser light sources with the same wavelength.

Assuming that n laser beams are transmitted, the distance between the transmitting apertures is d, receiving aperture diameter is D, and transmission distance is L. The geometric relationship between the transmitting apertures is shown in Fig. 2.2.

Let the instantaneous superimposed light intensity of n beams at a point (x, y) in the receiving plane be $I(x, y)$. The statistical properties of the light intensity I_k (k

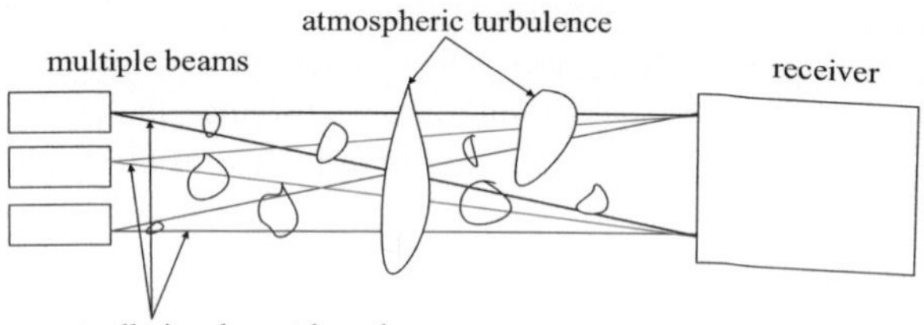

Fig. 2.1 Schematic diagram of multibeam emissions in a turbulent atmosphere

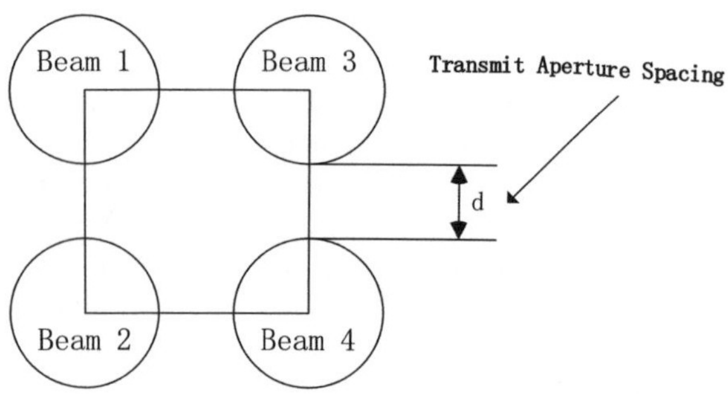

Fig. 2.2 Schematic diagram of multi-aperture emission geometry

= 1–n) of all laser beams at the receiving point obey the same negative exponential distribution. Let I_k (k = 1–n) be statistically independent and make the same contribution to $I(x, y)$; then,

$$I(x, y) = \sum_{k=1}^{n} I_k, \quad \langle I_k \rangle = \langle I \rangle / n. \tag{2.4}$$

The probability density function of $I(x, y)$ is

$$P_I(I) = \left(\frac{n}{\langle I \rangle} \right)^n \frac{I^{n-1} \exp(-nI/\langle I \rangle)}{\Gamma(n)} \quad I \geq 0. \tag{2.5}$$

For finite-aperture reception, the optical power detected by the receiver is the integral of the superimposed optical intensities over the plane area of the receive aperture. The integrated light intensity at the receiving plane is

$$I_s = \iint_R W_r(x, y) I(x, y) \mathrm{d}x\mathrm{d}y. \tag{2.6}$$

In the formula, $W_r(x, y)$ is the aperture weight function:

$$W_r(x, y) = \begin{cases} 1 & \sqrt{x^2 + y^2} \leq D/2 \\ 0 & \sqrt{x^2 + y^2} > D/2. \end{cases} \tag{2.7}$$

There is a certain coherence between the light intensities of each beam. When the distance d between the emission apertures increases, the coherence between I_k decreases, and the coherence increases continuously as it propagates in space; however, the turbulence causes the beam coherence to decrease.

If the coherence between the beams is strong, light and dark fringes will appear on the receiving surface, which will increase the flicker and fluctuation of the received light intensity. For a laser-communication system, the coherence between the transmitted beams should be minimized. Therefore, the distance between the transmitting apertures should be $d \geq \sqrt{\lambda L}$.

2.1.2 Light-Intensity Distribution of Multi-beam Emissions

Using multi-beam emissions, the logarithmic amplitude at the receiving end is obtained by adding more independent scattering elements than a single beam [11]. Assuming that the receiving aperture is sufficiently large and the aperture contains m mutually incoherent speckle elements, the integrated light contains a strong probability-density function [12]:

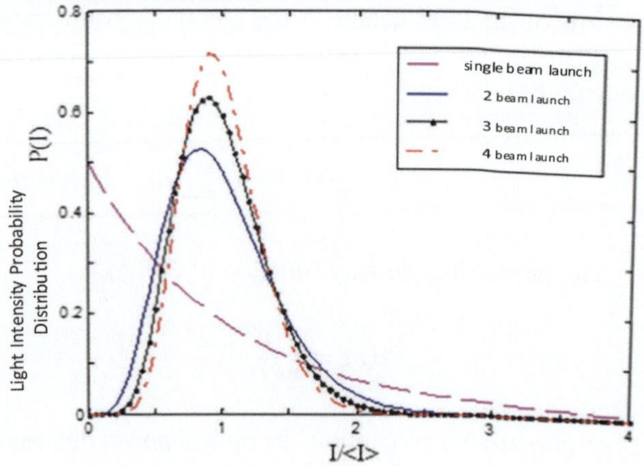

Fig. 2.3 Probability distribution of light intensity for different numbers of emitted beams, m = 1

$$P_{I_s}(I_s) = \left(\frac{nm}{\langle I_s \rangle}\right)^{nm} \frac{I_s^{nm-1} \exp\left(-\frac{nmI_s}{\langle I_s \rangle}\right)}{\Gamma(nm)} \quad I_s \geq 0, \tag{2.8}$$

where m is the number of mutually incoherent speckle elements [12]:

$$m \approx \begin{cases} (D_r/\rho_0)^2 & D_r \geq \rho_0 \\ 1 & otherwise. \end{cases} \tag{2.9}$$

Figure 2.3 shows the probability distribution of the light intensity emitted by different beams when the atmospheric refractive index is $C_n^2 = 3 \times 10^{-15}$. It can be seen that as the number of beams increases, the probability distribution of the light intensity gradually changes from a negative exponential distribution to a normal distribution. In the figure, speckle element m is 1. $n = 1$ represents the case of a single-beam emission.

Owing to the reduction in the coherence of each beam caused by turbulence, the number m of mutually incoherent speckle elements in the receiving aperture D_r is much greater than 1 after a long-distance laser transmission.

Figure 2.4 shows the probability distribution of the light intensity corresponding to different numbers of emitted beams when the atmospheric refractive index is $C_n^2 = 3 \times 10^{-15}$ and m is greater than 1. Compared with Fig. 2.3, it can be seen that with the increase of the number of speckle elements, the peak point of the light-intensity probability-density function is closer to the mean value, and the light-intensity probability curve is closer to a normal distribution.

Figure 2.5 shows the probability-distribution curve of the light intensity when four beams are emitted and different apertures are selected for reception. It can be seen that when the aperture diameter increases, the light-intensity probability distribution tends to be normal, and the peak value of the light-intensity probability also increases. When the beam propagation is through strong turbulence, the probability density of the light intensity I is [13, 14]

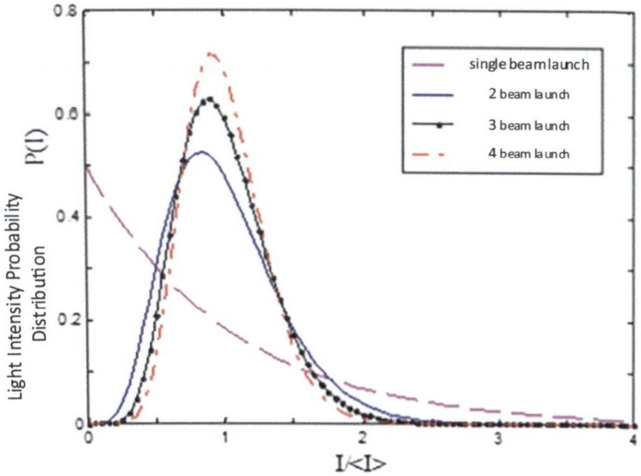

Fig. 2.4 Probability distribution of light intensity for different numbers of emitted beams, m > 1

$$P_{I_k}(I_k) = \frac{2}{\Gamma(m)} m^{(m+1)/2} I_k^{(m-1)/2} K_{m-1}[2(mI_k)^{1/2}]. \tag{2.10}$$

The light intensity of n beams at the receiving end has the same distribution.

$$I = I_1 + I_2 + \cdots + I_n = \sum_{k=0}^{n} I_k \tag{2.11}$$

The distribution I_k is

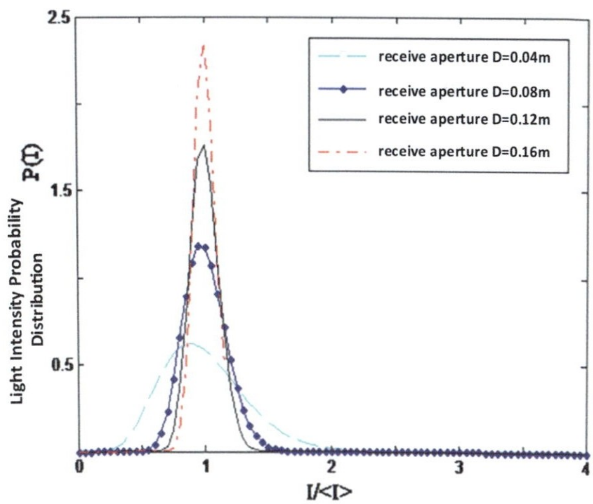

Fig. 2.5 Probability curve of light intensity when four beams transmit and receive with different aperture diameters

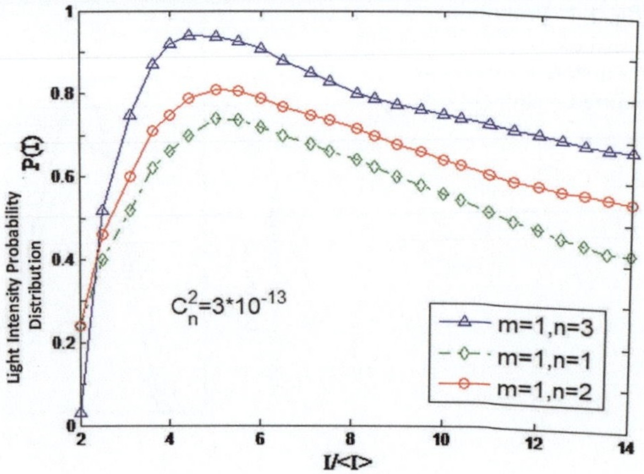

Fig. 2.6 Probability of light intensity when different numbers of beams are fired

$$\varphi_{I_k}(\omega) = \int_0^{\infty} \exp(j\omega I_k)p(I_k)dI_k = n^{(n-1/2)} \times (j\omega)^{-n} \times \exp(jn/\omega). \tag{2.12}$$

Therefore, in a strongly turbulent environment, the probability-density function of the light intensity emitted by multiple beams is

$$P_I(I) = \frac{1}{2\pi} \int_{-\infty}^{\infty} \exp(-j\omega I)\varphi_I(\omega)d\omega = m^{(m-1)n/2} \times I^{-mn/2} \times (mn)^{mn/2} \times K_{nm+1}[2(nmI)^{1/2}]. \tag{2.13}$$

Figure 2.6 shows the light-intensity probability curve under strongly turbulent conditions. Using a single beam, two beams, and three beams, it can be observed that with an increase in the number of emitted beams, the probability-distribution value of the light intensity gradually increases; however, the peak points are no longer concentrated with the increase in the light intensity.

2.1.3 Bit-Error Rate Characteristics of a Multi-beam Transmitting System

When a laser beam passes through a weak turbulence region, the probability-density distribution of the light-intensity fluctuation obeys a log-normal distribution, and when the beam is in a variance-saturated scintillation region, the light-intensity fluctuation obeys a negative exponential distribution. When the beam is transmitted in a strong turbulent region and the transmission distance is long, if the transmission distance L and the distance r between the emission lenses satisfy certain conditions, when the number of beams n increases, the atmospheric scintillation obeys the log-normal distribution.

The light-intensity flicker has a certain relationship with the transmission distance, number of emitted beams, and distance r between the emission apertures. The light-intensity flicker causes a noise current, which increases the bit-error rate of the system. Multibeam emissions can be used for the same signal transmission to ensure the reliability of the system. When a multibeam transmission is used, the logarithmic amplitude variance at the receiving end is [15]

$$\sigma_{n\chi}^2 = \frac{1}{n}(1-\gamma)\sigma_\chi^2 + \gamma\sigma_\chi^2. \tag{2.14}$$

In the above formula, n is the number of emitted beams, and r is the normalized correlation coefficient.

$$\gamma = \frac{\int_0^\infty J_0\left[(xk/L)^{1/2}\frac{1}{r}\right](1-\sin x/x)^{-4/3}\mathrm{d}x}{\int_0^\infty (1-\sin x/x)x^{-4/3}\mathrm{d}x} \tag{2.15}$$

Figure 2.7 shows the relationship between the light-intensity fluctuation variance and correlation coefficient when different beams are emitted. It can be seen that the larger the correlation coefficient, the larger the light-intensity fluctuation variance. When the correlation coefficient is 1, each light beam becomes perfectly coherent; however, there is no perfectly coherent light in practice. The figure shows the relationship between the light-intensity fluctuation variance and correlation coefficient when two, three, and four beams of light are used (Fig. 2.8).

The distance between each transmitted-beam aperture is an important design parameter. To avoid interference, the distance between any two emission apertures should be greater than the coherent wavelength γ_0 of the emitted laser light. The coherent wavelength of the emitted light source is $\gamma_0 = \sqrt{\lambda L}$. From the definition

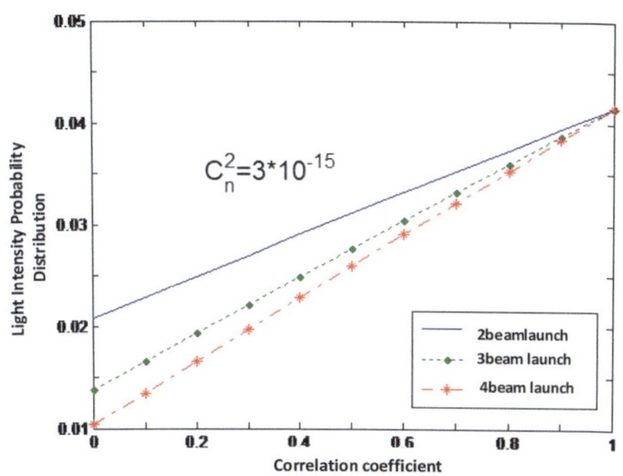

Fig. 2.7 Relationship between the light-intensity fluctuation variance and correlation coefficient when two, three, and four beams are used

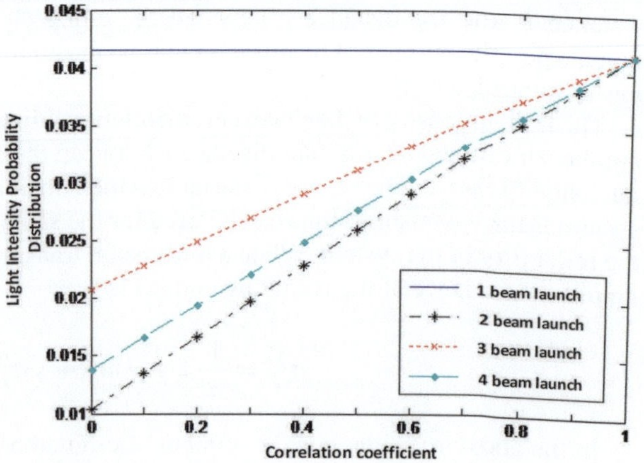

Fig. 2.8 Relationship between the light-intensity fluctuation variance and correlation coefficient when different numbers of beams are emitted

of the bit-error rate and Eq. (2.14), the bit-error rate of a multibeam transmission system can be obtained:

$$BER = \frac{1}{2}\left[erfc\left(\frac{4}{\sqrt{2}\frac{1}{n}(1-\gamma)\sigma_{\ln\chi}^{2} + \gamma\sigma_{\ln\chi}}\right)\right]. \quad (2.16)$$

Figures 2.9 and 2.10 show the variation in the system bit-error rate with the transmission distance. With an increase in the transmission distance, the bit-error rate also increases, and the increase in the transmitted beam reduces the bit-error rate of the system. The performance of the communication system is effectively improved when multiple beams are used for transmission, and certain transmission conditions are met.

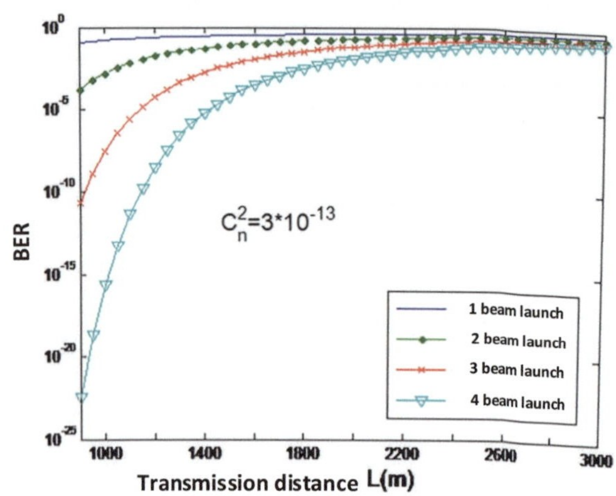

Fig. 2.9 Variation of the bit-error rate of different numbers of transmitted beams with transmission distance under strong turbulence

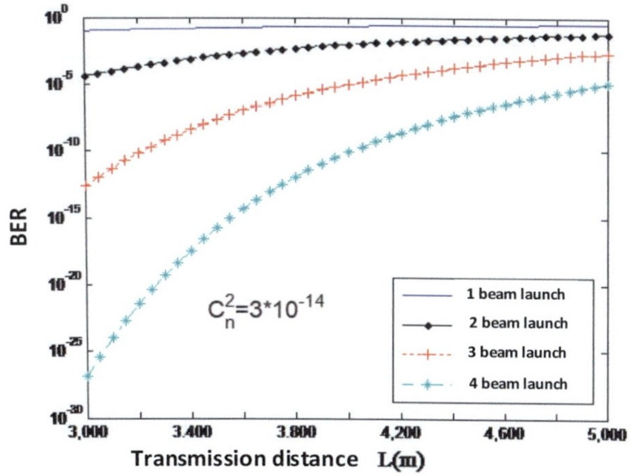

Fig. 2.10 Variation of the bit-error rate of different numbers of transmitted beams with transmission distance under moderate turbulence

Figure 2.11 shows the variation of the bit-error rate of a multibeam emission with the correlation coefficient. With the increase of the correlation coefficient, the bit-error rate also increases. Figure 2.12 shows the relationship between the bit-error rate, correlation coefficient, and transmission distance when three beams of light are emitted. On the one hand, it shows that with the increase of the transmission distance, the bit-error rate increases; on the other hand, it shows that the bit-error rate increases with the increase of the correlation coefficient.

Under the same turbulence environment, when the emitted beam increases, the light-intensity probability curve tends to be more normal, and the bit-error rate performance is also improved. With an increase in turbulence, the light-intensity probability curve no longer obeys a normal distribution, and multibeam emissions can improve the effect of turbulence on light-intensity fluctuations.

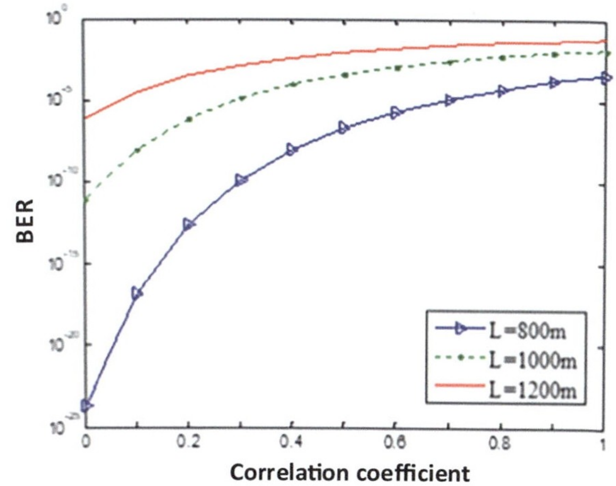

Fig. 2.11 Curve of the bit-error rate of a multibeam transmission as a function of the correlation coefficient

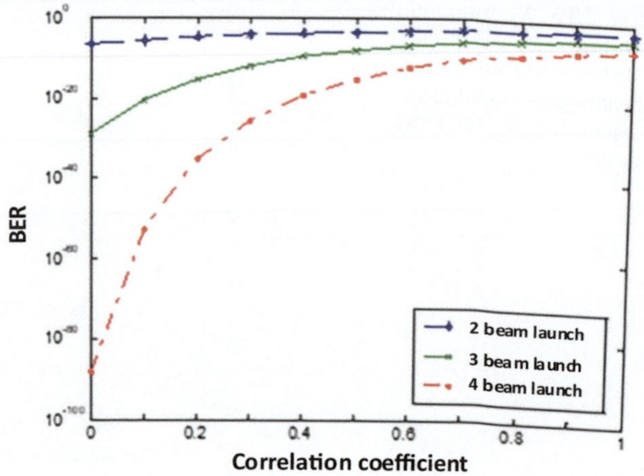

Fig. 2.12 Relationship between the bit-error rate of a three-beam transmission and the correlation coefficient and transmission distance

As the correlation coefficient increases, the variance in the light-intensity fluctuation also increases. However, when the emission beam increases, the variance of the light-intensity fluctuation gradually decreases, and when the correlation coefficient $r = 1$, the effect of a multibeam emission is the same as that of a single-beam emission. The results show that the bit-error rate performance of a multibeam transmission is much better than that of a single-beam transmission [16, 17].

2.2 Multi-aperture Reception

Multi-aperture receiving has an effect similar to that of large-aperture receiving, and can produce an aperture-smoothing effect on the light-intensity scintillation. Aperture smoothing (aperture averaging) can reduce the amplitude of the light-intensity flicker, thereby reducing the bit-error rate performance of the receiving system.

Similar to multibeam transmission, multi-aperture receiving technology also has the effect of space diversity. Its basic idea is to obtain signals from different paths at the receiving end, combine these signals appropriately to form a total received signal, and then detect the signal. Through the diversity function, the channel fading can be suppressed and the performance of the receiving system can be improved. Multi-aperture receiving technology can produce an aperture-smoothing effect on the fluctuation of the optical signal, which is an effective method of suppressing the influence of atmospheric turbulence.

2.2.1 *Equivalent Averaging Factor for Multi-aperture Receivers*

Figure 2.13 shows the geometric configuration of a multi-aperture receiver. r_1, r_2, and r_3 represent the center-position vectors of the three receiving apertures, and p represents a position vector on the sub-receiving apertures. The receive apertures in Fig. 2.13 may be larger.

Figure 2.14 shows a schematic diagram of the structure of the three-aperture receiving system.

Assuming that the shape and size of the N sub-apertures are the same, the total instantaneous luminous flux received by the receiving system including the N apertures is

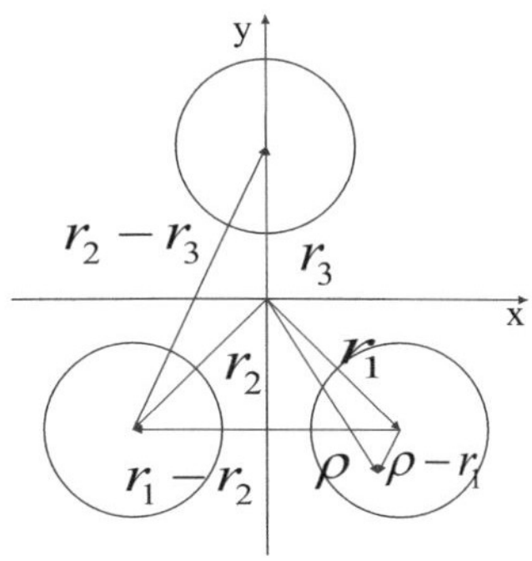

Fig. 2.13 Schematic diagram of three-aperture receiving geometry

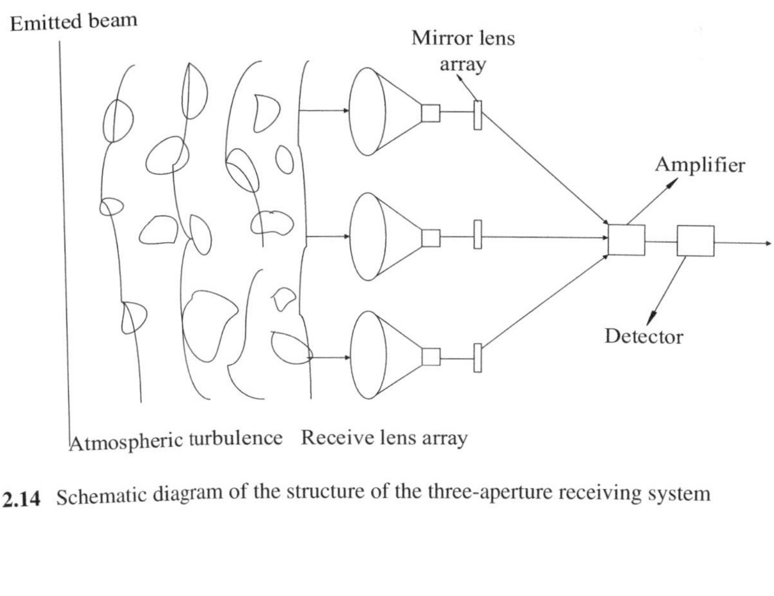

Fig. 2.14 Schematic diagram of the structure of the three-aperture receiving system

$$P = \sum_{i=1}^{N} \int_{S_i} T(D, \rho - r_i) I(\rho) \mathrm{d}\rho. \tag{2.17}$$

The aperture function for a circular aperture is

$$T(D, \rho') = \begin{cases} 1 & |\rho'| \leq D/2 \\ 0 & |\rho'| > D/2. \end{cases} \tag{2.18}$$

Substituting variables for Eq. (2.17), we can obtain

$$P = \int_{-\infty}^{\infty} T(D, \mathrm{R}) \sum_{i=1}^{N} I(\mathrm{R} + r_i) \mathrm{dR}. \tag{2.19}$$

The average total light intensity (i.e., the first moment of P) received by N sub-apertures can be written as

$$\langle P \rangle = \int_{-\infty}^{\infty} T(D, \mathrm{R}) \sum_{i=1}^{N} \langle I(\mathrm{R} + r_i) \rangle \mathrm{dR}. \tag{2.20}$$

From Eqs. (2.19) and (2.20), we obtain.

$$\begin{aligned} \sigma_P^2 = & \int_S \int_{S'} T(D, R') T(D, \mathrm{R}) \sum_{i=1}^{N} \sum_{j=1}^{N} \langle I(\mathrm{R} + r_i) \times I(R' + r_j) \rangle \mathrm{dR}\mathrm{d}R' \\ & - \int_S \int_{S'} T(D, R') T(D, \mathrm{R}) \sum_{i=1}^{N} \sum_{j=1}^{N} \left[\langle I(\mathrm{R} + r_i) \rangle \times \langle I(R' + r_j) \rangle \right] \mathrm{dR}\mathrm{d}R'. \end{aligned} \tag{2.21}$$

If we substitute $x = \mathrm{R} - R'$ in the above formula, it can be transformed into

$$\sigma_P^2 = \sum_{i=1}^{N} \sum_{j=1}^{N} \int_S \int_{S'} T(D, R) T(D, R - x) C_I(x + r_i - r_j) \mathrm{dR}\mathrm{d}x. \tag{2.22}$$

Using the definition of the aperture-averaging factor of an aperture receiving system [18], the formula for the average factor of a multi-aperture receiving system can be obtained:

$$A = \frac{\sigma_P^2}{\left(\frac{N\pi D^2}{4} \right)^2 C_I(0)}. \tag{2.23}$$

From Eqs. (2.22) and (2.23), we obtain

$$A = \frac{16}{(N\pi D^2)^2} \sum_{i=1}^{N} \sum_{j=1}^{N} \int_S \int_{S'} T(D, \mathrm{R}) T(D, \mathrm{R} - x) C_I(|x + r_i - r_j|) \mathrm{dRd}x, \quad (2.24)$$

where $C_I(\cdot)$ is the spatial covariance function of the light intensity, which can be defined as

$$\begin{aligned} C_I(|x + r_i - r_j|) &= C_I(R + r_i, R' + r_j) \\ &= \langle I(R + r_i) I(R' + r_j) \rangle - \langle I(R + r_i) \rangle \langle I(R' + r_j) \rangle. \end{aligned} \quad (2.25)$$

$K(x)$ is the modulation-transfer function of the receive aperture, as follows:

$$K(x, D) = \begin{cases} \frac{D^2}{2} \left\{ \cos^{-1}\left(\frac{x}{D}\right) - \left(\frac{x}{D}\right) \left[1 - \left(\frac{x}{D}\right)^2 \right]^{1/2} \right\} & x \le D \\ 0 & x > D. \end{cases} \quad (2.26)$$

Then, the expression for the aperture-averaging factor of a multi-aperture receiving system can be obtained as follows:

$$A = \frac{16}{(N\pi D^2)^2} \sum_{i=1}^{N} \sum_{j=1}^{N} C_P(|r_i - r_j|) \quad (2.27)$$

$$C_P(|\rho + x|) = \int_0^D K(x) C_I(|x + \rho|) / C_I(0) \mathrm{d}x. \quad (2.28)$$

Based on the large-scale fluctuation modulation theory proposed by Andrews [19], the light-intensity fluctuation variance function can be written as

$$C_I(\rho + x) = \exp[C_{\ln x}(\rho + x) + C_{\ln y}(\rho + x)] - 1. \quad (2.29)$$

In the case of plane-wave propagation, the large-scale log-intensity covariance can be expressed as [18]

$$\begin{aligned} C_{\ln x}(\rho + x) = 1.06\sigma_l^2 &\int_0^1 \int_0^\infty \eta^{-11/6} \exp(-\eta/\eta_x) J_0\left((\rho + \rho')\sqrt{k\eta/L}\right) \\ &\times (1 - \cos \eta\xi) \mathrm{d}\eta \mathrm{d}\xi \\ &\approx 0.15\sigma_l^2 \eta_x^{7/6} {}_1F_1(7/6; 1; -0.25k(\rho + x)^2 \eta_x/L). \end{aligned} \quad (2.30)$$

The small-scale log-intensity covariance can be written as [18]

$$
\begin{aligned}
C_{\ln y}(\rho + x) &= 1.06\sigma_l^2 \int_0^1 \int_0^8 (\eta + \eta_y)^{-11/6} J_0((\rho + x)\sqrt{k\eta/L})(1 - \cos \eta\xi)\mathrm{d}\eta\mathrm{d}\xi \\
&\approx 1.265\sigma_l^2 \left(\frac{k(\rho + x)^2}{L\eta_y}\right)^{5/12} \mathrm{K}_{5/6}\left[\left(\frac{k(\rho + x)^2\eta_y}{L}\right)^{1/2}\right]
\end{aligned} \tag{2.31}
$$

$$
\begin{aligned}
C_I(\rho + x) = \exp\Big\{&0.15\sigma_l^2\eta_x^{7/6}\,{}_1F_1(7/6;\,1;\,-0.25k(\rho + x)^2\eta_x/L) \\
&+ 1.265\sigma_l^2\left(\frac{k(\rho + x)^2}{L\eta_y}\right)^{5/12} \mathrm{K}_{5/6}\left[\left(\frac{k(\rho + x)^2\eta_y}{L}\right)^{1/2}\right]\Big\} - 1.
\end{aligned} \tag{2.32}
$$

We substitute the above formula into Eq. (2.28). Then, from Eq. (2.27), the expression for the aperture-averaging factor of a multi-aperture receiving system can be obtained.

Figure 2.15 shows the aperture-averaging factor when receiving with three apertures in different turbulent environments. The transmission distance was 1000 m, laser wavelength was 810 nm, and number of receiving apertures was n. It can be seen that:

(1) With an increase in the pore diameter, the pore-size average factor decreases rapidly; however, when the pore size increases to a certain extent, the decrease in the pore-size average factor tends to be slight.
(2) For weak turbulence, the aperture-averaging factor of the system decreases rapidly with the increase in aperture, and is much less than 0.1. The aperture-smoothing effect is significant, and is also obvious for medium and strong turbulent flows.
(3) The performance of the three-aperture receiver is better than that of a single large-aperture receiver.

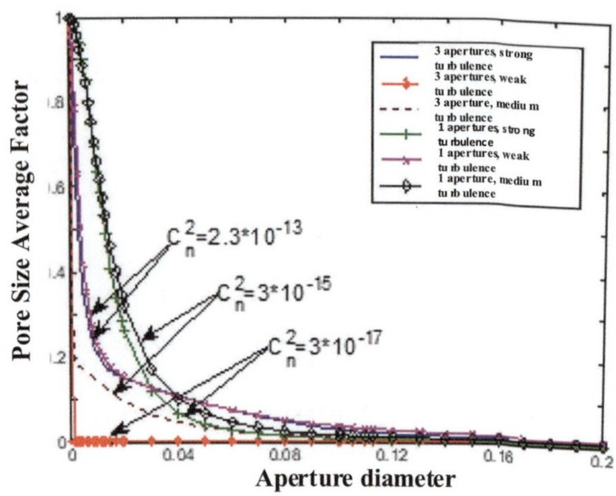

Fig. 2.15 Performance comparison of single-aperture and triple-aperture receiver systems under different turbulence conditions

The aperture averaging factor was investigated when the spacing between the respective receiving apertures was varied: 0.02, 0.03, 0.04, 0.05, and 0.06 m. As shown in Fig. 2.16, each curve gradually becomes smaller as the aperture spacing increases; however, the magnitude of the reduction becomes more gradual. When the aperture spacing increases to 0.08–0.1 m, the variation in the aperture-averaging factor is very small, and there is no need to increase the aperture spacing.

Figure 2.17 shows the calculation results of the aperture-averaging factor when received by two, three, four, and five apertures, when the receiving aperture and aperture spacing are fixed values under strong turbulence. It can be observed that with an increase in the number of receiving apertures, the aperture-averaging factor decreases.

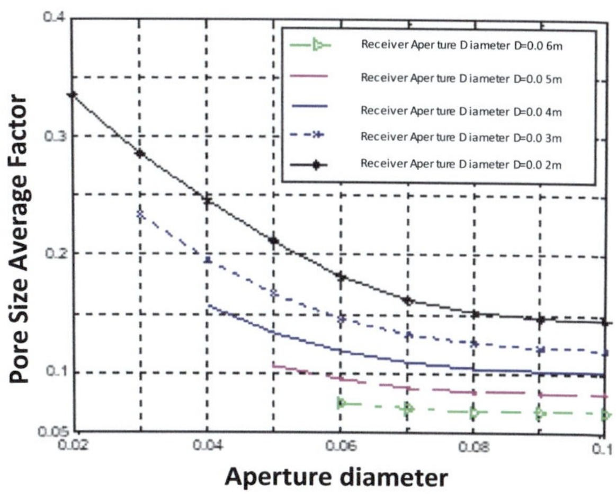

Fig. 2.16 Relationship between the aperture-averaging factor for different receiver apertures as a function of aperture spacing

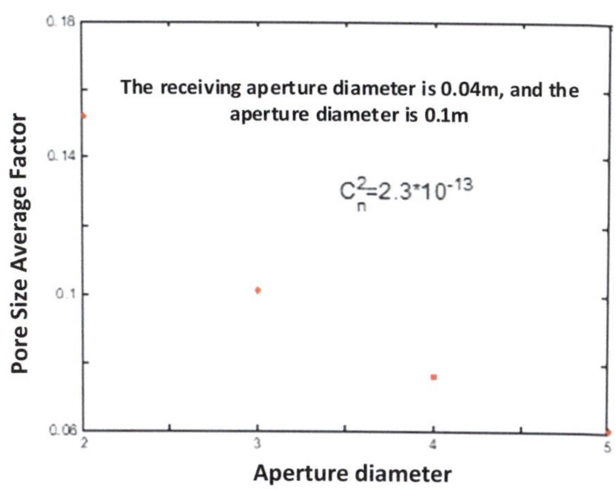

Fig. 2.17 Aperture-averaging factor values for different numbers of receive apertures

Figure 2.18 shows the average factor curve of apertures corresponding to different numbers of receiving apertures under strong turbulence conditions. The aperture spacing is 0.4 m. The curve in the figure shows that the larger the number of apertures, the smaller the aperture-averaging factor. When the aperture diameter is 0.12 m, the aperture-averaging factor tends to be flat, and the effect of increasing the aperture area is insignificant.

Figure 2.19 shows the average factor curve of apertures corresponding to different numbers of receiving apertures under moderate turbulence conditions, and the aperture spacing is 0.4 m. Compared with a strongly turbulent environment, the spatial averaging factor is much smaller under moderate turbulence.

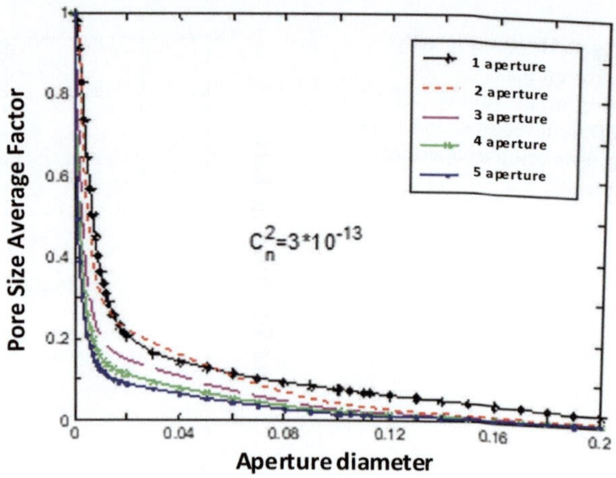

Fig. 2.18 Aperture-averaging factor curves for different numbers of receive apertures in strong turbulence conditions

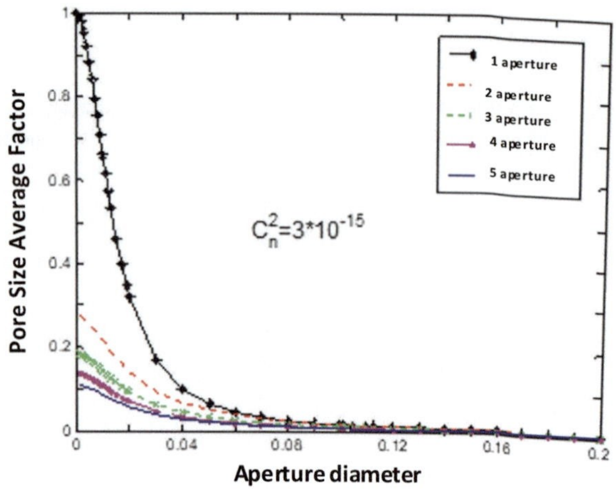

Fig. 2.19 Aperture-averaging factor curves for different numbers of receiver apertures in moderate turbulence conditions

2.2.2 Analysis of Bit-Error Rate Performance

The bit-error rate (BER) of the system depends on the average received power, scintillation over the aperture, and received noise (mainly thermal and shot), which vary with atmospheric turbulence.

After the optical signal passes through atmospheric turbulence, the output signal-to-noise ratio (SNR) of the detector is [20]

$$\langle SNR \rangle = \frac{i_s}{\sigma_N} = \frac{SNR_0}{\sqrt{\frac{P_{S0}}{\langle P_S \rangle} + A\sigma_I^2 SNR_0^2}}. \tag{2.33}$$

For a pulse-position modulation (PPM) system, when there is no turbulence effect, the BER of the system is as follows [21, 22]:

$$P_{e\_PPM} = \frac{1}{2} erfc\left(\frac{1}{2\sqrt{2}}\sqrt{\frac{L}{2}\log_2 LSNR}\right). \tag{2.34}$$

Under the influence of turbulence, the bit-error rate of a PPM system averages the probability-density function of the light-intensity fluctuation; that is, the probability-density function of the light-intensity fluctuation from a weak to strong turbulent region is modeled by the gamma–gamma distribution [19]:

$$P(I) = \frac{2(\alpha\beta)^{(\alpha+\beta)/2}}{\Gamma(\alpha)\Gamma(\beta)} I^{(\alpha+\beta)/2-1} K_{\alpha-\beta}[2(\alpha\beta I)^{1/2}], \quad I > 0 \tag{2.35}$$

$$\sigma_{\ln x}^2(D) = \frac{0.16\sigma_1^2 \eta_x^{7/6}}{(1 + d^2\eta_x/4)^{7/6}}. \tag{2.36}$$

It is related to the receiving aperture size:

$$\sigma_{\ln y}^2(D) = \frac{1.27\sigma_1^2 \eta_y^{-5/6}}{1 + 0.3d^2\eta_y}. \tag{2.37}$$

From Eqs. (2.34) and (2.35), the general expression of the bit-error rate of the PPM signal can be obtained as

$$P_e = \frac{(\alpha\beta)^{(\alpha+\beta)/2}}{\Gamma(\alpha)\Gamma(\beta)} \cdot \int_0^\infty erfc\left(\frac{1}{2\sqrt{2}}\sqrt{\frac{L}{2}\log_2 L\langle SNR \rangle s}\right) \times s^{(\alpha+\beta)/2-1} \cdot K_{(\alpha-\beta)}\left(2\sqrt{\alpha\beta s}\right) ds. \tag{2.38}$$

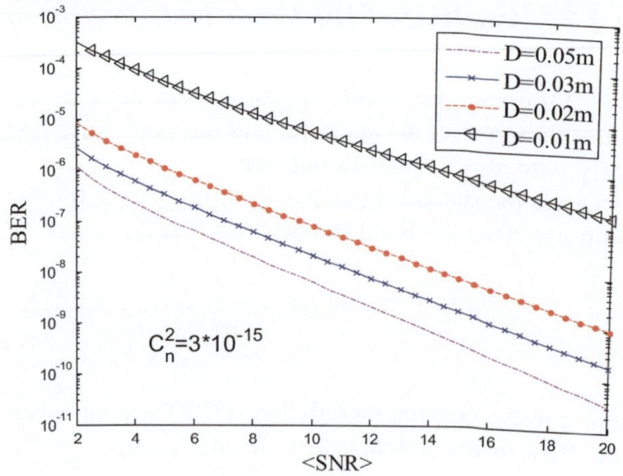

Fig. 2.20 Bit-error rate and signal-to-noise ratio of different receiving apertures

Substituting Eq. (2.33) into the above equation, the bit-error rate of a multi-aperture receiving system in a turbulent atmosphere can be deduced as follows:

$$P_e = \frac{(\alpha\beta)^{(\alpha+\beta)/2}}{\Gamma(\alpha)\Gamma(\beta)} \cdot \int_0^{\infty} erfc\left(\frac{1}{2\sqrt{2}}\sqrt{\frac{L}{2}\log_2 L \frac{SNR_0}{\sqrt{\frac{P_{S0}}{\langle P_S\rangle} + A\sigma_I^2 SNR_0^2}} s}\right) \times s^{(\alpha+\beta)/2-1} \cdot K_{(\alpha-\beta)}\left(2\sqrt{\alpha\beta s}\right)\mathrm{d}s. \quad (2.39)$$

Figure 2.20 shows the BER curve of a 256-PPM signal transmission when receiving with three apertures. The investigated receiving-aperture sizes are 0.01 m, 0.02 m, 0.03 m, and 0.05 m, and the atmospheric refractive-index structure constant is 2.3×10^{-13} $m^{-2/3}$; with the increase in the receiving aperture, the bit-error rate gradually decreases.

As can be seen from Fig. 2.21, when the number of receivers increases, the bit-error rate curve also decreases; however, the decrease is relatively shallow, and the bit-error rate decreases to a small order of magnitude with an increase in the signal-to-noise ratio (ordinate in the figure; logarithmic scale).

Figure 2.22 shows the bit-error rate curve of the 256-PPM transmission signal. Comparing Figs. 2.21 and 2.22, it can be seen that the bit-error rate decreases rapidly as the turbulence intensity decreases.

Figure 2.23 shows the relationship between the signal-to-noise ratio and bit-error rate when using 8-PPM, 16-PPM, and 256-PPM signal transmissions. It can be seen that when the number of time slots of the PPM signal increases, the bit-error rate of the received signal decreases significantly.

The analysis results show that in the same turbulent environment, when the receiving area is the same, the performance of multi-aperture receiving is better than that of single-aperture receiving. When receiving with the same number of

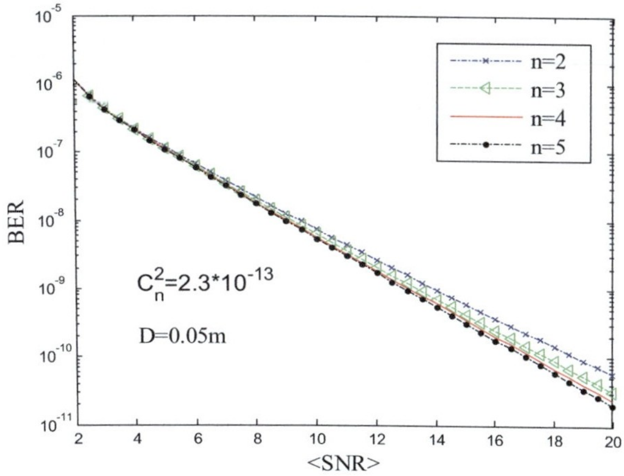

Fig. 2.21 Bit-error rate received with different numbers of receiving apertures

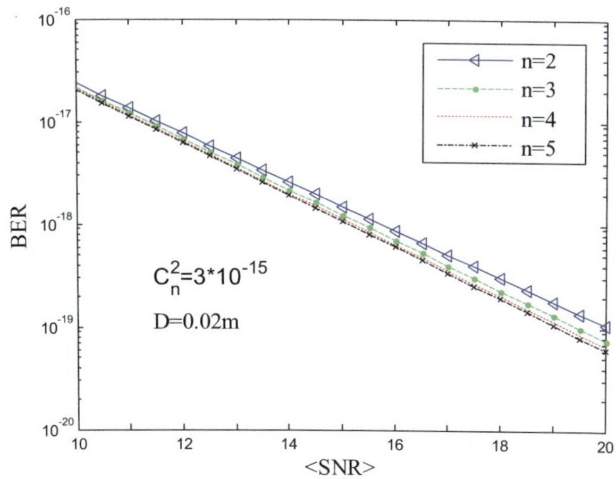

Fig. 2.22 BER curves of different numbers of receiving apertures

apertures, the aperture-averaging factor gradually decreases with the weakening of the turbulence. When it decreases to a certain extent, it tends to flatten, and continuing to increase the aperture area will not further suppress turbulence.

The bit-error rate formula of a multi-aperture receiving system using PPM modulation in a turbulent environment was deduced, the bit-error rate characteristics of receiving systems with different numbers and different receiving areas were analyzed, and the bit-error rate under different turbulent conditions was analyzed, as well as the relationship with the signal-to-noise ratio. Finally, the bit-error rate characteristics of PPM signals with different data-bit lengths are compared, which provides a reference for the design of wireless laser receiving systems [23].

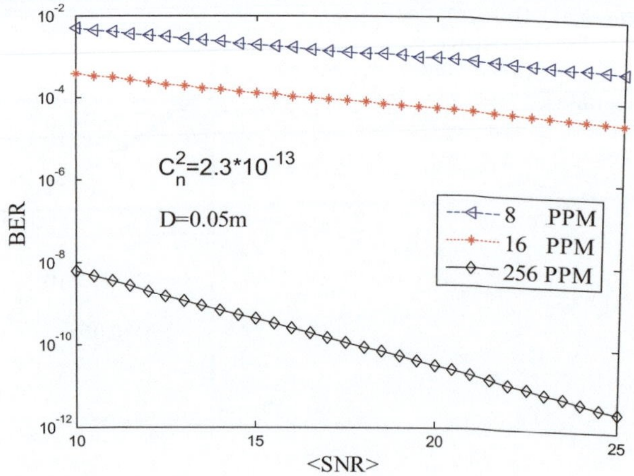

Fig. 2.23 BER curves of different PPM transmission signals at the receiving end

2.3 Multi-beam Emission/Multi-aperture Receiving Experiments

This section introduces related experiments on multibeam emission and multibeam reception. The main purpose of the experiment was to measure and analyze the light-intensity variance at the receiving end and to compare the light-intensity variance of multi-transmission and single-reception with single transmission and single reception, and a single beam at the transmitter.

When light is emitted, the receiving end detects the light-intensity variance of single-, double-, and triple-aperture reception, to compare the performance of multibeam and single-beam transmissions, and the transmission performance of multi-aperture and single-aperture reception.

2.3.1 Composition and Experiment on a Multibeam Emission System

Figure 2.24 shows the composition of the experimental system for multibeam emission. The entire system consists of two parts: transmitter and receiver. In this experiment, the transmitter used two laser beams with the same wavelength, the laser used green light with a 532-nm wavelength, and the receiver used a single aperture. The diameter of the receiving lens is 0.1 m.

The laser is emitted through a Cassegrain telescope and the emitted beam is transmitted through the atmosphere. At the receiving end, it converges into a beam of light through the receiving lens, and then is detected by the detector. A data-acquisition card connected to the computer acquires the measured data, which are used to calculate the light-intensity variance.

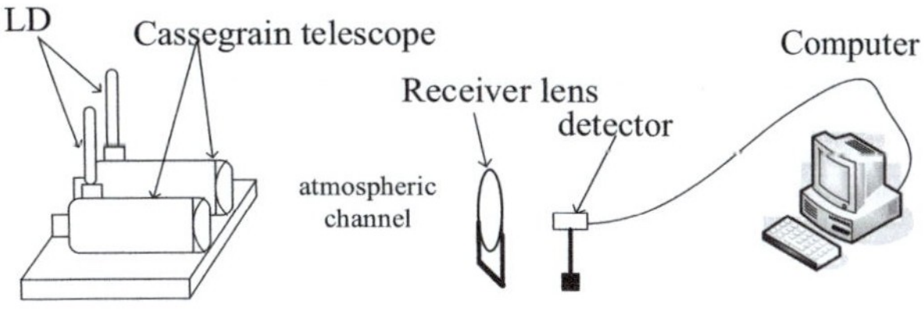

Fig. 2.24 Composition of the multibeam-emission experimental system

The experiment was conducted at 9:00 PM on November 24, 2011 between the fifth and sixth teaching floors at Xi'an University of Technology. During the experiment, the weather was sunny and breezy.

Because the receiving end was a single receiver, only one of the receiving apertures was used in the experiment. In this experiment, the light-intensity data of one and two emitted beams of light were collected to compare the performance of single-transmission and multi-transmission systems.

The key points in the experiment were the alignment of the beam and maintaining its stability. Because this was a long-distance experiment, the laser beam took a long time to align. Additionally, the laser beam needed to be relatively stable at the receiving end for data acquisition. Otherwise, the measured data exhibited increased jitter.

Figure 2.25 shows the sampling voltage measured when a single beam transmits a single aperture and receives it. The sampling frequency was 2 MHz, and 3000 data points were acquired in the experiment. The image shows discrete points. It can be seen from the figure that most of the data are concentrated around 47, and a small number of data points are slightly off-center, which is caused by the flickering of the light intensity.

Figure 2.26 shows the sampled voltage values when two beams of light were used for transmission and a single lens was used at the receiving end. It can be seen

Fig. 2.25 Measured voltage values for single-transmit / single-receive

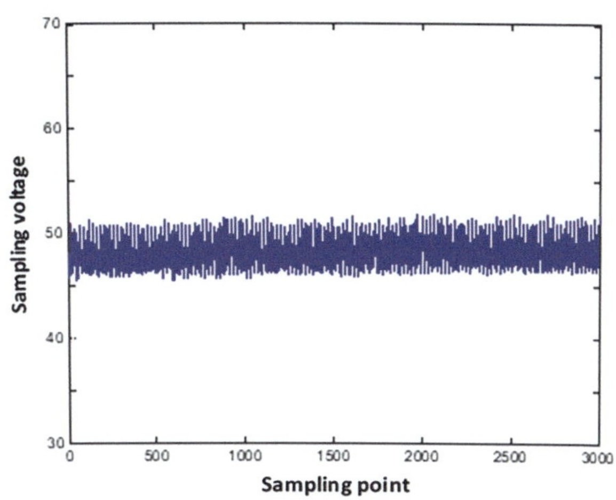

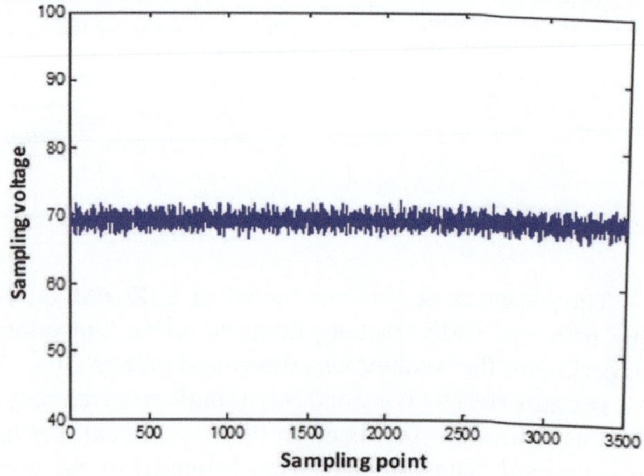

Fig. 2.26 Measured voltage values for dual-beam transmit / single-aperture receive

from the figure that, compared with single-beam transmission, the voltage fluctuation when using two beams of light is reduced and the signal is stabilized.

According to theoretical calculations, the light-intensity variance values of single-transmission–single-reception and dual-transmission–single-reception are 3.4001 and 2.3800, respectively, and the calculated variances of the experimentally measured data are 1.8815 and 0.8852, respectively. Theoretical and experimental measurement data show that the variance of the light intensity emitted by a double beam is smaller than that of a single-beam, and using a multibeam emission system can reduce the light-intensity fluctuation, which is beneficial for improving the performance of communication systems.

2.3.2 *Multi-aperture Receiving-System Composition and Related Experiments*

Figure 2.27 shows a schematic diagram of a multi-aperture receiving system. In the figure, the transmitting end is mainly composed of a Cassegrain telescope and laser. The receiving end is composed of a receiving lens with three apertures, beam-converging lens array, detector, and a computer for data acquisition and processing. The distance between the three lenses is 0.1 m, and the diameter of the lens is 0.1 m.

When the laser beam at the transmitting end reaches the surface of the receiving lens and passes through the lens, three beams of light appear. The three beams of light must be combined into one beam by a lens and fed into a mirror array that converges the beams. However, the optical path is more complicated. The key point is to keep the emitted beam stable; otherwise, the beams are difficult to combine.

This experiment was carried out at 9:00 pm on October 25, 2011, when the weather was cloudy. The laser used in the experiment was green light with a 532-nm wavelength, and the sampling frequency of the data-acquisition card at the receiving end

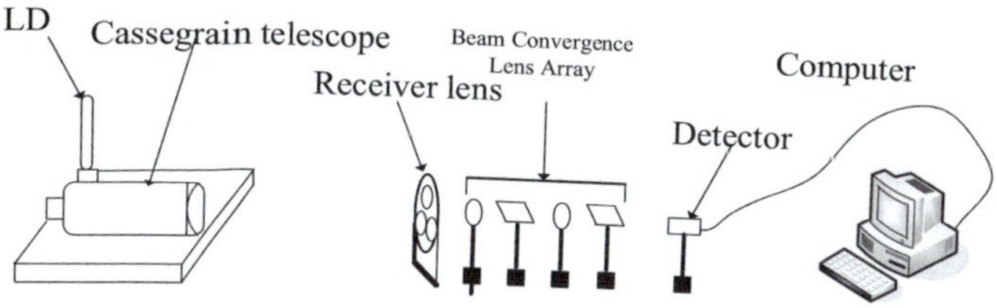

Fig. 2.27 Structure diagram of a multi-aperture receiving system

was 40 MHz. When the beam passed through the three-aperture receiving lens, data were received by the detector and collected by the computer.

During the experiment, the data received by one, two, and three apertures were measured in two groups. Figure 2.28 shows one of the sampling voltage values when a single aperture was used for reception. Figure 2.29 shows one of the sampled voltage values during the two-aperture reception. Figure 2.30 shows one of the sampled voltage values when three apertures were received.

Comparing single-aperture, two-aperture, and three-aperture receiving, it can be observed that as the number of apertures increases, the measured data jitter decreases, and the overall stability improves. The variance was calculated according to the data measured in this experiment, and the two groups of data in each case were calculated.

The measured variances of single-aperture reception were 5.9763 and 5.717, and the theoretically calculated variance was 4.8638. When receiving with two apertures, the calculated variances of the experimental data were 2.8183 and 2.8181 and the theoretically calculated value was 2.0372. When using three-aperture receiving, the two groups of variances calculated from the experimental data were 0.1042 and 0.1050, while the theoretically calculated variances was 0.5998.

Overall, it can be concluded that the light-intensity variance gradually decreases with an increase in the number of apertures, indicating that multi-aperture receiving

Fig. 2.28 Sampling-voltage values for single-aperture reception

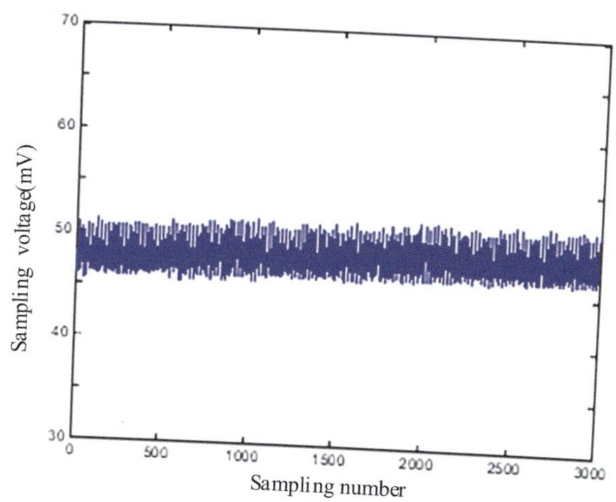

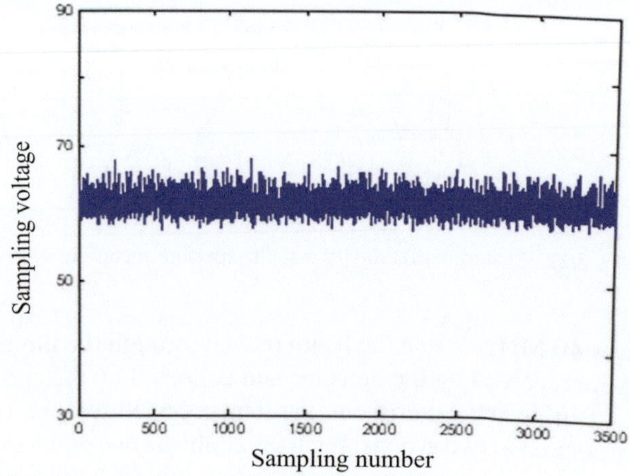

Fig. 2.29 Sampling-voltage values for two-aperture reception

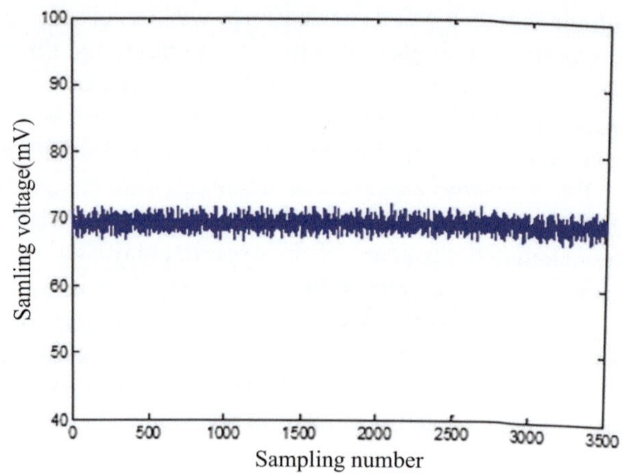

Fig. 2.30 Sampling-voltage values for three-aperture reception

can more effectively suppress the influence of turbulence and improve the laser-transmission performance.

2.4 Summary

We conducted related experiments on multibeam emission and multi-aperture receiving, calculated the light-intensity variance of the data measured by the experiment, and compared the light-intensity fluctuation variance of different numbers of light beams when transmitting, and the light-intensity variance of receiving with different numbers of receiving apertures. We concluded that multibeam transmitting and multi-aperture receiving systems could suppress the turbulence effect more effectively than single-beam transmitting and single-aperture receiving systems, and improve the performance of wireless laser-communication systems.

References

1. Churnside JH (1991) Aperture averaging of optical scintillations in the turbulent atmosphere. Appl Opt 30(15):1982–1994
2. Wang H, Ke X (2008) MIMO atmospheric laser communication and its key technologies. Laser J 29(3):52–54
3. Wang H, Ke X (2008) Free-space optical communication based on vertical layered space-time coding. China Laser 35(6):874–878
4. Andrews LC, Phillips RL, Hopen CY (2000) Aperture averaging of optical scintillations: power fluctuaions and the temporal spectrum. Waves Random Media 10(1):53–70
5. Ma D, Wei J, Zhuang Z (2004) Multi-beam transmission performance analysis and channel modeling in atmospheric laser communication. J Opt 24(8):1020–1024
6. Minch JR, Gervais DR, Townsend DJ (2006) Adaptive transceivers for mobile free-space optical communications. In: IEEE Military communications conference, 1–5
7. Haas SM, Shapiro JH, Tarokh V (2002) Space-time codes for wireless optical communications EURASIP J Appl Signal Process 3:211–220
8. Ma D, Zhuang Z (2002) Research on atmospheric scintillation in space laser communication. China Laser A29(3):277–280
9. Wei Z, Chen H (2010) Design of receiving aperture for wireless optical communication based on aperture smoothing effect. Electron Des Eng 2010-04
10. Andrews LC, Phillips RL (1985) I-K distribution as a universal propagation model of laser beams in atmospheric turbulence. Opt Soc Am 89(2):160–163
11. Li D (2011) Research on modeling multi-beam transmission channel under the atmospheric turbulence. JCIT: J Convergence Inf Technol 6(2):183–190
12. Tyson RK (2002) Bit-error rate for free-space adaptive optics laser communications. J Opt Soc Am A 19:753–758
13. Dashen R (1979) Path integrals for wave in random media. Math Phys 20(5):894–920
14. Andrew LC, Halash MA, Hopen CY (2001) Theory of optical scintillation: Gaussian-beam wave model. Wave Random Media 11(2):271–291
15. Ke X, Song P, Pei G (2011) Research on multi-aperture receiving technology in wireless laser communication. J Opt 31(12):22–28
16. Chen D, Ke X (2012) Research on diversity receiving technology of subcarrier modulation wireless optical communication. J Commun 33(08):128–133
17. Goodman JW (1985) Statistical optics. Wiley, p 222
18. Zhang Y (2002) Light propagation and imaging in random media. National Defense Industry Press, Beijing, pp 228–229
19. Vetelino FS, Young C, Andrews L (2007) Aperture averaging effects on the probability density of irradiance fluctuations in moderate-to-strong turbulence. Appl Opt 46(11):2099–2108
20. Andrews LC, Phillips RL, Hopen CY (2001) Laser beam scintillation with applications. SPIE Optical Engineering Press, pp 211–212
21. Lee EA, Messerschmitt DG (1994) Digital communication. Kluwer Academic Publishers, Boston, pp 178–266
22. Barry JR (1997) Wireless infrared communication. Proc IEEE 85(2):265–298
23. Ke X, Chen J, Pei G (2012) Research on multi-beam emission technology in wireless laser communication. Optoelectron Eng 39(07):1–7

Chapter 3
Channel Model and Channel Capacity

When lasers are transmitted in atmospheric channels, they are affected by atmospheric turbulence, rain, fog, and haze, which increase the bit-error rate (BER) of the communication system and, in severe cases, lead to communication interruptions. To effectively suppress the effect of atmospheric turbulence, researchers have explored the application of optical multiple-input multiple-output (MIMO) technology to atmospheric laser communications [1–8].

Channel capacity is a measure of the ability of a given channel to send messages from a specified source, under specific constraints. It is usually expressed as the maximum bit rate that can be achieved by the channel, when the appropriate coding is used and the error probability is within an acceptable range. In this chapter, the channel model of an optical MIMO system is discussed, and its channel capacity is analyzed using different modulation methods.

3.1 Channel Model of an Optical MIMO System

Optical MIMO technology uses multiple transmit and receive apertures for communication at the transmitter and receiver ends, respectively. This can exponentially increase the capacity and spectrum utilization of the communication system without an additional increase in bandwidth [9, 10]. MIMO technology can effectively suppress the atmospheric-turbulence effect by obtaining a diversity gain to attenuate the light-intensity fluctuation of the received signal [5, 11–13].

Because some coherence exists between the beams, the mutual distance between multiple receivers affects the beam coherence. The greater the distance between the receivers, the smaller the coherence between the beams. For wireless laser-communication systems, if the coherence between the beams is strong, bright and dark stripes will appear on the detector receiving surface, resulting in an increase in the received light-intensity fluctuations, and preventing the system from meeting

X. Ke, *Space-Time Coding Theory in Optical Wireless MIMO Systems*,
Optical Wireless Communication Theory and Technology,
https://doi.org/10.1007/978-981-97-6449-5_3

the requirements of normal communication. Therefore, it is necessary to reduce the coherence between the received beams as much as possible.

Coherence increases as it propagates in space; however, turbulence causes beam coherence to decrease [14]. According to Taylor's frozen-field theory, the turbulent structure remains constant for a short period, and the change with time is only a translation of the entire turbulent pattern.

Assuming that a narrowband beam is transmitted over a long distance, the refractive-index undulations along the propagation direction are superimposed and partially cancelled out; this is much less than the refractive-index undulations perpendicular to the transmission direction. If $l_0 < \sqrt{\lambda Z} < L_0$, the atmospheric-turbulence coherence length can be approximately defined as $\rho_0 \approx \sqrt{\lambda Z}$, and the atmospheric-turbulence coherence time (intensity undulation–dependent time) can be defined as $\tau_0 = \rho_0 / u_\perp$, where λ denotes the optical wavelength, l_0 and L_0 are the inner and outer turbulence scales, respectively, Z denotes the transmission distance, and $u_\perp$ denotes the component of the mean wind speed (omitting wind-speed fluctuation) perpendicular to the beam-transmission direction. If the transverse coherence length of the receiver is greater than $\rho_0 \approx \sqrt{\lambda Z}$, the received light intensity will satisfy the incoherence condition, which requires the distance between the receivers $D_r \geq \sqrt{\lambda Z}$.

To enable the receiver to effectively receive the light signal, but also to reduce the distance between the individual receivers as much as possible, A cannot exceed the received light coverage of the regional range; that is, to meet $D_r \leq Z\theta$, where θ is the divergence angle of the light beam. To ensure that the received light intensity will satisfy the incoherence condition, D_r should satisfy $\sqrt{\lambda Z} \leq D_r \leq Z\theta$.

For example, when $\lambda = 1.55$ μm, $Z = 10$ km, and $\theta = 50$ μrad, then the distance between receivers should be 12.45–50 cm. If the distance between each transmitting and receiving aperture is much greater than the atmospheric coherence length (coherence length), the signals received on each receiving aperture can be considered independent of each other.

Because the atmospheric-channel correlation time is much longer than the symbol period of wireless laser communications (generally less than 0.01 μs), the channel parameters can be considered to remain constant during the symbol period (i.e., the channel gain h_{ji} can be approximated as a constant within $0 \leq t \leq T$).

3.1.1 *Channel Model of an Optical MIMO System*

The channel model for an optical MIMO communication system with M lasers and N detectors is shown in Figure 3.1. The signal sent by the ith laser is denoted as x_i and the photocurrent received by the jth detector is denoted as y_j. h_{ji} denotes the optical-intensity fading coefficient from the ith laser to the jth detector. It is a constant in a characteristic period and approximately obeys a lognormal distribution. Then,

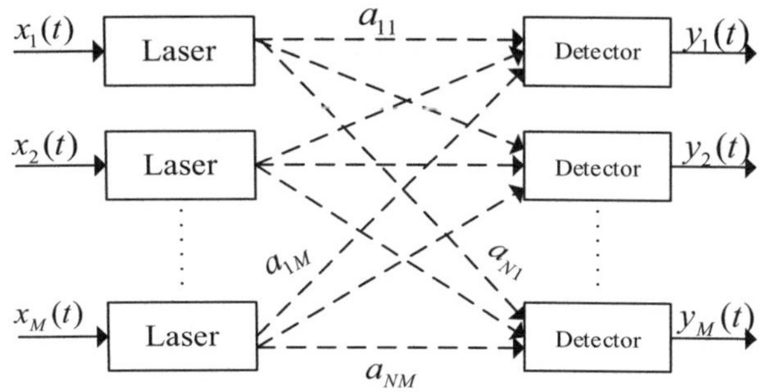

Fig. 3.1 Channel model of a MIMO optical communication system

$$y_j(t) = \eta \sum_{i=1}^{M} h_{ji} x_i(t) + n_{0,j}(t) \tag{3.1}$$

or appears in matrix form as

$$y(t) = \eta H x(t) + n_0(t) \tag{3.2}$$

where η denotes the photoelectric conversion efficiency. We define $x(t) = (x_1(t), x_2(t), \ldots, x_M(t))^T$ as the input signal matrix, $y(t) = (y_1(t), y_2(t), \ldots, y_N(t))^T$ as the output signal matrix, and $y(t) = (y_1(t), y_2(t), \ldots, y_N(t))^T$ as the noise matrix. Each component contains Gaussian white noise with mean m_x and variance N_0, and they are statistically independent of each other. H is the channel gain matrix.

$$H = \begin{bmatrix} h_{11} & h_{12} & \cdots & h_{1M} \\ h_{21} & h_{22} & \cdots & h_{2M} \\ \vdots & \vdots & \vdots & \vdots \\ h_{N1} & h_{N2} & \cdots & h_{NM} \end{bmatrix} = \begin{bmatrix} a_{11}^2 & a_{12}^2 & \ldots & a_{1M}^2 \\ a_{21}^2 & a_{22}^2 & \ldots & a_{2M}^2 \\ \ldots & \ldots & \ldots & \ldots \\ a_{N1}^2 & a_{N2}^2 & \ldots & a_{NM}^2 \end{bmatrix} \tag{3.3}$$

In wireless laser-communication systems, on–off keying (OOK) is a typical intensity modulation/direct detection (IM/DD) method, and its channel model is given by Eq. (3.3).

3.1.2 Channel Model of MIMO with PPM Modulation

Compared to OOK modulation, pulse-position modulation (PPM) has high optical-power utilization and orthogonality [15]. For PPM modulation, assuming that the period of the modulated symbol is T, a PPM symbol is divided into Q time slots of length T_s. Because the transmitted symbols are represented by the number of time slots in which a pulse appears, Q modulated symbols exist for Q-PPM modulation.

When modulated symbols are sent with equal probability, each symbol represents $\log_2^Q$ bits of information, and the throughput of PPM modulation can be increased by increasing the value of Q. When the symbol period is constant, an increase in Q implies a decrease in the time-slot length, which reduces the background noise in each time slot [16–18].

In the following analysis, we assume the following:

(1) The total system power E_s is assumed to be constant and the average power of each laser is E_s/M;
(2) The channel is set to be a flat, slow fading channel;
(3) The signals are set to be independent of each other;
(4) The characters will be sent with equal probability and each character represents $\log_2^Q$ bits of information; the Poisson counting model is used for the detector;
(5) The character period of Q-PPM modulation is set to T, which is divided into Q time slots with a time-slot length of $T_s = T/Q$.

In Q-PPM modulation, common photoelectric detection methods are the Poisson [19] and Gaussian mechanisms [20].

3.1.2.1 Channel Model of the Poisson Mechanism

In atmospheric laser communication using PPM modulation, the number of photons received at the receiver typically obeys the Poisson counting model. That is, the probability of receiving k photons in one optical-pulse time slot is

$$P(k) = \frac{e^{-\lambda}\lambda^k}{k!}, \tag{3.4}$$

where λ denotes the average count energy of each optical pulse. For an $M \times N$ MIMO system, the number $y^{(n)}$ of photons received at the nth detector can be expressed as

$$y^{(n)} = \frac{n_s}{M}\sum_{m=1}^{M} x^{(m)} h_{nm} + n_b n_0, \tag{3.5}$$

where $x^{(m)}$ denotes the signal sent to the mth laser, and h_{nm} denotes the light-intensity fading coefficient from the mth laser to the nth detector.

$n_s = \frac{\eta P_r T}{hf} = \frac{\eta E_s}{hf}$ denotes the average counting rate of the received photons when there is no fading. η denotes the quantum-conversion efficiency, h denotes Planck's constant, f denotes the frequency of the optical carrier, P_r denotes the average power received at each detector, and E_s denotes the total power of the system. n_b denotes the average count rate generated by noise (which includes the dark current), that is, $n_b = \frac{\eta P_b T}{hf}$, where P_b denotes the average noise power of the detector.

Without loss of generality, we represent the "*on*" time slot by 1 and the "*off*" time slot by 0. The received signal $y^{(n)}(j = 1, 2)$ is comprised of "*on*" and "*off*" time slots. Let $\lambda_{on,n}$ be the photon count for the "*on*" time slot and M be the photon count for the "*off*" time slot. Because n_0 is an $1 \times Q$ vector with one element, Eq. (3.5) can be transformed into

$$\begin{cases} \lambda_{on,n} = \frac{n_s}{M} \sum_{n=1}^{M} h_{nm} + n_b \\ \lambda_{off,n} = n_b. \end{cases} \tag{3.6}$$

3.1.2.2 Channel Model of the Gaussian Mechanism

The channel model for the PPM-modulated Gaussian mechanism is given by Eq. (3.2). Equation (3.2) can be transformed as

$$\begin{cases} I_{on,n} = \eta \sum_{n=1}^{M} h_{nm} + n_0 \\ I_{off,n} = n_0, \end{cases} \tag{3.7}$$

where $I_{on,n}$ is the photocurrent in the "*on*" time slot and $I_{off,n}$ is the photocurrent in the "*off*" time slot.

3.2 Channel Capacity of Optical MIMO Systems

In 1948, Shannon expressed the channel capacity of single-input single-output (SISO) systems; that is, the channel capacity C can be expressed as

$$C = \log_2(1 + SNR) \quad \text{bit/s/Hz}, \tag{3.8}$$

where SNR is the signal-to-noise ratio at the receiving end. This formula provides the upper limit of the information-transmission rate that can be achieved over a reliable communication link.

For MIMO systems in radio-frequency (RF) communication, Foschini and Telatar studied the channel capacity and obtained an equation for it [21, 22]. For wireless laser-communication systems, prior research [23, 24] has provided performance bounds for the channel capacity of optical MIMO systems, and we use an approach similar to RF communication to study the channel capacity of the OOK and PPM modulation methods.

3.2.1 Channel Coefficients to Determine the Channel Capacity of a MIMO System

Using the channel model of Eq. (3.2), according to information theory [25], the channel capacity of an optical MIMO system can be defined as

$$C = \max_{p(x)} I(x; y), \tag{3.9}$$

where $p(x)$ is the probability-density function of vector x, and $I(x; r)$ is the mutual-information quantity between vectors x and r.

$$I(x; y) = H(y) - H(y|x), \tag{3.10}$$

where $H(n)$ and $H(y)$ are the entropies of noise vector n and received-signal vector r, respectively, and $H(y|x)$ is the conditional entropy of received signal y, given that the transmitted signal is x. According to [26], the channel capacity of an optical MIMO system can be obtained as

$$C = \max_{R_{xx}, tr(R_{xx})=M} \log_2 \det\left[I_N + \left(\frac{\eta^2 E_s^2}{MN_0}\right) HR_{xx}H^H\right], \tag{3.11}$$

where $\det(\cdot)$ denotes the matrix determinant; R_{xx} and R_{yy} are the covariance matrices of transmit-signal matrix x and the receive-signal matrix r, respectively, $tr(R_{xx})$ is the trace of R_{xx}, and I_N is a unitary array of order N.

Assuming that knowledge about the channel is not available at the transmitter, $R_{xx} = I_M$. Then, Eq. (3.11) can be simplified as

$$C = \log_2 \det\left[I_N + \left(\frac{\eta^2 E_s^2}{MN_0}\right) HH^H\right]. \tag{3.12}$$

According to the theory of singular-value decomposition, the singular-value decomposition of channel matrix A is $A = UDV^H$, where U and V are the unitary matrices of $N \times N$ and $M \times M$, that is, $UU^H = I$ and $VV^H = I$, respectively. D is the nonnegative diagonal matrix of $M \times M$, and its diagonal element λ_i is the positive eigenvalue of matrix HH^H. According to singular-value decomposition theory and matrix properties $\det\left(I + UDU^H\right) = \det(I + D)$, the channel capacity of the system can be obtained as

$$C = \sum_{i=1}^{M} \log_2\left(I_N + \frac{\eta^2 E_s^2}{MN_0}\lambda_i\right). \tag{3.13}$$

As shown in the above equation, the channel capacity of an optical MIMO system is mainly determined by the singular values of H or by the eigenvalues of HH^H.

These eigenvalues determine the gain of an independent parallel channel. The channel capacity of the type in Eq. (3.13) can be achieved using spatially multiplexed Bell-Labs layered space-time (BLAST) codes, which are not satisfied by space-time block codes, so there is a loss in channel capacity.

3.2.2 *Channel Capacity of Random-Time MIMO Systems with Channel Coefficients*

In an actual wireless fading environment, owing to the existence of channel time variability, multipath effects, and other factors, the fading coefficient of a channel is no longer determined, but becomes a random variable; thus, the channel capacity C of an optical MIMO system is also a random variable. Therefore, when analyzing the channel capacity of optical MIMO, we introduce two statistical properties, namely, the all-state ephemeral capacity and outage-probability capacity, to characterize the channel capacity of the system. Their physical significance is given as follows.

3.2.2.1 Ergodic Capacity

The ergodic capacity C_{avg} of a MIMO channel is the average channel capacity of the channel matrix H. In an ergodic channel, the ergodic capacity is the error-free transmission rate that can be achieved using an asymptotically optimal codeword. Thus, the all-state ephemeral capacity is the Shannon capacity of the channel. It is generally used to measure the channel-capacity performance in the overall sense of the system, which can be obtained by statistically averaging Eq. (3.13):

$$C_{avg} = E_A[C] \tag{3.14}$$

where $E_A[\cdot]$ denotes the instantaneous channel capacity C, averaged with respect to the channel-fading coefficient.

To specifically analyze the channel capacity of an optical MIMO system, the following Monte Carlo method was used to analyze the system channel capacity with different numbers of transmitting and receiving apertures and compare it with the capacity of a SISO system. The simulation conditions are as follows

(1) We set the total transmit-power constant $E_s = 1$ to take a normalized value;
(2) A system using an avalanche photodiode (APD) detector, with photoelectric conversion efficiency $N = 0.5$;
(3) Light-intensity flicker obeys a lognormal distribution and the scintillation index $S.I. = 0.6, 1.0$;
(4) Aperture numbers are 1×1, 1×2, 1×4, 2×2, 2×3, 4×4, and 4×5 systems.

To analyze the effect of the number of transceiver apertures, noise, and atmospheric scintillation on the system channel capacity, the average system channel capacities when the scintillation index is 0.6 and 1.0, are analyzed using the Monte Carlo method, as shown in Figs. 3.2 and 3.3.

Figure 3.2 shows the variation curves of the average channel capacity (C_{avg}) and signal-to-noise ratio of the system when the scintillation index is 0.6. The following can be seen from Fig. 3.2:

(1) In the case where the number of receive apertures is greater than or equal to the number of transmit apertures, the channel capacity of the system increases almost exponentially with the number of transmit apertures, in the order of the

Fig. 3.2 Variation curve of the average channel capacity versus signal-to-noise ratio for scintillation index (*S.I.*) = 0.6

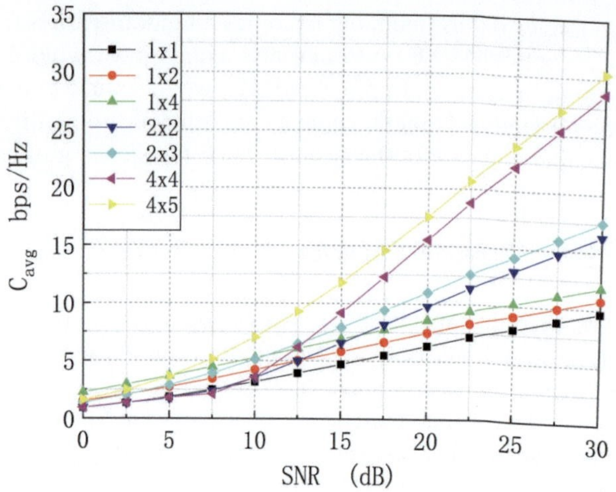

Fig. 3.3 Variation curve of average channel capacity with SNR for $S.I. = 1.0$

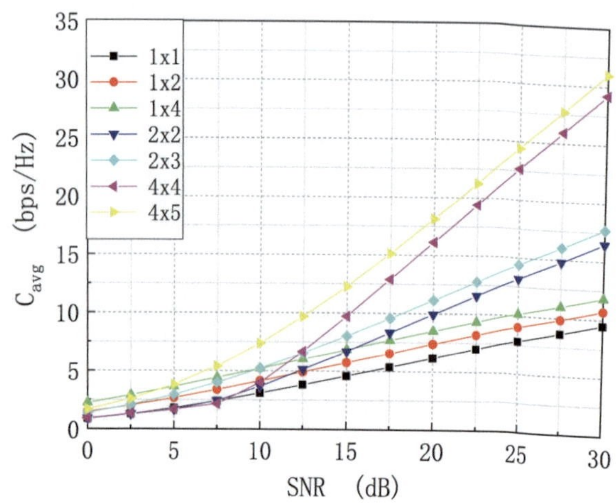

number of transmit apertures from 1 to 2 and then to 4, reflecting the advantage of using multiplexing technology at the transmitting end.

(2) The channel capacity increases with the number of receive apertures in the order of apertures 1×1, 1×2 and 1×4, 2×2 and 2×3, and 4×4 and 4×5, with the same number of transmit apertures. This means that increasing the number of receive apertures is equivalent to increasing the receive aperture, which reflects the effect of the receive diversity. Therefore, the channel capacity of the system gradually increases as the number of apertures increases. In particular, when the number of receive apertures is greater than or equal to the number of transmit apertures, the system capacity increases almost proportionally to the number of transmit apertures.

(3) When the total transmit power is limited, that is, the total transmit energy is normalized, the increase in the channel capacity with the number of transmit apertures is not very obvious when the signal-to-noise ratio is small; however, the trend of the channel capacity with the number of transmit apertures becomes more obvious when the signal-to-noise ratio is large.

Figure 3.4 shows the variation curve of the average channel capacity of the system with the signal-to-noise ratio when the scintillation index is 1.0. This figure reflects the same law as Fig. 3.3.

To reflect the variation in channel capacity with the number of transceiver apertures when the number of apertures is certain, we analyzed the variation law of the channel capacity with the number of transceiver apertures when the signal-to-noise ratio was certain. The results are shown in Figs. 3.5 and 3.6.

Figure 3.4 shows the variation in the average channel capacity with the number of transmitting apertures when $N = 1$ and $S.I. = 0.6$. As can be seen from the figure, when the number of received apertures is certain, the channel capacity increases

Fig. 3.4 Relationship between the average channel capacity and the number of transmitting apertures when $S.I. = 0.6$ and $N = 1$

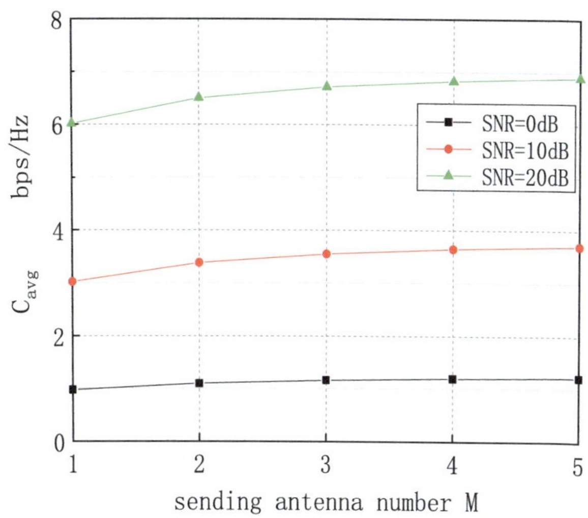

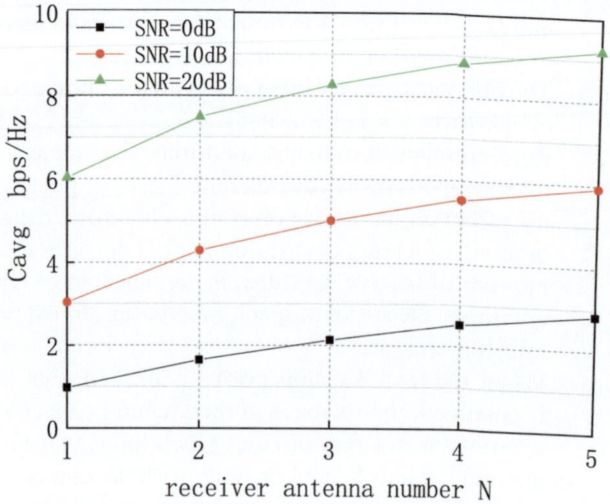

Fig. 3.5 Relationship between the average channel capacity and number of receiving apertures when $S.I. = 0.6$ and $M = 1$

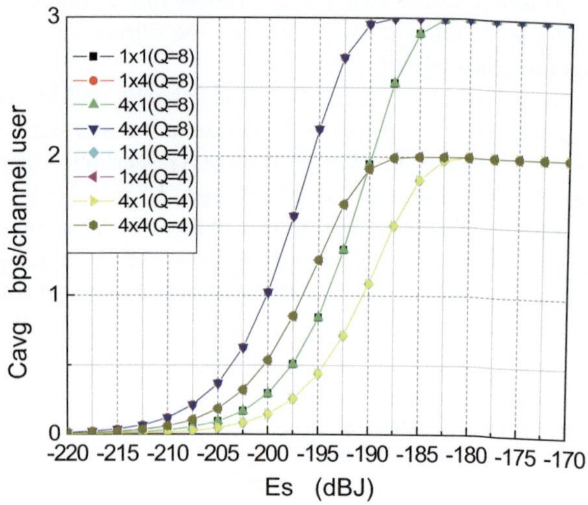

Fig. 3.6 Change curve of C_{avg} and E_s without fading or background radiation

with the number of transmitting apertures; however, the magnitude of the increase is smaller, that is, there is a "flat-layer" effect.

Figure 3.5 shows the variation in the average channel capacity with the number of received apertures when $M = 1$ and $S.I. = 0.6$. As shown in the figure, when the number of transmitting apertures is certain, the channel capacity increases significantly with an increase in the number of receiving apertures.

3.2.2.2 Interruption Capacity

The interruption capacity (outage capacity) is the rate at which a system can transmit information with a high probability of assurance. This analytically quantifies the performance (capacity) level that can be guaranteed at a certain level of reliability. It is defined as follows.

P_{out} is the probability that the channel capacity of a system is less than a certain fixed value, C_{out}; that is,

$$P_{out} = P_r\{C < C_{out}\}, \tag{3.15}$$

where P_{out} is the interruption probability and C_{out} is the interruption capacity. It is important to emphasize that the interruption probability and interruption capacity correspond. That is, the system is guaranteed to have a capacity higher than the fixed-value C_{out} with a probability $1 - P_{out}$, or the capacity is lower than the fixed-value C_{out} with a probability P_{out}, for all implementations of the system.

3.2.3 *Channel Capacity of a Typical MIMO System*

One characteristic of MIMO technology is that it obtains a large multiplexing or diversity gain, which can effectively improve the system performance. Depending on the number of apertures at the transmitting and receiving ends, MIMO can also include single-input multiple-output (SIMO) and multiple-input single-output (MISO) systems, as opposed to traditional single-input single-output (SISO) systems. MISO, SIMO, and SISO can be considered as special cases of MIMO.

3.2.3.1 Channel Capacity of a Single-Input Single-Output System

A single-input single-output (SISO) system uses one antenna to transmit data and one antenna to receive data. For a given channel, the power of the transmitted signal is *P*, the received signal-to-noise ratio is *SNR*, and the normalized capacity of the channel can be expressed according to Shannon's formula [27]:

$$C = \log_2(1 + SNR). \tag{3.16}$$

From Eq. (3.16), it can be seen that the channel capacity grows logarithmically with the SNR. In regions with high SNR, the channel capacity can increase by 1 bit/s/Hz for every 3-dB increase in SNR. Actual wireless channels are generally time-varying, and if their channel fading is denoted by *h*, the channel capacity can be expressed as

$$C = \log_2\left(1 + SNR|h|^2\right). \tag{3.17}$$

3.2.3.2 Channel Capacity of a Single-Input Multiple-Output System

A single-input multiple-output (SIMO) system transmits data using one antenna at the transmitter side and receives data using N antennas at the receiver side. Thus, the receiver side can obtain N different feedback fading samples of the signal. The channel-characteristic matrix H can be expressed as

$$H = [h_{11}, h_{21}, \ldots, h_{N1}]^H. \tag{3.18}$$

where h_{i1} denotes the channel-fading coefficient from the transmitting antenna to the ith receiving antenna, satisfying

$$H^H H = \sum_{i=1}^{N} |h_{i1}|^2. \tag{3.19}$$

For channel-fading determination, the channel capacity of the system can be expressed as [28]

$$C = \log_2(1 + NSNR) = \log_2\left(1 + \sum_{i=1}^{N} |h_{i1}|^2 SNR\right). \tag{3.20}$$

This channel capacity corresponds to the maximum linear combination at the receiver [29].

$$|h_{11}|^2 = |h_{21}|^2 = \cdots = |h_{N1}|^2 = 1 \tag{3.21}$$

When the elements of the channel matrix are used as the norm in Eq. (3.21), the capacity in Eq. (3.18) can be written as

$$C = \log_2(1 + N \cdot SNR). \tag{3.22}$$

From the above equation, it can be seen that compared to the channel capacity of the SIMO system, the received signal-to-noise ratio of the SIMO system is N times that of the SISO system, and the channel capacity increases logarithmically with an increase in the number of received antennas.

3.2.3.3 Channel Capacity of a Multiple-Input Single-Output System

A multiple-input single-output (MISO) system is a communication system in which M antennas are used at the transmitter end to transmit data, and one antenna is used at the receiver end to receive data. The channel-characteristic matrix of a MISO system can be expressed as

$$H = [h_{11}, h_{21}, \ldots, h_{1M}]^H, \tag{3.23}$$

where h_{1j} denotes the channel fading coefficient from the jth transmitting antenna to the receiving end, which satisfies

$$HH^H = \sum_{j=1}^{M} |h_{1j}|^2 \tag{3.24}$$

An equal-power transmission scheme is generally adopted when the channel-state information is unknown. Assuming that the total transmitted power is P, the power at each transmit antenna is P/M, and thus, the channel capacity of the MISO system can be expressed as [30]

$$C = \log_2\left(1 + HH^H \frac{SNR}{M}\right) = \log_2\left(1 + \sum_{j=1}^{M} |h_{1j}|^2 \frac{SNR}{M}\right). \tag{3.25}$$

The total transmit power is evenly distributed to M transmit antennas, and the transmitted power at each transmit antenna becomes the original $1/M$. If the channel coefficient is normalized according to Eq. (3.21), the channel capacity of the MISO system is

$$C = \log_2(1 + SNR). \tag{3.26}$$

It can be seen from the above formula that the received signal-to-noise ratio is the same as that of the system; therefore, the channel capacity will not increase with the increase in the number of transmitting antennas.

3.2.3.4 Channel Capacity of a MIMO System

A multiple-input multiple-output (MIMO) system is a communication system in which the transmitter sends data with M antennas and the receiver receives data with N antennas. The channel capacity of the MIMO system was analyzed according to the channel model given in the previous section. The elements in channel matrix H satisfy $h_{i,j} \sim C_N(0, 1)$, that is, the Rayleigh-fading channel.

From the relationship between the received information and conditional information entropy, we obtain

$$\begin{aligned} I(x; y) &= H(y) - H(y|\mathrm{x}) \\ &= H(y) - H((Hx + \eta)|\mathrm{x}) \\ &= H(y) - H(\eta|\mathrm{x}) \\ &= H(y) - H(\eta) \end{aligned}, \tag{3.27}$$

where $H(\cdot)$ represents information entropy. The information entropy of the received signal y is expressed as

$$H(y) \leq \log_2\left[\det\left(\pi e Q_{vy}\right)\right], \tag{3.28}$$

where $\det(\cdot)$ denotes the determinant operation of the square matrix. If and only if y obeys the complex cyclic-symmetric Gaussian distribution, the equal sign holds and the information entropy reaches the maximum. The entropy of the noise vector satisfies the following equation:

$$H(\eta) \leq \log_2\left[\det\left(\pi e Q_{\eta\eta}\right)\right] = \log_2\left[\pi e \sigma^2\right] \tag{3.29}$$

From Eq. (3.12), the covariance matrix of the received signal y is

$$HH^H = U \Lambda U^H C_{out}. \tag{3.30}$$

When the channel-state information at the transmitter is unknown, if an input signal x with a given transmission power P obeys the complex cyclic-symmetric Gaussian distribution and the allocated transmission power is equal, that is, when $Q_{xx} = I_M \cdot P/M$ is satisfied, the system channel capacity can reach the maximum value [31]. The expression for the channel capacity can be obtained from Eq. (3.13):

$$\begin{aligned} C &= \log_2\left\{\det\left(Q_{vy} Q_{\eta\eta}^{-1}\right)\right\} \\ &= \log_2\left\{\det\left[\left(HQ_{xx}H^H + Q_{\eta\eta}\right)Q_{\eta\eta}^{-1}\right]\right\} \\ &= \log_2 \det\left(I_N + HQ_{xx}H^H/\sigma^2\right) \\ &= \log_2 \det\left(I_N + P/M \cdot I_M HH^H/\sigma^2\right) \\ &= \log_2 \det\left(I_N + \frac{SNR}{M} HH^H\right) \end{aligned} \tag{3.31}$$

If k is used to represent the rank of the channel matrix H, that is, $rank(H) = k$, it can be seen from matrix theory that the number of nonzero eigenvalues of H is equal to the rank k of matrix H. For an $N \times M$-dimensional channel-characteristic matrix H, the maximum 0value of the rank is the smaller of M and N: $\max(k) = \min(M, N)$; that is, there are at most $\min(M, N)$ nonzero eigenvalues.

We perform eigenvalue decomposition for $N \times N$-dimensional matrix HH^H:

$$HH^H = U \Lambda U^H, \tag{3.32}$$

where Λ represents the diagonal matrix with the characteristic value of HH^H as the main diagonal element:

$$\Lambda = \text{diag}\{\lambda_1, \lambda_2, \ldots, \lambda_N\} \quad (\lambda_1 \geq 0). \tag{3.33}$$

U represents a unitary matrix composed of eigenvectors corresponding to the eigenvalues of the above matrix, which satisfies

$$U^H U = UU^H = I_N. \tag{3.34}$$

Substituting Eq. (3.30) into Eq. (3.31), the channel capacity can be expressed as

$$C = \log_2 \det\left(I_N + \frac{SNR}{M} U \Lambda U^H \right). \tag{3.35}$$

Because $\det(I_M + AB) = \det(I_N + BA)$, Eq. (3.34) can be simplified as

$$C = \log_2 \det\left(I_N + \frac{SNR}{M} \Lambda \right). \tag{3.36}$$

The above formula is adjusted to obtain

$$C = \sum_{i=1}^{k} \log_2\left(I_N + \frac{SNR}{M} \lambda_i \right). \tag{3.37}$$

Equation (3.37) shows that MIMO channels can be theoretically equivalent to k independent parallel subchannels, and the eigenvalue of matrix HH can represent the power gain of each subchannel. It can be seen that the channel capacity of the MIMO system has a significant relationship with the rank k of the channel matrix.

The number of independent subchannels can be obtained from the matrix rank. Assuming that a large number of scattering media are present in the wireless-propagation environment, generally, as long as the distance between antennas is sufficiently large, the fading of signals received by any two receiving antennas can be considered to be independent of each other; that is, they reach the receiving end through independent transmission paths. Therefore, Eq. (3.35) can be explained by the fact that a MIMO system uses multipath propagation to convert adverse factors into favorable factors; thus, the signals received by different receiving antennas are uncorrelated [32].

In a quasi-static fading channel, the channel matrix changes randomly after several symbol periods, which is a non-ergodic channel. Regardless of the target transmission rate, reliable communication of the channel at this data-transmission rate cannot be guaranteed [33]. At this point, the channel capacity must be represented by the interrupt capacity C_{out}.

Channel interrupts are used to describe the properties of non-ergodic channels. When the mutual information of the channel cannot support the desired transmission rate, the channel is interrupted [34], which can be expressed as

$$\{H : I(x; y|\mathrm{H} = \mathrm{H}) \prec R\}. \tag{3.38}$$

Assuming that the probability of a channel interruption is $q\%$, the interrupt capacity C_{out} is defined as

$$P(I \prec C_{out}) = q\%(100 - q)\%. \tag{3.39}$$

Equation (3.39) indicates that when the information transmission rate is C_{out}, the channel can reliably transmit with a probability of $(100 - q)\%$.

Considering these situations, the channel capacity of an optical MIMO system is significantly higher than that of a classical system. However, the increased channel capacity can be used to improve the transmission rate of the system, and the reliability of the communication system can be improved by adding information redundancy, while keeping the information rate unchanged. A reasonable compromise can also be made between the two.

3.2.4 *Channel Capacity of a MIMO System Using PPM Modulation*

The channel capacity reflects the effectiveness of the communication system and is the maximum transmission rate that can achieve error-free transmission in a given channel. Therefore, we studied the channel capacity of an optical MIMO system under the Poisson distribution for PPM modulation. Because the channel capacity of PPM-modulated optical MIMO under background radiation is complex, we focus on the channel capacity of PPM-modulated MIMO without background radiation. The channel capacity of PPM was studied with and without fading.

When PPM modulation is adopted, the channel capacity and mutual information still satisfy Eqs. (3.9) and (3.10), respectively. At this time, $H(y)$ and $H(y|x)$ are changed. According to the characteristics of PPM modulation, $H(y)$ and $H(y|x)$ are, respectively,

$$H(y) = \log_2^O \tag{3.40}$$

$$H(y|\mathrm{x}) = \log_2^Q \cdot \exp\left(-\frac{\eta E_s \sum_{m-1}^{M} \sum_{n=1}^{N} a_{nm}^2}{hfM}\right). \tag{3.41}$$

Then, the channel-capacity formula becomes

$$C = \log_2^Q\left[1 - \exp\left(-\frac{\eta E_s \sum_{m-1}^{M}\sum_{n=1}^{N} a_{nm}^2}{hfM}\right)\right]. \tag{3.42}$$

The average channel capacity can be expressed as

$$C_{\text{avg}} = E_H[C] = \log_2^Q\left[1 - \prod_{m=1}^{M}\prod_{n=1}^{N}\int_0^{\infty} f_A(a_{nm})\cdot\exp\left(-\frac{\eta E_s a_{nm}^2}{hfM}\right)da_{nm}\right]. \tag{3.43}$$

It can be seen from Eq. (3.43) that the channel capacity at this time is related to the number of time slots Q, number of transmit apertures M, number of receive apertures η, transmission power E_s, photoelectric conversion efficiency N, and channel-fading parameter a. To better explain their relationship, we considered two cases.

3.2.4.1 No Background Radiation or Fading

When there is no background radiation or fading, $\sum_{m-1}^{M}\sum_{n=1}^{N} a_{nm}^2 = MN$. The average channel capacity can be obtained from Eq. (3.43):

$$C_{avg} = \log_2^Q\left[1 - \exp\left(-\frac{\eta N E_s}{hf}\right)\right]. \tag{3.44}$$

It can be seen that the average channel capacity is independent of the number of transmit apertures M but is related to the number of time slots Q, the number of receive apertures N, and the transmission power E_s. To better illustrate the relationship between them, we simulated the average channel capacity under the condition of limited average power; the results are shown in Fig. 3.6.

Figure 3.6 shows the curves of C_{avg} and E_s for different MIMO systems. At this time, $\eta = 0.5$ and $\lambda = 1.550$ μm. The following can be deduced from the figure:

(1) Although the average channel capacity increases with the increase in transmission power E_s, when E_s is greater than a certain value E_0, where $E_0 = -180$ dBJ, the channel capacity will reach the saturation point and will not increase with the increase in transmission power.
(2) The channel capacity of the system increases with an increase in the number of receiving apertures n;
(3) The channel capacity of the system does not change with an increase in the number of transmit apertures M;
(4) The channel capacity of the system increases with an increase in the number of slots Q.

(2) Fading, but no Background Radiation

When there is fading, but no background radiation, the average channel capacity is expressed as in Eq. (3.24).

For lognormal fading, the average channel capacity is

$$C_{\text{avg}} = E_H[C] = \log_2^Q\left[1 - \prod_{m=1}^{M}\prod_{n=1}^{N}\int_0^{\infty}\frac{1}{\left(2\pi\sigma_x^2\right)^{1/2}a_{nm}}\exp\left(-\frac{\log_e^{a_{mn}} - \mu_x}{2\sigma_x^2}\right)\cdot\exp\left(-\frac{\eta E_s a_{nm}^2}{hfM}\right)da_{nm}\right]. \tag{3.45}$$

To ensure the incoherent transmission of light beams in optical MIMO systems, the distance D_r between any two receiving apertures should satisfy $\sqrt{\lambda Z} \leq D_r \leq Z\theta$, where θ is the divergence angle of the beam, λ is the wavelength of the light, and Z is the transmission distance. When $\lambda = 1550$ nm and $Z = 1$ km, then $D_r = 4$ cm. In other words, incoherent transmission can be realized, as long as the distance between the detectors is $D_r \geq 4$ cm. Because the distance between detectors is only a few tens of centimeters, it is considered that the fading coefficients are approximately equal, and the average channel capacity expressed in Eq. (3.45) becomes

$$C_{\text{avg}} = \log_2^Q\left\{1 - \left[\int_0^{\infty}\exp\left(-\frac{\eta E_s a^2}{hfM}\right)\cdot\frac{1}{\left(2\pi\sigma_x^2\right)^{1/2}a}\exp\left(-\frac{\log_e^a - \mu_x}{2\sigma_x^2}\right)da\right]^{NM}\right\}. \tag{3.46}$$

As can be seen from Eq. (3.46), it is difficult to obtain the closed form of the average channel capacity.

Rayleigh distribution can also be used to describe the fading in atmospheric laser communication. The fluctuation characteristics of light intensity subject to the Rayleigh distribution have been studied [28]. The advantages of Rayleigh distribution are as follows:

(1) The Rayleigh distribution is more suitable than a lognormal distribution for describing strong fluctuations, that is, deep fading;
(2) In Rayleigh distributed fading, MIMO systems can obtain a more significant diversity gain; that is, the bit-error rate curve becomes steeper;
(3) The Rayleigh distribution is easier to calculate; that is, it is easier to obtain the closed form of the MIMO-system channel capacity and bit-error rate.

To compare the characteristics of the two distributions, we use the same assumption, that is, $E\left[A^2\right] = 1$. In the Rayleigh distribution, the light intensity $I = A^2$ is a unilateral exponential random variable. The probability-density function corresponding to amplitude A is

$$f_A(a) = 2ae^{-a^2} \quad a > 0 \tag{3.47}$$

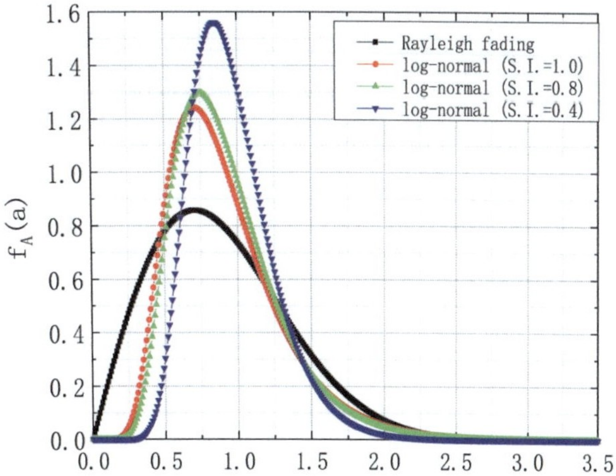

Fig. 3.7 Probability-density functions of the Rayleigh distribution and lognormal distribution

The corresponding scintillation index $S.I. = 1.0$. To better explain the light-intensity fluctuation characteristics under the two distribution forms, we studied the probability-density functions of the two distribution forms, and the results are shown in Fig. 3.7.

It can be seen from Fig. 3.7 that although the two distribution forms are different, the greater the scintillation index, the closer the probability-density functions of the two distribution forms become. In particular, when the scintillation index is $S.I. = 1$, in the definition domain of $a \in (1, 3.5)$, the value of a in the two distributions is very close. Therefore, the Rayleigh distribution is helpful for describing the closed form and performance bound of the MIMO-system channel capacity and bit-error rate. Under the same assumptions, the expression of the average channel capacity under Rayleigh fading is

$$C_{avg} = \log_2^Q\left[1 - \left(1 + \frac{\eta E_s}{hfM}\right)\right]^{-NM}. \tag{3.48}$$

As shown, the average channel capacity is related to the number of time slots Q, number of transmit apertures M, number of receive apertures N, transmit power E_s, and channel fading parameters a. Similarly, to illustrate the relationship between them, the average channel capacity and interrupt channel capacity are simulated under the condition of limited average power. The results are shown in Figs. 3.8, 3.9, 3.10 and 3.11.

Figure 3.8 shows the change curve of C_{avg} and E_s when there is lognormal fading, but no background radiation. At this time, $S.I. = 1$. The following can be seen from the figure:

(1) The average channel capacity increases with the increase in transmission power E_s; however, when $E_s \geq -175$ dBJ, the channel capacity will reach the saturation point and will not increase with the increase in transmission power.

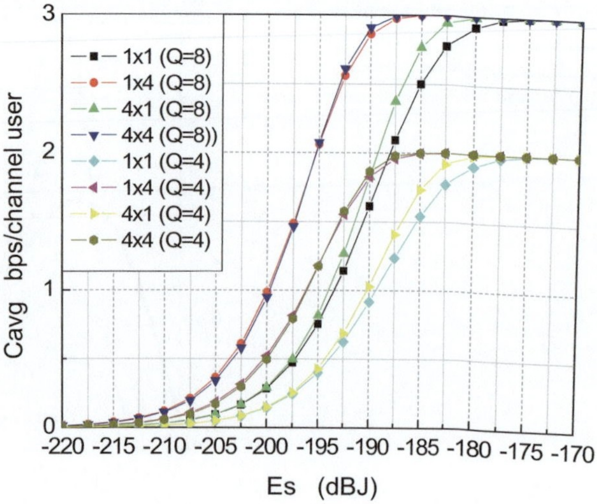

Fig. 3.8 Change curve of C_{avg} and E_s with lognormal fading and no background radiation

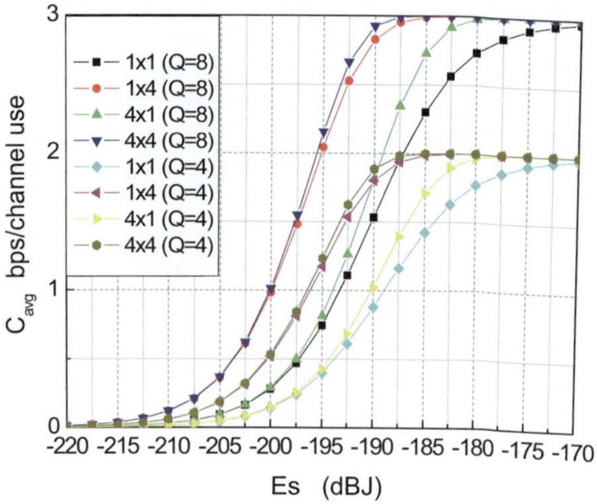

Fig. 3.9 Change curve of C_{avg} and E_s with Rayleigh fading and no background radiation

(2) The channel capacity of the system increases obviously with the increase of the number of receiving apertures *n*;
(3) The channel capacity of the system increases only slightly with the increase of the number of transmit apertures *M*;
(4) The system channel capacity increases obviously with the increase of the number of slots *Q*.

Figure 3.9 shows the change curves of C_{avg} and E_s when there is Rayleigh fading and no background radiation. It can be seen from the figure that the pattern reflected this time is very similar to that in Fig. 3.8. The differences are as follows:

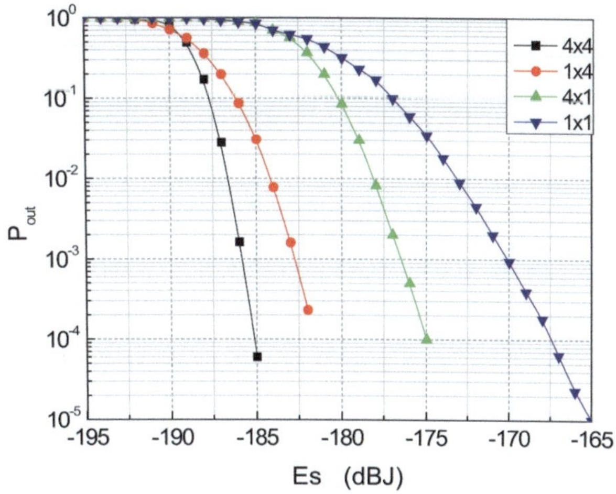

Fig. 3.10 Change curve of outage probability and E_s with lognormal fading and no background radiation

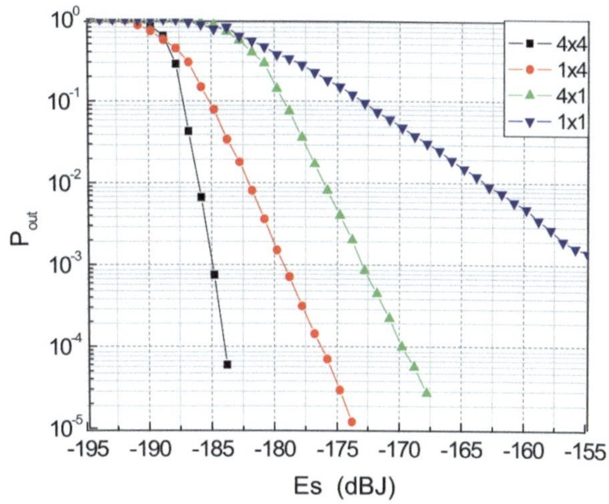

Fig. 3.11 Change curve of outage probability and E_s with Rayleigh fading and no background radiation

(1) In the Rayleigh distribution, the benefit of the system channel capacity from the increase in the number of transmit apertures M is slightly greater than that in the lognormal distribution;
(2) When the channel capacity reaches the saturation point, the transmission power required is different in the Rayleigh distribution, $E_s \geq -170\,\text{dBJ}$.

Figure 3.10 shows the change curve of the outage probability and E_s when there is lognormal fading and no background radiation. At this time, $\eta = 0.5$, $\lambda = 1.550\,\mu\text{m}$, and $S.I. = 1.0$. The following can be seen from the figure:

(1) The interrupt probability decreases significantly with the increase in transmission power E_s;

(2) The outage probability of the system decreases with the increase in the number of transmit apertures M and the number of receive apertures N; the effect of the increase in the number of receive apertures is more obvious. This is because the gain in diversity is equivalent to an increase in the receiving aperture.

Figure 3.11 shows the change curve of the outage probability and E_s in the case with Rayleigh fading and no background radiation. It can be observed from the figure that the law reflected at this time is very similar to that in Fig. 3.10. The difference between them is that the curve change in the outage probability in Figure 3-11 is relatively smooth, and the improvement in power obtained under the same outage probability is greater.

3.3 Comparison of the Performance of OOK Modulation and PPM Modulation

3.3.1 Performance When OOK Is Modulated with PPM

Performance simulations were performed for 1×1, 1×2, 2×2, and 2×3 systems to compare the variation curves of the bit-error rate (BER) with the signal-to-noise ratio (SNR) for different modulation methods of PPM and OOK. The number of code elements for the simulations in Figs. 3.13 and 3.14 is $N = 10000$, and the number of code elements for the simulations in Figs. 3.15 and 3.16 is $N = 15000$, with the parameters set to a channel-fading factor of 0.6 and an optical-conversion efficiency of 0.7.

Comparing Fig. 3.12 with Figs. 3.14 and 3.13 with Fig. 3.15, it can be seen that the higher the number of simulation code elements, the smoother and more statistically significant the curve, and the better it shows the performance characteristics of the two modulation methods in the free-space optical (FSO) MIMO system.

The BER of the system under PPM modulation is lower than that of the system under OOK modulation for a certain signal-to-noise ratio, as can be seen from the curves in Figs. 3.14 and 3.15. PPM is an orthogonal modulation method. Although OOK modulation occupies less bandwidth, it has a higher average transmit power compared to PPM modulation. PPM modulation is suitable for communication systems with restricted pulse-repetition intervals [35]. Therefore, it is more suitable than OOK modulation for FSO MIMO communication systems.

It is also clear from the graph that the BER is lower when the receive diversity is greater than the transmit diversity, than when both diversities are equal. This demonstrates the superiority of MIMO systems for multiterminal transceivers. It can be observed that PPM modulation is more suitable than OOK modulation for free-space laser-communication MIMO systems.

Fig. 3.12 BER curve for PPM modulation (N = 10,000)

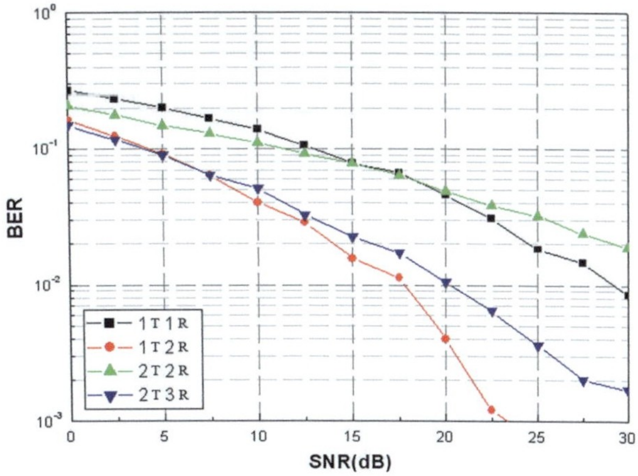

Fig. 3.13 BER curve for OOK modulation (N = 10,000)

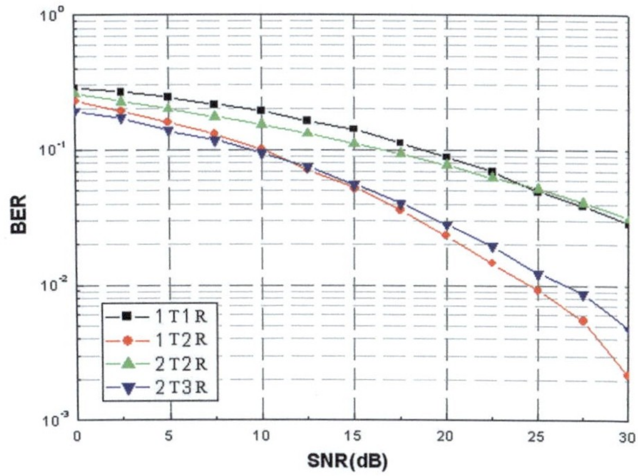

Fig. 3.14 BER curve for PPM modulation (N = 15,000)

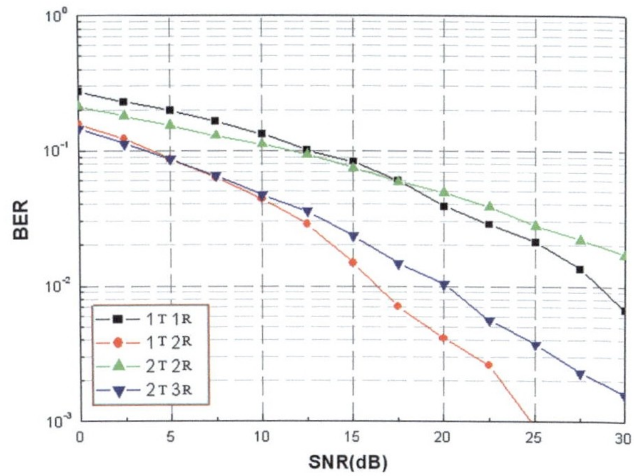

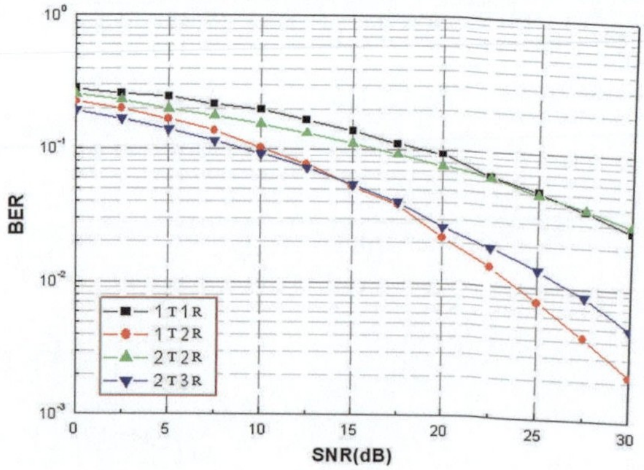

Fig. 3.15 BER curve for OOK modulation (N = 15,000)

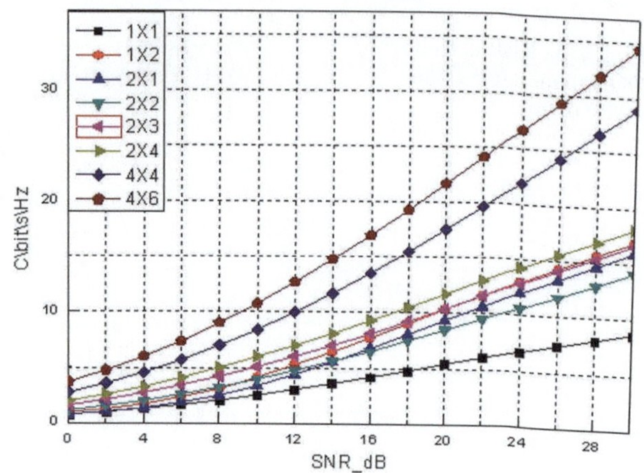

Fig. 3.16 Variation curve of channel capacity versus signal-to-noise ratio

3.3.2 *Wireless Optical MIMO Channel-Capacity Simulation*

The variation in channel capacity with the number of transmitting and receiving antennas and noise was simulated using the Monte Carlo method. The conditions for the simulation were as follows: the number of samples was 10,000, numbers of transmitting and receiving antennas were 1×1, 1×2, 2×1, 2×2, 2×3, 2×4, 4×4, and 4×6, and the scintillation index S.I. was set to 0.6.

The channel capacity curve with SNR derived from the simulation conditions is shown in Fig. 3.16, and Table 3.1 shows the SNR (dB) required to obtain different systems with a channel capacity of 7 bit/s/Hz. We conclude the following from the simulation plots and tables.

(1) The channel capacity of the FSO MIMO system increases sharply with an increase in the SNR and number of antenna pairs, indicating that the system is very effective in increasing the channel capacity. For example, by comparing

Table 3.1 SNR required to obtain a system capacity of 7 bit/s/Hz for different systems

Wireless communication systems	Signal-to-noise ratio (SNR_dB)
1 × 1 (SISO)	24.5
1 × 2 (SIMO)	16
2 × 1 (MISO)	15
2 × 2 (MIMO)	13.5
2 × 3 (MIMO)	12
2 × 4 (MIMO)	11
4 × 4 (MIMO)	8
4 × 6 (MIMO)	5

2 × 2, 2 × 4, 4 × 4, and 4 × 6 configurations, the channel capacities of the system are 7.5, 10, 15, and 20 bit/s/Hz, respectively, at a signal-to-noise ratio of 18 dB.

(2) By comparing the channel capacity curves of 1 × 1, 1 × 2, and 2 × 1, it is shown that both the transmit-diversity and receive-diversity techniques can increase the channel capacity; that is, the two techniques serve the same purpose. When the product of the number of transmitting and receiving antennas is a certain value, the effect of the receive diversity technology is more obvious than that of the transmit diversity technology, in terms of increasing the channel capacity.

(3) To obtain the same channel capacity, a MIMO system requires a much lower signal-to-noise ratio than the SISO, SIMO, and MISO systems. Table 1 gives the required signal-to-noise ratio for each system to achieve a channel capacity of 7 bit/s/Hz. Comparing the four curves, 1 × 1, 1 × 2, 2 × 1, and 4 × 4, we can see that the MIMO system capacity is 16.5 dB higher than that of the SISO system, 8 dB higher than that of the SIMO system, and 7 dB higher than that of the MISO system.

In summary, with the increase in the number of transmitting and receiving antennas, the channel capacity of the system increases almost exponentially, indicating that the MIMO system has a significant increase in channel capacity compared to the SISO system. This proves that the channel capacity can be significantly increased, without increasing the spectrum resources and transmitting power, when the MIMO technology is used in wireless optical communication, as proposed in this paper. Thus, the information-transmission efficiency can be improved by increasing the channel capacity. Similarly, increasing the information redundancy while maintaining a constant transmission rate can also improve the system performance.

3.4 Space-Time Coding Design Guidelines

The space-time coding design guidelines are established from the perspective of spatial-diversity gain and coding gain and are set differently depending on whether the fading coefficients of different channels are independent or correlated.

First, we assume that the signal to be sent is $c = x_1^1, x_1^2, \ldots x_1^N, x_2^1, x_2^2, \ldots x_2^N, \ldots x_p^1, x_p^2, \ldots x_p^N,$ and the codeword obtained at the receiver using the maximum-likelihood judgment is $e = e_1^1, e_1^2, \ldots e_1^N, e_2^1, e_2^2, \ldots e_2^N, \ldots, e_p^1, e_p^2, \ldots, e_p^N$. Thus, we have the following definition.

Definition 3.1 [36] We define the codeword difference matrix $B(c, e)$ as

$$B(c,e) = \begin{bmatrix} e_1^1 - x_1^1 & e_2^1 - x_2^1 & \ldots & e_p^1 - x_p^1 \\ e_1^2 - x_1^2 & e_2^2 - x_2^2 & \ldots & e_p^2 - x_p^2 \\ e_1^3 - x_1^3 & e_2^3 - x_2^3 & \ldots & e_p^3 - x_p^3 \\ \vdots & \vdots & \ddots & \vdots \\ e_1^N - x_1^N & e_2^N - x_2^N & \ldots & e_P^N - x_P^N \end{bmatrix} \tag{3.49}$$

Definition 3.2 [36, 37] We define the codeword distance matrix as follows:

$$A(c, e) = B(c, e) \bullet B^H(c, e) \tag{3.50}$$

In the above equation, $(\cdot)^H$ is the conjugate transpose of the matrix. Because $A(c, e) = A^H(c, e)$ is an $N_t \times N_t$ matrix and the eigenvalues of the matrix $A(c, e)$ are nonnegative real numbers, it is a nonnegative definite Hermitian matrix. If the rank of $A(c, e)$ is denoted by r and is the number of transmitting antennas in a slow-fading channel, then $A(c, e)$ will have r nonzero eigenvalues $\lambda_i (i = 1, \ldots, r)$. δ_H is defined as the number of distinct space-time code words in c and e, which is the space-time symbol-by-symbol Hamming distance between the two code words and is equal to the number of transmitting antennas in a fast-fading channel.

Based on the relationship between the symbol period of the signal and the correlation time of a time-varying channel under time-varying channel conditions, the channel can be divided into fast-fading and slow-fading channels.

3.4.1 Coding Design Guidelines for Slow Raleigh-Fading Channels

If the coherence time of a time-varying channel is significantly longer than the symbol period of the signal, such a time-varying channel is called a slow-fading channel. Under such conditions, the time-varying channel-fading coefficient is static and essentially unchanged over multiple consecutive symbol periods.

Definition 3.3 [36, 37] In a slow-fading channel, the product of the minimum rank number and the number of receiving antennas, rN_R, is referred to as the minimum diversity. The coding gain can be expressed as

$$G_c = \frac{\left(\prod_{i=1}^{r} \lambda_i\right)^{1/r}}{d_u^2} \tag{3.51}$$

In the above equation, d_u^2 is the square of the Euclidean distance when there is relatively no coding.

The minimum error probability is possible when the product $\prod_{i=1}^{r} \lambda_i$ of the smallest nonzero eigenvalues of the matrix $A(c, e)$ of the codeword group with the smallest rank takes the maximum value.

Theorem 3.1 [36, 37] When $rN_R < 4$, we draw the following conclusions:

(1) The minimum rank of the matrix $A(c, e)$ of all codeword groups is maximized.
(2) The product $\prod_{i=1}^{r} \lambda_i$ of the smallest nonzero eigenvalues of the matrix $A(c, e)$ of a codeword group with minimum rank takes the maximum value.

Theorem 3.2 [36, 37] When $rN_R \geq 4$, we draw the following conclusions:

(1) Ensure that the minimum rank of all codeword group matrices $A(c, e)$ satisfies $rN_R \geq 4$.
(2) The minimum trace $\sum_{i=1}^{r} \lambda_i$ of all different codeword group matrices $A(c, e)$ obtains its maximum value.

3.4.2 *Coding-Design Guidelines for Fast Raleigh-Fading Channels*

If the coherence time of a time-varying channel is less than the symbol period of the signal, the time-varying channel is called a fast-fading channel. Similar to frequency-selective fading, a fast-fading channel is also known as time-selective fading.

Definition 3.4 In a fast-fading channel, the product of the minimum null-time symbol-by-symbol Hamming distance and the number of receiving antennas, $\delta_H N_R$, is referred to as the minimum diversity. The coding gain is defined as follows:

$$G_c = \frac{\left(d_p^2\right)^{1/\delta_H}}{d_u^2} \tag{3.52}$$

where d_p^2 is the product of the squares of the Euclidean distances between two null-time symbol sequences.

$$d_p^2 = \prod_{t \in \rho(c_t.e_t)} |c_t - e_t|^2 \tag{3.53}$$

where $\rho(c_t, e_t)$ is the desirable time point at $|c_t - e_t| \neq 0$.

Theorem 3.3 [36, 37] When $\delta_H N_R < 4$, we draw the following conclusions:

(1) The minimum space-time symbol-by-symbol Hamming distance δ_H between all different codeword pairs is maximum.
(2) The minimum product distance d_p^2 is the greatest along the path of the codeword pair with the minimum symbol-by-symbol Hamming distance δ_H.

Theorem 3.4 [36, 37] When $\delta_H N_R \geq 4$, we draw the following conclusions:

(1) The product of the minimum null-time symbol-by-symbol Hamming distance and number of receiving antennas $\delta_H N_R \geq 4$.
(2) The minimum Euclidean distance is maximized for all code word pairs.

From Theorem 3.2, it is clear that the number of diversities in a space-time coding system largely determines the choice of coding-design criteria for a fading channel. For a fading channel, the total diversity refers to the product of the number of receive diversities (i.e., the number of receive antennas) and the number of transmit diversities given in the coding scheme.

When the total diversity is small, the determinant with the maximum minimum rank should be chosen for slow-fading channels, and the coding with the maximum minimum symbol-by-symbol Hamming distance and product distance should be chosen for fast-fading channels to obtain the maximum diversity and coding gains. In this case, the coding performance is mainly determined by the diversity gain, which is decisive when the total number of diversities is large.

References

1. Cao Q, Brandt-Pearce M, Wilson SG (2007) Free space optical MIMO system using PPM modulation and a single optical Amplifier. Commun Network China, 508–512
2. Cvijetic N, Wilson Stephen G, Brandt-Pearce M (2008) Performance bounds for free-space optical MIMO systems with APD receivers in atmospheric turbulence. IEEE J Sel Areas Commun 26(3):3–11
3. Letzepis N, Holland I, Cowley W (2008) The Gaussian free space optical MIMO channel with Q-ary pluse position modulation. IEEE Trans Wireless Commun 7(5):1744–1753
4. Chatzidiamantis, N. D., Uysal, M., Tsiftsis, T. A., & Karagiannidis, G. K. (2009) EM-based maximum-likelihood sequence detection for MIMO optical wireless systems. In: IEEE ICC 2009, 1–5
5. Zeinab H, Jarir F, Mohsen K (2009) MIMO free space optical communications in turbid and turbulent atmosphere. J Commun 4(8):524–532

6. Mohamed MDA, Dabbo A, Hranilovic S (2008) MIMO optical wireless channels using Halftoning. IEEE 2008:1360–1364
7. Riediger MLB, Schober R, Lampe L (2008) Multiple-symbol detection for photon-counting MIMO free-space optical communications. IEEE Trans Wireless Commun 7(12):5369–5379
8. Wilson SG, Brandt-Pearce M, Cao Q, Baedke M et al (2005) Optical repetition MIMO transmission with multipulse PPM. IEEE J Sel Areas Commun 23(9):1901–1910
9. Hofmann WH, Zhu, NH, Görblich et al (2009) 1.55 μm VCSEL arrays for optical multiple-input multiple-output (MIMO). In: International quantum electronics conference (IQEC), 2009, 1–2
10. Cvijetic N, Wilson SG, Brandt-Pearce M (2007) Receiver optimization in turbulent free-space optical MIMO channels with APDs and Q-ary PPM. IEEE Photonics Technol Lett 19(2):103–105
11. Vilnrotter V, Lau CW, Srinivasan M et al (2005) Optical array receiver for communication through atmospheric turbulence. IEEE J Lightwave Technol 23(4):1664–1675
12. Brandt-Pearce M, Wilson S, Cao Q et al (2004) Code design for optical MIMO systems over fading channels. In: Proceedings of the 38th Asilomar conference on signals, systems & computers, Monterey, CA, 2004, pp 871–875
13. Djordjevic IB, Denic S, Anguita J, Vasic B, Neifeld MA (2008) LDPC-coded MIMO optical communication over the atmospheric turbulence channel. J Lightwave Technol 26(5):478–487
14. Chen J (2007) Channel and performance study of laser atmospheric transmission. Huazhong University of Science and Technology, Wuhan
15. Ke X, Chen J (2012) Comparison of the performance of pulse position modulation for wireless laser communication classes. Laser Technol 36(1):67–76
16. Qin L, Ke X (2007) Capacity analysis of optical MPPM channels without background noise. Opt Eng 34(7):107–110
17. Qin L, Ke X (2007) A two-pulse MPPM coding and mapping method. J Xi'an Univ Technol 23(3):269–272
18. Qin L, Du Y, Ke X, Yang L (2008) Implementation of MPPM compiled code system with three pulses in atmospheric laser communication. Semicond optoelectron 29(3):403–406
19. Wilson SG, Brandt-Pearce M, Cao Q et al (2005) Free-space optical MIMO transmission with Q-ary PPM. IEEE Trans Commun 53(8):1402–1412
20. Letzepis N, Holland I, Cowley W (2008) The Gaussian free space optical MIMO channel with Q-ary pulse position modulation. IEEE Trans Wireless Commun 7(5):1744–1753
21. Telatar E (1995) Capacity of multi-antennas Gaussian channels. AT&T Bell Labs. Internal Tech Memo Eur Trans Telecomm 1999 10(11):585–595
22. Foschini GJ, Gans M (1998) On limits of wireless communication in a fading environment when using multiple-element antennas. Wireless Pers Commun 3:311–335
23. Haas SM et al (2001) Capacity of wireless optical communication. IEEE J Sel Areas Commun 21(8):1346–1357
24. Chakraborty K (2005) Capacity of the MIMO optical fading channel. In: Information theory, 2005 ISIT 2005:530–534
25. Tao C, Tao C (1999) Optical information theory. Science Press, Beijing
26. Paulraj A, Nabar R, Gore D (2003) Introduction to space-time wireless communications. Cambridge University Press, Now York
27. Cover TM, Thomas JA (1991) Elements of information theory. Wiley
28. Wilson SG, Brandt-Pearce M, Cao Q, Baedke M (2005) Optical repetition MIMO transmission with multipulse PPM. IEEE J Sel Areas Commun 23(9):1901–1910
29. Pauiraj A, Nabar R, Gore D (2003) Introduction to space-time wireless eommunieations. Cambridge University Press, Cambridge
30. Vucetic B, Yuan J (2004) Space-time coding techniques (Tran by Wang X). Machinery Industry Press, Beijing
31. Goldsmith A (2007) Wireless communication (Translated by Yang H, Li W, Guo W et al). People's Post and Telecommunications Publishing House, Beijing

32. Winters JH (1987) On the capacity of radio communication systems with diversity in a Rayleigh fading environment. IEEE J Sel Areas Commun 5(5):868–875
33. Shelf T (2007) Research on diversity multiplexing tradeoffs under correlated channels. M.S. Thesis. Dalian University of Technology, Dalian
34. Ozarow L, Shamai S, Wyner A (1994) Information-theoretic considerations in cellular mobile radio. IEEE Trans Veh Technol 43:357–375
35. Alamout SM (1998) A simple transmit diversity technique for wireless communications. IEEE J Sel Area Commun 16:1451–1457
36. Branka V, Yuan J, Wang X et al (2004) Space-time coding technology. Mechanical Industry Press, 2004.07
37. Tarokh V, Seshadri N (1998) Space-time codes for high data rate wireless communication: performance criterion and code construction. IEEE Trans Inf Theory 44:744–765

Chapter 4
Orthogonal Space–Time Block Codes

Space–time coding (STC) is the combination of spatial and temporal transmission signals, which is essentially combining spatial and temporal two-dimensional processing. Space–time coding introduces temporal and spatial correlation in the signals transmitted by different antennas to provide diversity and coding gains for the receiver without sacrificing bandwidth.

In space, the spatial diversity of multiple transmitting and receiving antennas is used to improve the capacity and speed of wireless communication systems. In terms of time, different signals are transmitted in different time slots using the same antenna, so that the receiver can obtain receive diversity. In this manner, diversity and coding gains can be obtained simultaneously, thus, achieving a high transmission rate and reliable transmission of information.

Constructing space–time coding involves seeking a reasonable compromise between decoding complexity, optimizing system performance, and maximizing the information rate. Orthogonal space–time coding is a coding method to improve system performance by increasing information redundancy. It overcomes channel fading by generating spatial and temporal correlations between multiple transmitting antennas and transmitting signals for each period [1]. Orthogonal space–time block codes (OSTBC) are orthogonal codes designed to construct space–time codes.

The design principle of a codeword is to satisfy the orthogonality between each transmitting antenna. When receiving, the maximum-likelihood detection algorithm is used for decoding. Because of the orthogonality between codewords, only simple linear processing is needed at the receiving end. In this chapter, starting from classical Alamouti space–time coding [1] and improved Alamouti space–time coding [2], combined with the characteristics of pulse-position modulation (PPM) technology, we discuss a space–time coding method suitable for wireless laser communication.

X. Ke, *Space-Time Coding Theory in Optical Wireless MIMO Systems*,
Optical Wireless Communication Theory and Technology,
https://doi.org/10.1007/978-981-97-6449-5_4

4.1 Alamouti Space–Time Coding

The earliest space–time coding (STC) structure was proposed by Alamouti [1], and has been widely valued. A space–time block code (STBC) is an encoding form, based on transmission diversity. Trokh [3] extended the two-antenna transmitting-diversity scheme proposed by Alamouti to multiple transmitting antennas and proposed a space–time block code using orthogonal design theory. Such orthogonal space–time block codes can obtain a full set gain and are relatively simple to decode.

Because atmospheric laser communication is very different from traditional radio-frequency (RF) communication, space–time coding must be modified to accommodate it. Therefore, Simon and Vilnrotter et al. improved the classical Alamouti code [2] and proposed an orthogonal space–time coding scheme suitable for intensity modulation/direct detection (IM/DD) optical communication. Alamouti space–time coding is not only easy to implement but also can obtain a full-score set gain.

4.1.1 *Coding Alamouti Codes*

The transmit-diversity scheme proposed by Alamouti is shown in Fig. 4.1 [1]. In Alamouti space–time coding, the binary information bits sent by the source are firstly modulated (constellation mapping); then, the modulated symbols x_1 and x_2 are sent into the encoder, and encoded in the following manner:

$$X = \begin{bmatrix} x_1 & x_2 \\ -x_2^* & x_1^* \end{bmatrix} \tag{4.1}$$

The encoded symbols are sent from the two antennas respectively, that is, in the transmitting cycle of the first symbol, symbols x_1 and x_2 are sent simultaneously from transmitting antennas 1 and 2. In the second symbol-transmission period, symbols $-x_1^*$ and x_1^* are sent simultaneously from transmitting antennas 1 and 2, respectively. x_1^* is the complex conjugate of x_1. This method encodes in both the space and

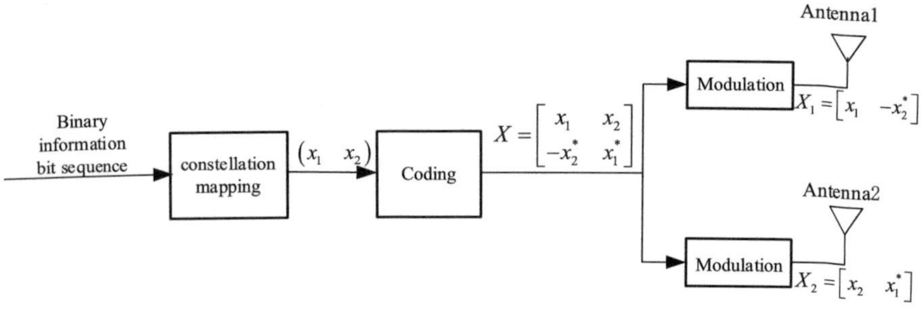

Fig. 4.1 Alamouti transmit-diversity space–time coding scheme

time domains. X_1 and X_2 are the symbols sent from transmitting antennas 1 and 2, respectively; then,

$$X_1 = \left(x_1 - x_2^*\right) \tag{4.2}$$

$$X_2 = \left(x_2 x_1^*\right) \tag{4.3}$$

$$X_1 X_2^T = x_1 x_2^* - x_1 x_2^* = 0 \tag{4.4}$$

The coding matrix has the following characteristics:

$$XX^T = \begin{bmatrix} |x_1|^2 + |x_2|^2 & 0 \\ 0 & |x_1|^2 + |x_2|^2 \end{bmatrix} = \left(|x_1|^2 + |x_2|^2\right) I_2 \tag{4.5}$$

where I_2 is the identity matrix of 2×2.

4.1.2 Alamouti-Code Decoding

Figure 4.2 shows the Alamouti space–time coding receiver with a receiving antenna. It is assumed that the channel is fast fading; that is, the fading coefficient remains constant for two consecutive symbol-transmission cycles. Suppose $h_1(t)$ represents the channel-fading coefficient from transmitting antenna 1 to the receiving antenna at time t, $h_2(t)$ represents the channel-fading coefficient from transmitting antenna 2 to the receiving antenna at time t, and both of them are assumed to follow the Rayleigh distribution.

Then, at the receiving end, the received signal in two consecutive symbolic cycles can be expressed as

$$r_1 = h_1 x_1 + h_2 x_2 + n_{01} \tag{4.6}$$

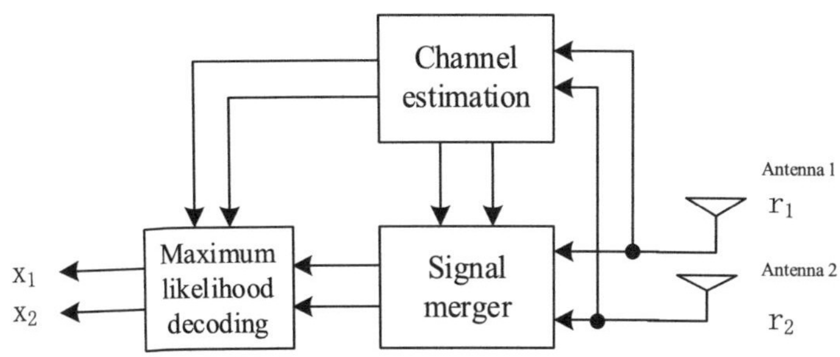

Fig. 4.2 Alamouti space–time block-code receiver

$$r_2 = -h_1x_2^* + h_2x_1^* + n_{02} \tag{4.7}$$

where r_1 and r_2 represent the signal received by the receiving antenna at moments t and $t + T$, respectively. n_{01} and n_{02} represent the independent complex Gaussian white noise of the receiving antenna at moments t and $t + T$, respectively. It is assumed that the receiver knows the fading coefficients h_1 and h_2 of the channel, and uses the maximum-likelihood decoding method. That is, it finds a pair of symbols $(\hat{x}_1, \hat{x}_2)$ from the constellation to minimize the Euclidean distance of the following equation:

$$\begin{aligned} &d^2(r_1, h_1\hat{x}_1 + h_2\hat{x}_2) + d^2\left(r_2, -h_1\hat{x}_2^* + h_2\hat{x}_1^*\right) \\ &= \left|r_1 - h_1\hat{x}_1 - h_2\hat{x}_2'\right|^2 + \left|r_2 + h_1\hat{x}_2^* - h_2\hat{x}_1^*\right|^2 \to \min \end{aligned} \tag{4.8}$$

We substitute Eqs. (4.6) and (4.7) into Eq. (4.8); then, the above equation can become

$$\begin{aligned} (\hat{x}_1, \hat{x}_2) = \arg\ \min&\left(|h_1|^2 + |h_2|^2 - 1\right)\left(|\hat{x}_1|^2 + |\hat{x}_2|^2\right) + d^2(\tilde{x}_1, \hat{x}_1) \\ &+ d^2(\tilde{x}_2, \hat{x}_2) \quad (\hat{x}_1, \hat{x}_2) \in C \end{aligned} \tag{4.9}$$

where C is the set of all possible transmitted symbols, and $\tilde{x}_1, \tilde{x}_2$ are the signals obtained by combining the channel-fading coefficient and the received signal:

$$\tilde{x}_1 = h_1^*r_1 + h_2r_2^* = \left(|h_1|^2 + |h_2|^2\right)x_1 + h_1^*n_{01} + h_2n_{02}^* \tag{4.10}$$

$$\tilde{x}_2 = h_2^*r_1 - h_1r_2^* = \left(|h_1|^2 + |h_2|^2\right)x_2 - h_1n_{02}^* + h_2n_{01} \tag{4.11}$$

The combined signals $\tilde{x}_1$ and $\tilde{x}_2$ are functions of x_1 and x_2, respectively, when the receiver has obtained the channel-fading coefficients h_1 and h_2. Therefore, Eq. (4.9) can be decomposed into two independent decoding algorithms, namely,

$$\hat{x}_1 = \arg\ \min\left(|h_1|^2 + |h_2|^2 - 1\right)\ |\hat{x}_1|^2 + d^2(\tilde{x}_1, \hat{x}_1), \quad \hat{x}_1 \in S \tag{4.12}$$

$$\hat{x}_2 = \arg\ \min\left(|h_1|^2 + |h_2|^2 - 1\right)\ |\hat{x}_2|^2 + d^2(\tilde{x}_2, \hat{x}_2), \quad \hat{x}_2 \in S \tag{4.13}$$

where S is the modulation mapping constellation. If the quadrature amplitude modulation (QAM) or M-ary phase-shift keying (M-PSK) constellations are used, $(|h_1|^2 + |h_2|^2 - 1)|\hat{x}_i|^2$ $(i = 1, 2)$ is constant for all signals, given the signal-fading coefficient. Therefore, decision Eqs. (4.12) and (4.13) can be simplified as

$$\hat{x}_1 = \arg\ \min\ d^2(\tilde{x}_1, \hat{x}_1), \quad \hat{x}_1 \in S \tag{4.14}$$

$$\hat{x}_2 = \arg\ \min\ d^2(\tilde{x}_2, \hat{x}_2), \quad \hat{x}_2 \in S \tag{4.15}$$

As can be seen from Eqs. (4.14) and (4.15), maximum-likelihood decoding can be completed, as long as the codeword with the smallest metric value is found, and the computational complexity is not high when the maximum-likelihood decoding algorithm is adopted.

4.2 Alamouti Space–Time Coding in Optical Communication

In the traditional Alamouti code, negative and complex numbers exist; however, complex conjugate and negative numbers cannot be represented by "*on*" and "*off*". Because intensity modulation is generally adopted in atmospheric laser communication, researchers [2] improved the traditional Alamouti code and proposed a space–time coding method suitable for the IM/DD of turbulent atmospheric channels. The improved Alamouti code uses two transmitting antennas and one receiving antenna and has error performance similar to the traditional Alamouti code.

4.2.1 *Improving the Coding of Alamouti Code*

Simon and Vilnrotter [2] proposed an improved Alamouti code. The encoding method operates in the real-number field and avoids negative and complex signals by complementing the symbols. The improved coding matrix is [2]

$$X = \begin{bmatrix} x_1 & \overline{x}_2 \\ x_2 & x_1 \end{bmatrix} \tag{4.16}$$

where $\overline{x}_2$ represents the complement of symbol x_2. In this coding method, the key is obtaining complement $\overline{x}_2$. Therefore, for different modulation modes, the complement form of a symbol x, namely $\overline{x}$, is given below. Its complement is defined as

$$\overline{x}_i = A - x_i \tag{4.17}$$

where A represents the pulse amplitude and is a constant related to the transmitted light intensity. As can be seen from the above equation, taking the complement means reversing the state of the transmitting optical signals "*on*" and "*off*". For on–off keying (OOK) modulation, assume that the signal is

$$\begin{cases} s_1(t) = 0 \quad 0 < t < T \\ s_2(t) = A \quad 0 < t \le T \end{cases} \tag{4.18}$$

where T represents the character period. Then, its complement is

$$\begin{cases} \overline{s}_1(t) = A\ 0 < t < T \\ \overline{s}_2(t) = 0\ 0 < t \leq T \end{cases} \tag{4.19}$$

For binary PPM modulation, it is assumed that the transmitted signal form is

$$\begin{aligned} s_1(t) &= \begin{cases} 0\ 0 < t < \frac{T}{2} \\ A\ \frac{T}{2} \leq t \leq T \end{cases} \\ s_2(t) &= \begin{cases} A\ 0 < t < \frac{T}{2} \\ 0\ \frac{T}{2} \leq t \leq T \end{cases} \end{aligned} \tag{4.20}$$

Then, its corresponding complement form is

$$\begin{aligned} \overline{s}_1(t) &= \begin{cases} A\ 0 < t < \frac{T}{2} \\ 0\ \frac{T}{2} \leq t \leq T \end{cases} \\ \overline{s}_2(t) &= \begin{cases} 0\ 0 < t < \frac{T}{2} \\ A\ \frac{T}{2} \leq t \leq T \end{cases} \end{aligned} \tag{4.21}$$

4.2.2 Improved Alamouti-Code Decoding

According to the characteristics of the improved Alamouti code [2] and the optical-MIMO channel model, for a 2 × 1 atmospheric laser-communication system, the numbers r_1 and r_2 respectively represent the signals received by the detector at moments t and $t + T_s$:

$$\begin{cases} r_1 = \eta(h_1 C_1 + h_2 C_2) + n_1 \\ r_2 = \eta(h_1 \overline{C}_2 + h_2 C_1) + n_2 \end{cases} \tag{4.22}$$

where n_1 and n_2 represent Gaussian white noise with a mean value of 0 and variance of N_0. Using the same method as the traditional Alamouti coding scheme to merge signals,

$$\begin{cases} \tilde{x}_1 = h_1 r_1 + h_2 r_2 = \eta(h_1^2 + h_2^2)C_1 + \eta h_1 h_2 A + h_1 n_1 + h_2 n_2 \\ \tilde{x}_2 = h_2 r_2 - h_1 r_1 = \eta(h_1^2 + h_2^2)C_2 - \eta h_1 h_2 A + h_2 n_1 + h_1 n_2 \end{cases} \tag{4.23}$$

It can be seen from Eq. (4.23) that $\tilde{x}_1$ and $\tilde{x}_2$ are related to $\eta h_1 h_2 A$, in addition to the codeword sent by themselves. When $\eta h_1 h_2 A$ is constant in the two adjacent character periods, Eq. (4.23) or the deformed formula can be used for signal decision, namely

$$\begin{cases} \tilde{x}_1 - \eta h_1 h_2 A = h_1 r_1 + h_2 r_2 = \eta(h_1^2 + h_2^2)C_1 + h_1 n_1 + h_2 n_2 \\ \tilde{x}_2 + \eta h_1 h_2 A = h_2 r_2 - h_1 r_1 = \eta(h_1^2 + h_2^2)C_2 + h_2 n_1 + h_1 n_2 \end{cases} \tag{4.24}$$

We extend this coding method to find a space–time coding scheme suitable for arbitrary Q-PPM modulation. Suppose that any Q-PPM modulated signal is

$$s_i(t) = \begin{cases} A & \frac{iT}{Q} \leq t \leq \frac{(i+1)T}{Q} \\ 0 & \text{other} \end{cases} \tag{4.25}$$

Similar to 2-PPM modulation, the complement form $s_i(t)$ is

$$\overline{s}_i(t) = \begin{cases} 0 & \frac{iT}{Q} \leq t \leq \frac{(i+1)T}{Q} \\ A & \text{other} \end{cases} \tag{4.26}$$

As can be seen from Eq. (4.26), although the complement form $\overline{s}_i(t)$ of $s_i(t)$ can be obtained for any Q-PPM modulation, we notice that the signal $\overline{s}_i(t)$ in other time slots is A, which means that the number of "*on*" time slots sent is increased; that is, the number of "*on*" time slots is Q-1, and the total power on each symbol will be changed.

4.3 Orthogonal Space–Time Coding in Bit-Complement Form

To adopt space–time coding in IM/DD atmospheric laser communication, Simon and Vilnrotter [2] improved the Alamouti code. Their code has the same performance as the Alamouti code, but it is only suitable for OOK and 2-PPM modulation; it is inapplicable to any Q-PPM modulation ($Q > 2$). To obtain a code suitable for any Q-PPM modulation ($Q > 2$) space–time block code, we modified the improved Alamouti code.

4.3.1 Bit-Complement Orthogonal Space–Time–Code Coding Method

In this study, we combine space–time coding with modulation technology and construct a coding method that uses the relationship between binary bits (i.e., the complementary relationship between binary bits "**0**" and "**1**") to first carry out a bit complement before Q-PPM modulation.

Bit-complement orthogonal space–time coding (BCOSTBC) is modified again, based on the improved Alamouti code. It has the same coding matrix as the improved

Table 4.1 Codewords corresponding to the bit-complement orthogonal space–time codes in 4-PPM modulation

Quad code element	Binary bit		
		x	$\overline{x}$
0	**0 0**	1000	0001
0	**0 1**	0100	0010
0	**1 0**	0010	0100
0	**1 1**	0001	1000

Alamouti code; the main difference is that the method of obtaining complement $\overline{x}_i$ is different. The specific method is as follows: At time $t + T$, bit $k = \log_2 Q$ corresponding to x_2 is first supplemented, and then Q-PPM modulation is carried out to obtain the modulated symbol $\overline{x}_i$. This ensures that the total power on each symbol remains constant when sending the modulated $\overline{x}_i$ symbol.

In Q-PPM modulation, because the transmitted signal x_i is composed of "*on*" and "*off*" in a timeslot, without loss of generality, the "*on*" timeslot is represented by 1 and the "*off*" timeslot is represented by 0. Then, the Q-PPM modulation symbol can be represented by a sequence of "**0**" and "**1**". Taking 4-PPM modulation as an example, the codewords encoded by the bit-complement orthogonal space–time code are shown in Table 4.1.

4.3.2 *Decoding Bit-Complement Orthogonal Space–Time Coding*

For an $M \times N$ MIMO system (M represents the number of transmitting antennas and N represents the number of receiving antennas), after considering the influence of thermal noise, the signal $r^{(n)}$ received on the nth detector can be expressed as

$$r^{(n)} = \frac{\eta I_s}{M} \sum_{m=1}^{M} x^{(m)} a_{nm}^2 + \eta I_b + v_n \tag{4.27}$$

where $x^{(m)}$ is the signal sent from the mth laser; η is the photoelectric conversion efficiency; I_s represents the maximum received light intensity without fading; a_{nm} represents the path-amplitude fading gain from the nth laser to the Z th detector; I_b represents the intensity of light produced by background radiation; v_n is Gaussian white noise with a mean value of 0 and variance of N_0, indicating the noise generated by the detector and its attached electronic circuit.

It is assumed that the receiver can accurately estimate the intensity-fading coefficient of the channel. Taking the 2×2 system as an example, the decoding process of bit-complement orthogonal space–time coding is illustrated. $r_1^{(n)}$ and r_2^n represent the signals received by the nth detector at moments t and $t + T$, respectively:

$$\begin{cases} r_1^{(n)} = \frac{\eta I_s}{M}\left(a_{n1}^2 x_1 + a_{n2}^2 x_2\right) + \eta I_b + v_{n1} \\ r_2^{(n)} = \frac{\eta I_s}{M}\left(a_{n1}^2 \bar{x}_2 + a_{n2}^2 x_1\right) + \eta I_b + v_{n2} \end{cases} \tag{4.28}$$

In BCOSTBC code, the signal can be obtained by equal-gain merging:

$$\begin{cases} \tilde{x}_1 = \sum\limits_{n=1}^{N}\left[a_{n1}^2 r_1^{(n)} + a_{n2}^2 r_2^{(n)}\right] \\ \tilde{x}_2 = \sum\limits_{n=1}^{N}\left[a_{n2}^2 r_1^{(n)} + a_{n1}^2 \bar{r}_2^{(n)}\right] \end{cases} \tag{4.29}$$

$$\begin{cases} \tilde{x}_1 = \frac{\eta I_s}{M}\left(\sum\limits_{n=1}^{N}\sum\limits_{m=1}^{M} a_{nm}^4 x_1 + \sum\limits_{n=1}^{N}\prod\limits_{m=1}^{M} a_{nm}^2 x_2 + \sum\limits_{n=1}^{N}\prod\limits_{m=1}^{M} a_{nm}^2 \bar{x}_2\right) + \sum\limits_{n=1}^{N}\left(a_{n1}^2 v_{n1} + a_{n2}^2 v_{n2}\right) \\ \quad + \eta I_b \sum\limits_{n=1}^{N}\sum\limits_{m=1}^{M} a_{nm}^2 \\ \tilde{x}_2 = \frac{\eta I_s}{M}\left(\sum\limits_{n=1}^{N}\sum\limits_{m=1}^{M} a_{nm}^4 x_2 + \sum\limits_{n=1}^{N}\prod\limits_{m=1}^{M} a_{nm}^2 x_1 + \sum\limits_{n=1}^{N}\prod\limits_{m=1}^{M} a_{nm}^2 \bar{x}_1\right) + \sum\limits_{n=1}^{N}\left(a_{n2}^2 v_{n1} + a_{n1}^2 v_{n2}\right) \\ \quad + \eta I_b \sum\limits_{n=1}^{N}\sum\limits_{m=1}^{M} a_{nm}^2 \end{cases} \tag{4.30}$$

According to Eq. (4.30), $\tilde{x}_1$ and $\tilde{x}_2$ are only related to x_1 and x_2, respectively, and are unrelated to the other symbols. Therefore, the signal can be detected using this item. Without loss of generality, the combined signal $\tilde{x}_j (j = 1, 2)$ also consists of "*on*" and "*off*" time slots. If $I_{on,i}$ is the current in the "*on*" slot and I_{off} is the current in the "*off*" slot, then $I_{on,i}$ and I_{off} are, respectively,

$$\begin{cases} I_{on,j} = \frac{\eta I_s}{M}\sum\limits_{n=1}^{N}\sum\limits_{m=1}^{M} a_{nm}^4 + \sum\limits_{n=1}^{N}\sum\limits_{m=1}^{M} a_{nm}^2 v_{nm} + \eta I_b \sum\limits_{n=1}^{N}\sum\limits_{m=1}^{M} a_{nm}^2 \\ I_{off,j} = \sum\limits_{n=1}^{N}\sum\limits_{m=1}^{M} a_{nm}^2 v_{nm} + \eta I_b \sum\limits_{n=1}^{N}\sum\limits_{m=1}^{M} a_{nm}^2 \end{cases} \tag{4.31}$$

It is assumed that the background radiation received by any detector is uniform and the received signal is first processed by a filter. In this case, the current signal generated by the background radiation can be eliminated after passing through the integrator. To simplify the form of the output signal, we define the electric energy on each symbol of any detector in the absence of turbulence as $E_s = \frac{(\eta I_s)^2 T}{Q}$. Then, the output of the detection signal through the integrator can be simplified as

$$\begin{cases} z_{on,n} = \frac{\sqrt{E_s}}{M}\sum\limits_{n=1}^{N}\sum\limits_{m=1}^{M} a_{nm}^4 + \sum\limits_{n=1}^{N}\sum\limits_{m=1}^{M} a_{nm}^2 v_n \\ z_{off,n} = \sum\limits_{n=1}^{N}\sum\limits_{m=1}^{M} a_{nm}^2 v_n \end{cases} \tag{4.32}$$

Let z_{nq} be an implementation of the n detector at slot q. The received observation set can be expressed as $Z = \{z_{nq}, n = 1, \ldots, N; q = 0, \ldots, Q-1\}$.On a Q-PPM modulation symbol, Q^j_{on} and Q^j_{off} represent the "*on*" and "*off*" time-slot sets of the j modulation symbol, respectively.

According to Bayesian theory, the logarithmic-likelihood ratio can be defined as

$$\Lambda(r) = \log \frac{P_r(r = on|Z, a)}{P_r(r = off|Z, a)} = \log \frac{\sum_{q \in Q^j_{on}} P_r(Z|X = q, a) P_r(X = q)}{\sum_{q \in Q^j_{off}} P_r(Z|X = q, a) P_r(X = q)} \tag{4.33}$$

where $P_r(X = q)$ represents the prior probability at $X = q$. Assuming that all symbols are transmitted with equal probability, Eq. (4.33) can be rewritten as

$$\Lambda(r) = \log \frac{\sum_{q \in Q^j_{on}} \exp[\log P_r(Z|X = q, a)]}{\sum_{q \in Q^j_{off}} \exp[\log P_r(Z|X = q, a)]} \tag{4.34}$$

In a Gaussian distribution, the logarithmic-probability distribution is zero:

$$\begin{aligned} \log P_r(Z|X = i, a) = &-\frac{NQ}{2}\log N_0 - \frac{NQ}{2}\log 2\pi - \frac{1}{2N_0}\sum_{n=1}^{N}\left(z_{ni} - \tilde{a}\sqrt{E_s}\right)^2 \\ &- \frac{1}{2N_0}\sum_{n=1}^{N}\sum_{\substack{q=1 \\ q \neq i}}^{Q} z_{nq}^2 \end{aligned} \tag{4.35}$$

where $\tilde{a} = \frac{1}{M}\sum_{n=1}^{N}\sum_{m=1}^{M} a^4_{nm}$. By substituting Eq. (4.35) into Eq. (4.34), the logarithmic likelihood ratio of the BCOSTBC code is

$$\Lambda(r) = \log \frac{\sum_{q \in Q^j_{on}} \exp\left[\frac{\sqrt{E_s}}{N_0}\sum_{n=1}^{N} z_{nq}\tilde{a}\right]}{\sum_{q \in Q^j_{off}} \exp\left[\frac{\sqrt{E_s}}{N_0}\sum_{n=1}^{N} z_{nq}\tilde{a}\right]} \tag{4.36}$$

When the signal-to-noise ratio (SNR) is large, it can be approximated by $\log \sum_{i=1}^{N} \exp(a_i) \approx \max(a_1, \ldots, a_N)$; then, Eq. (4.36) can be simplified as

$$\Lambda(r) \approx \frac{\sqrt{E_s}}{N_0}\left(\max_{q \in Q^j_{on}} \sum_{n=1}^{N} z_{nq}\tilde{a} - \max_{q \in Q^j_{on}} \sum_{n=1}^{N} z_{nq}\tilde{a}\right) \tag{4.37}$$

Then, the maximum likelihood criterion of the BCOSTBC code is $\hat{x}_{nq} = \begin{cases} 1 & \Lambda(r) \geq 0 \\ 0 & \Lambda(r) < 0 \end{cases}$.

4.3.3 Bit-Complement Orthogonal Space–Time–Coding Error Performance

Without loss of generality, we assume that each channel between the laser and detector is independent of the others. Then, the symbol error rate P_s of the BCOSTBC code can be expressed as

$$P_s = 1 - P \tag{4.38}$$

where P represents the average correct probability of the Q-PPM symbol received by the jth detector. When the length of the Q-PPM modulation symbol is fixed and a character has only a Q-1 space–time slot, then the average correct probability P of the jth detector receiving the Q-PPM symbol can be expressed as

$$P = \int (1 - p(\mathit{off}\,|on, a)) \cdot (1 - p(on|\mathit{off}\,, a))^{Q-1} \cdot f_{A^2}(a)da \tag{4.39}$$

where $f_{A^2}(a)$ represents the probability-density function $\sum_{n=1}^{N}\sum_{m=1}^{M} a_{nm}^4 a_{nm}^4$, which obeys the $\chi^2(t)$ distribution. The probability-density function of the $\chi^2(t)$ distribution is zero:

$$f_{A^2}(a) = \begin{cases} \frac{1}{2^{\frac{t}{2}}\Gamma(\frac{t}{2})} x^{\frac{t}{2}-1} e^{-\frac{a}{2}} & a > \mu_x \\ 0 & a \le \mu_x \end{cases} \tag{4.40}$$

Because the probability-density function of the $\chi^2(t)$ distribution is difficult to use for calculation, the convexity of the bit-error rate (BER) function is also taken into account; i.e., $\mathrm{E}[P(a)] \ge P[\mathrm{E}(a)]$, where $\mathrm{E}[\,\cdot\,]$ represents the mean value. If $\frac{1}{NM}\sum_{n=1}^{N}\sum_{m=1}^{M} a_{nm}^4$ is considered as the average of a^4, then the average probability of being right can be simplified.

$p(on|\mathit{off}\,, a)$ is the probability that no information pulse is sent, and misjudged as an information pulse when the light-intensity fluctuation is a^2. $p(\mathit{off}\,|on\,, a)$ is the probability that the sent information pulse is not detected when the light-intensity fluctuation is a^2. $p(on|\mathit{off}\,, a)$ and $p(\mathit{off}\,|on\,, a)$ are

$$p(on|\mathit{off}\,, a) = \frac{1}{\sqrt{2\pi N_0 \sum_{n=1}^{N}\sum_{m=1}^{M} a_{nm}^4}} \int_b^{\infty} \exp\left(-\frac{r_j^2}{2N_0 \sum_{n=1}^{N}\sum_{m=1}^{M} a_{nm}^4}\right) dr_j \tag{4.41}$$

$$p(\mathit{off}\,|on, a) = \frac{1}{\sqrt{2\pi N_0 \sum_{n=1}^{N}\sum_{m=1}^{M} a_{nm}^4}} \int_{-\infty}^{b} \exp\left(-\frac{\left(r_j - \sqrt{E_s}\tilde{a}\right)^2}{2\sum_{n=1}^{N}\sum_{m=1}^{M} a_{nm}^4 N_0}\right) dr_j \tag{4.42}$$

where b represents the decision threshold, and the best decision threshold is $b* = \frac{1}{2}\sqrt{E_s}\tilde{a}$. Then, Eqs. (4.41) and (4.42) can be transformed into

$$p(on|off\ , a) = \frac{1}{2}\left[1 - erf\left(\frac{\sqrt{E_s}\tilde{a}}{2\sqrt{2N_0\sum_{n=1}^{N}\sum_{m=1}^{M}a_{nm}^4}}\right)\right] \tag{4.43}$$

$$p(off\,|on, a) = \frac{1}{2}\left[1 + erf\left(\frac{\sqrt{E_s}\tilde{a}}{2\sqrt{2N_0\sum_{n=1}^{N}\sum_{m=1}^{M}a_{nm}^4}}\right)\right] \tag{4.44}$$

If the SNR is defined as $SNR = \frac{E_s}{N_0}$ without fading, then the SNR per bit is $\frac{E_b}{N_0} = \frac{SNR}{R\log_2^Q}$, where R represents the transmission rate. Equations (4.43) and (4.44) are substituted into Eq. (4.39) and denoted by SNR:

$$P = \frac{1}{2Q}\int erfc\left(\frac{\sqrt{SNR}\tilde{a}}{2\sqrt{2\sum_{n=1}^{N}\sum_{m=1}^{M}a_{nm}^4}}\right)\left[1 + erf\left(\frac{\sqrt{SNR}\tilde{a}}{2\sqrt{2\sum_{n=1}^{N}\sum_{m=1}^{M}a_{nm}^4}}\right)\right]^{Q-1} \cdot f_A(a)da \tag{4.45}$$

Because the BCOSTBC coding method applies to optical MIMO technology, the distance between detectors is only a few tens of cm apart, and its attenuation coefficients can be considered to be approximately equal. Then, the average sign-correct probability of the ith detector can be further simplified as

$$P = \frac{1}{2^Q}\int erfc\left(\sqrt{\frac{N}{8M}\cdot SNR\cdot a^2}\right)\left[1 + erf\left(\sqrt{\frac{N}{8M}\cdot SNR\cdot a^2}\right)\right]^{Q-1} \cdot f_A(a)da \tag{4.46}$$

Substituting Eq. (4.46) into Eq. (4.38), the sign-error probability of the BCOSTBC code is

$$P_s = 1 - \frac{1}{2^Q}\int erfc\left(\sqrt{\frac{N}{8M}\cdot SNR\cdot a^2}\right)\left[1 + erf\left(\sqrt{\frac{N}{8M}\cdot SNR\cdot a^2}\right)\right]^{Q-1} \cdot f_A(a)da \tag{4.47}$$

The error ratio can also be obtained according to the relationship between the error ratio and the sign-error probability:

$$BER = \frac{Q}{2(Q-1)}P_s \tag{4.48}$$

4.3.4 Simulation Analysis

The error performance of 1×1, 1×2, 2×1, and 2×2 systems was simulated by the Monte Carlo method where the total power of the system E_s is unchanged and the channel parameters are known. Simulation parameters $\eta = 0.5$ and $R = 1$. Figures 4.3, 4.4, 4.5 and 4.6 show the simulation results.

Figure 4.3 shows the BCOSTBC bit-error rate at 4-PPM modulation with scintillation index $S.I. = 0.6$. As shown in Fig. 4.3, the bit-error performance of the BCOSTBC system is significantly better than that of the 1×1 system. At $P_s = 10^{-2}$, compared with the 1×1 system, the error performance of the 2×1 and 2×2 systems, after adopting BCOSTBC codes, is improved by 8 dB and 18 dB, respectively. The

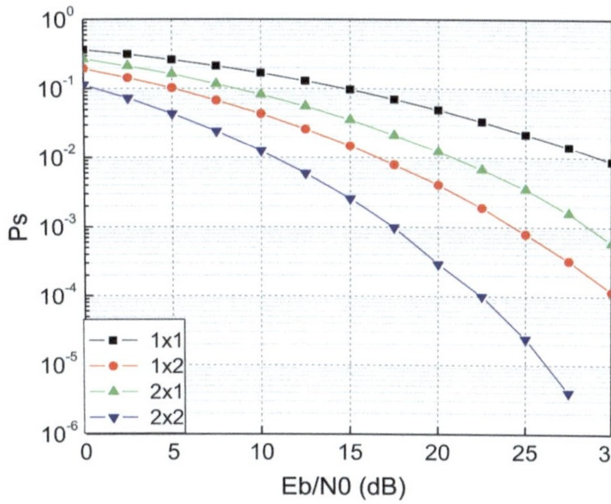

Fig. 4.3 Error rate of BCOSTBC code at 4-PPM with scintillation index (*S.I.*) = 0.6

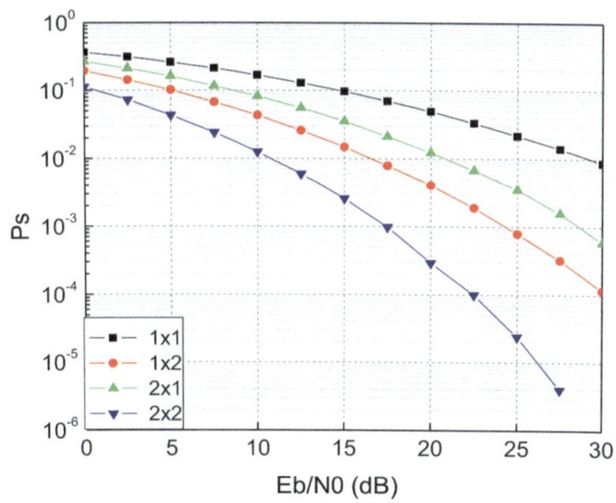

Fig. 4.4 Error rate of BCOSTBC code at 4-PPM modulation with *S.I.* = 1.0

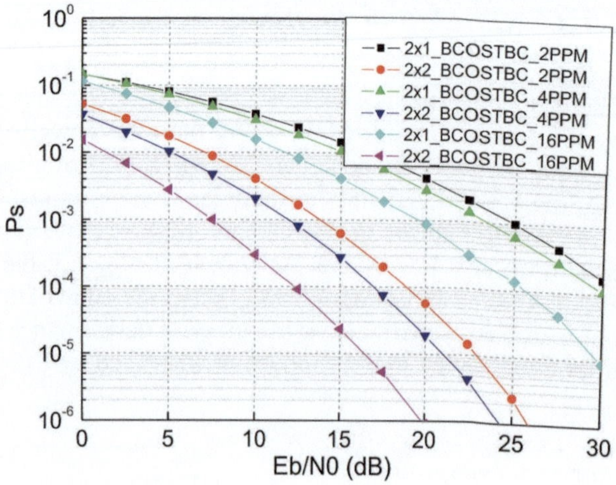

Fig. 4.5 Error rate of BCOSTBC codes\ under different modulations when $S.I. = 0.6$

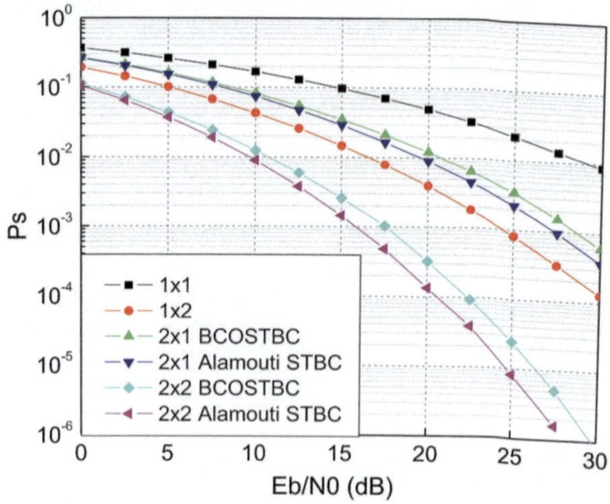

Fig. 4.6 Error-rate comparison between BCOSTBC code and improved Alamouti code at 4-PPM when $S.I. = 0.6$

results show that increasing the number of lasers and detectors increases the diversity gain and improves the system performance. It can be seen that the coding method is suitable for any $M * N$ atmospheric laser-communication system.

Figure 4.4 shows the BCOSTBC bit-error rate at 4-PPM modulation with $S.I. = 1.0$. As can be seen from the figure, the rule presented at this time is the same as that shown in Fig. 4.3. The difference lies in the improvement of the system's bit-error performance after adopting BCOSTBC coding.

Figure 4.5 shows the error rate of BCOSTBC codes in different modulation modes at $S.I. = 0.6$. It can be seen from the figure that under different modulation modes, the error rate of the system gradually decreases with the increase of SNR. Moreover, in the same system, the error rate of the system decreases with the increase of Q; that is, the larger Q is, the smaller P_s is.

In a 2×1 system with BCOSTBC code, the error rates at 4-PPM and 16-PPM were improved by about sum, respectively, at $P_s = 10^{-3}$ hours, relative to 2-PPM modulation. In a 2×2 system using BCOSTBC code, the error rate improved by approximately 2 and 6.5 dB when using 4-PPM and 16-PPM, respectively, compared with that using 2-PPM. Therefore, this coding method is suitable for arbitrary-base PPM modulation.

Figure 4.6 shows the error-performance comparison between BCOSTBC code and improved Alamouti code when $S.I. = 0.6$. As can be seen from the figure, compared with the improved Alamouti code [2], the error performance of the BCOSTBC code is decreased. At $P_s = 10^{-3}$, the error performance of the 2×1 and 2×2 systems decreased by approximately 1 dB and 1.5 dB, respectively. However, the BCOSTBC code reduces the number of transmitted light pulses when it represents the complex conjugate of characters; that is, the number of transmitted light pulses is reduced by $Q - 2$, which means that the transmitted power on a single character cycle is reduced by $Q - 2$ times.

4.4 Pulse-Position Orthogonal Space–Time Coding

An effective method to improve the reliability of communication links is to provide redundancy for received signals. Space–time coding can provide redundancy in both the time and space domains. For this reason, a repetitive code-encoding scheme was proposed in the literature [4]; however, this method utilized only spatial diversity, not temporal diversity. Brandt-Pearce and Wilson et al. used PPM modulation technology [5] to design a coding scheme with equal Hamming weights and analyzed the error-performance limit; however, the signal detection of the system was difficult.

Simon and Vilnrotter proposed an improved Alamouti coding scheme [2], which had error performance similar to the traditional Alamouti code and achieved a transmission rate of 1. However, when Q-PPM modulation ($Q > 2$) is adopted in this method, the number of optical pulses needs to be increased to describe the $\overline{x}$ vector. This changes the transmission power of the signal over the character cycle. We combine the orthogonality of Alamouti code with the orthogonality of Q-PPM modulation and propose a coding scheme suitable for atmospheric laser communication. The scheme has a simple coding and decoding algorithm, similar to the traditional Alamouti code error performance, and also achieves a transmission rate of 1.

4.4.1 Coding Principle of Pulse-Position Orthogonal Space–Time Coding

In the traditional Alamouti space–time block-code encoding scheme [1], complex conjugation and negative numbers of characters are required. However, IM/DD atmospheric laser communication has only two optical signal states: "*on*" and "*off*," and it is impossible to represent complex conjugation and negative numbers of symbols with these two states. Therefore, it is necessary to find a suitable coding method for IM/DD optical communication.

To make full use of the orthogonality of Q-PPM modulation to assist space–time codes to realize their orthogonality, different pulse positions are used to represent complex conjugation and negative numbers of symbols. For example, at $Q = 4$, when $x \to [1\ 0\ 0\ 0]$, $x^* \to [0\ 0\ 0\ 1]$, $-x^* \to [0\ 0\ 1\ 0]$, and $-x \to [0\ 1\ 0\ 0]$, the coding matrix is transformed into a real matrix represented by the pulse positions. This can then be transmitted in IM/DD optical communication.

4.4.2 Decoding Principle of Pulse-Position Orthogonal Space–Time Coding

Assuming that the receiver can accurately estimate the optical-intensity fading coefficient of the channel, taking the 2×2 system as an example and adopting the photon-counting model, the channel model is given in Eq. (3.5) $y^{(n)} = \frac{n_s}{M}\sum_{m=1}^{M} x^{(m)}h_{nm} + n_b n_0$. We assume that the representation method of the Alamouti coding scheme is still adopted; that is, $r_1^{(n)}$ and $r_2^{(n)}$ respectively represent the signals received by the n detector at moments t and $t + T$; then,

$$\begin{cases} r_1^{(n)} = \frac{n_s}{M}(h_{n1}x_1 + h_{n2}x_2) + n_b n_{01}^n \\ r_2^{(n)} = \frac{n_s}{M}\left[h_{n1}\left(-x_2^*\right) + h_{n2}x_1^*\right] + n_b n_{02}^n \end{cases} \tag{4.49}$$

where $h_{nm} = a_{nm}^2$. When coding, we note that $\left(-x_2^*\right)^* = -x_2$. Assuming the equal-gain combination of signals,

$$\begin{cases} \tilde{x}_1 = \sum\limits_{n=1}^{N}\left[h_{n1}r_1^{(n)} + h_{n2}\left(r_2^{(n)}\right)^*\right] \\ \tilde{x}_2 = \sum\limits_{n=1}^{N}\left[h_{n2}r_1^{(n)} + h_{n1}\left(-\left(r_2^{(n)}\right)^*\right)\right] \end{cases} \tag{4.50}$$

We substitute Eq. (4.49) into Eq. (4.50):

$$\begin{cases}\tilde{x}_1 = \frac{n_s}{M}\sum_{n=1}^{N}\sum_{m=1}^{M} h_{nm}^2 x_1 + \frac{n_s}{M}\sum_{n=1}^{N}\prod_{m=1}^{M} h_{nm}x_2 + \frac{n_s}{M}\sum_{n=1}^{N}\prod_{m=1}^{M} h_{nm}(-x_2)\\ +n_b\sum_{n=1}^{N}\left[h_{n1}n_{01}^{n} + h_{n2}\left(n_{02}^{n}\right)^{*}\right]\\ \tilde{x}_2 = \frac{n_s}{M}\sum_{n=1}^{N}\sum_{m=1}^{M} h_{nm}^2 x_2 + \frac{n_s}{M}\sum_{n=1}^{N}\prod_{m=1}^{M} h_{nm}x_1 + \frac{n_s}{M}\sum_{n=1}^{N}\prod_{m=1}^{M} h_{nm}(-x_1)\\ +n_b\sum_{n=1}^{N}\left[h_{n2}n_{01}^{n} + h_{n1}\left(n_{02}^{n}\right)^{*}\right]\end{cases} \quad (4.51)$$

According to Eq. (4.51), $\tilde{x}_1$ and $\tilde{x}_2$ are related to $x_1, x_2, -x_2$, or $-x_1$, but are mostly related to the signal emitted by the corresponding laser; that is, the first term is the largest, so it can be used to detect the signal. The x and $-x$ items contained in another detector signal can also be used to recover the signal; however, the error code is large in this case. According to the channel model in Eq. (3.5), the combined signal of traditional Alamouti codes is

$$\begin{cases}\tilde{x}_1 = \frac{n_s}{M}\sum_{n=1}^{N}\sum_{m=1}^{M} h_{nm}^2 x_1 + n_b\sum_{n=1}^{N}\left[h_{n1}^{*}n_{01}^{n} + h_{n2}\left(n_{02}^{n}\right)^{*}\right]\\ \tilde{x}_2 = \frac{n_s}{M}\sum_{n=1}^{N}\sum_{m=1}^{M} h_{nm}^2 x_2 + n_b\sum_{n=1}^{N}\left[h_{n2}^{*}n_{01}^{n} + h_{n1}\left(n_{02}^{n}\right)^{*}\right]\end{cases} \quad (4.52)$$

When comparing Eqs. (4.51) and (4.52), the first term in the two equations is observed to be identical, indicating that the code has an error performance similar to that of the traditional Alamouti code. Because x_1, $-x_1$, x_2, and $-x_2$ represent different pulse positions, $x_1 = x_2$ or $x_1 = -x_2$ will exist. Let the probability of its existence in the symbol set be $1 - p$; then, the probability of $x_1 \neq x_2$ and $x_1 \neq -x_2$ is p, where $p = \frac{1}{2^{k-1}}$ is related to Q ($k = \log_2^Q$). Two cases should be discussed:

(1) $x_1 \neq x_2$ and $x_1 \neq -x_2$ (probability p). Then, the signal is Eq. (4.51).
(2) $x_1 = x_2$ or $x_1 = -x_2$ (probability $1 - p$). Then, Eq. (4.51) becomes

$$\begin{cases}\tilde{x}_1 = \frac{n_s}{M}\left[\left(\sum_{n=1}^{N}\sum_{m=1}^{M} h_{nm}^2 + \sum_{n=1}^{N}\prod_{m=1}^{M} h_{nm}\right)x_1 + \sum_{n=1}^{N}\prod_{m=1}^{M} h_{nm}(-x_2)\right]\\ \quad + n_b\sum_{n=1}^{N}\left[h_{n1}n_{01}^{n} + h_{n2}\left(n_{02}^{n}\right)^{*}\right]\\ \tilde{x}_2 = \frac{n_s}{M}\left[\left(\sum_{n=1}^{N}\sum_{m=1}^{M} h_{nm}^2 + \sum_{n=1}^{N}\prod_{m=1}^{M} h_{nm}\right)x_2 + \sum_{n=1}^{N}\prod_{m=1}^{M} h_{nm}(-x_1)\right]\\ \quad + n_b\sum_{n=1}^{N}\left[h_{n2}n_{01}^{n} + h_{n1}\left(n_{02}^{n}\right)^{*}\right]\end{cases} \quad (4.53)$$

or

$$\begin{cases} \tilde{x}_1 = \frac{n_s}{M}\left[\left(\sum_{n=1}^{N}\sum_{m=1}^{M}h_{nm}^2 + \sum_{n=1}^{N}\prod_{m=1}^{M}h_{nm}\right)x_1 + \sum_{n=1}^{N}\prod_{m=1}^{M}h_{nm}x_2\right] \\ \quad + n_b\sum_{n=1}^{N}\left[h_{n1}n_{01}^n + h_{n2}\left(n_{02}^n\right)^*\right] \\ \tilde{x}_2 = \frac{n_s}{M}\left[\left(\sum_{n=1}^{N}\sum_{m=1}^{M}h_{nm}^2 + \sum_{n=1}^{N}\prod_{m=1}^{M}h_{nm}\right)x_2 + \sum_{n=1}^{N}\prod_{m=1}^{M}h_{nm}x_1\right] \\ \quad + n_b\sum_{n=1}^{N}\left[h_{n2}n_{01}^n + h_{n1}\left(n_{02}^n\right)^*\right] \end{cases} \tag{4.54}$$

Without loss of generality, the "*on*" time slot is 1 and the "*off*" time slot is 0; then, the combined signal $\tilde{x}_j(j = 1, 2)$ consists of "*on*" and "*off*" time slots. Let $\lambda_{on,i}$ be the photon count in slot "*on*" and λ_{off} be the photon count in slot "*off*". Because $g(x)$ is a $1 \times Q$ vector with one element, the detection of the signal $\tilde{x}_j$ can be transformed into the detection of a time-slot signal "*on*"; that is,

$$\begin{cases} \lambda_{on,j} = \frac{n_s}{M}\sum_{n=1}^{N}\sum_{m=1}^{M}h_{nm}^2 + n_b\sum_{n=1}^{N}\sum_{m=1}^{M}h_{nm} \quad x_1 \neq x_2 \text{ and } \mathrm{x}_1 \neq -\mathrm{x}_2, \mathit{probobilityis}\mathrm{p} \\ \lambda_{on,j} = \frac{n_s}{M}\left(\sum_{n=1}^{N}\sum_{m=1}^{M}h_{nm}^2 + \sum_{n=1}^{N}\prod_{m=1}^{M}h_{nm}\right) \\ \qquad + n_b\sum_{n=1}^{N}\sum_{m=1}^{M}h_{nm} \quad x_1 = x_2 \text{ or } \mathrm{x}_1 = -\mathrm{x}_2, \mathit{probobilityis}1 - \mathrm{p} \\ \lambda_{off,j} = n_b\sum_{n=1}^{N}\sum_{m=1}^{M}h_{nm} \end{cases} \tag{4.55}$$

Suppose the received observation set is $Z = \{z_{nq}, n = 1, \ldots, N, q = 0, \ldots, Q - 1\}$, where z_{nq} represents the photon count at the qth time slot of the nth detector. Assuming that the signal sent by the sender is x_j, the maximum-likelihood decision of $\tilde{x}_j$ can be expressed as

$$\tilde{x}_j = \arg\max_{x_j} f\left(\mathrm{Z}\middle|x_j\right) \tag{4.56}$$

Assuming the Poisson counting model is used and z_{nq} is independent with N detectors and Q time slots, the random matrix $r(x)$ of $Z \times Q$ can be transformed into a continuous product of $Z \times Q$ individual elements over a. Then, the maximum-likelihood decision can be converted to

$$\tilde{x}_j = \arg\max_{x_j}\prod\prod_{q\in Q_{on}^{(j)}}\frac{\exp(-\lambda_{on,n})(\lambda_{on,n})^{z_{nq}}}{z_{nq}!}\prod_{q\in Q_{off}^{(j)}}\frac{\exp(-\lambda_{off,n})(\lambda_{off,n})^{z_{nq}}}{z_{nq}!} \tag{4.57}$$

where $Q_{on}^{(j)}$ represents the set of "*on*" time slots and $Q_{off}^{(j)}$ represents the set of "*off*" time slots for modulation symbols.

In Eq. (4.57), for any x_j, $z_{nq}!$ in the denominator is constant, so it can be eliminated:

$$\tilde{x}_j = \arg\max_{x_j} \prod \prod_{q\in Q_{on}^{(j)}} \exp(-\lambda_{on,n})(\lambda_{on,n})^{z_{nq}} \prod_{q\in Q_{off}^{(j)}} \exp(-\lambda_{off,n})(\lambda_{off,n})^{z_{nq}} \quad (4.58)$$

For all modulation symbols, $\exp(-(Q-1)\lambda_{off,n})$ and $\exp(-\lambda_{on,n})$ can be eliminated by assuming an equal number of "*on*" time slots:

$$\tilde{x}_j = \arg\max_{x_j} \prod \prod_{q\in Q_{on}^{(j)}} (\lambda_{on,n})^{z_{nq}} \prod_{q\in Q_{off}^{(j)}} (\lambda_{off,n})^{z_{nq}} \quad (4.59)$$

The logarithm of Eq. (4.59) is

$$\tilde{x}_j = \arg\max_{x_j} \sum_n \left[\sum_{q\in Q_{on}^{(j)}} \ln((\lambda_{on,n})^{z_{nq}}) + \sum_{q\in Q_{off}^{(j)}} \ln((\lambda_{off,n})^{z_{nq}}) \right] \quad (4.60)$$

Then, the above equation can be rewritten as

$$\tilde{x}_j = \arg\max_{x_j} \sum_n \left[\sum_{q\in Q_{on}^{(j)}} z_{nq} \ln(\lambda_{on,n}) + \sum_{q\in Q_{off}^{(j)}} z_{nq} \ln(\lambda_{off,n}) \right] \quad (4.61)$$

$$\tilde{x}_j = \arg\max_{x_j} \sum_n \left[\sum_{q\in Q_{on}^{(j)}} z_{nq} \ln(\lambda_{on,n}) + \sum_{all\ q} z_{nq} \ln(\lambda_{off,n}) - \sum_{q\in Q_{on}^{(j)}} z_{nq} \ln(\lambda_{off,n}) \right] \quad (4.62)$$

In all q time slots, $\sum_{all\ q} z_{nq} \ln(\lambda_{off,n})$ is a constant; then, Eq. (4.62) can be simplified as

$$\tilde{x}_j = \arg\max_{x_j} \sum \left[\sum_{q\in Q_{on}^{(i)}} z_{nq} \ln\left(\frac{\lambda_{on}}{\lambda_{off}}\right) \right] \quad (4.63)$$

It can be seen from Eq. (4.63) that the maximum-likelihood decision criterion is transformed into the decision of the weighted sum in the "*on*" time slot. To simplify the maximum-likelihood decision criterion, we study the maximum-likelihood decision criterion of orthogonal space–time coding with pulse position in four cases: no fading without background radiation, no fading with background radiation, fading without background radiation, and fading with background radiation.

4.4.3 No Background Radiation and No Fading

In this case, according to Eq. (4.61) and taking $x_1 \neq x_2 and x_1 \neq -x_2$ as examples; then,

$$\begin{aligned}\tilde{x}_j &= \arg\max_{x_j} \sum_n \left[\sum_{q\in Q_{on}^{(j)}} z_{nq} \ln(\lambda_{on,n}) + \sum_{q\in Q_{off}^{(j)}} z_{nq} \ln(\lambda_{off,n}) \right] \\ &= \arg\max_{x_j} \sum_n \left[\sum_{q\in Q_{on}^{(j)}} z_{nq} \ln(\frac{n_s}{M} \sum_{n=1}^{N}\sum_{m=1}^{M} h_{nm}^2 + n_b \sum_{n=1}^{N}\sum_{m=1}^{M} h_{nm}) \right. \\ &\left. + \sum_{q\in Q_{off}^{(j)}} z_{nq} \ln(n_b \sum_{n=1}^{N}\sum_{m=1}^{M} h_{nm}) \right]\end{aligned} \tag{4.64}$$

When there is no background radiation in the system, $n_b = 0$ and $z_{nq} = 0$ in all "*off*" time slots, owing to the

$$\lim_{x\to 0^+} x \times \log(x) = 0 \tag{4.65}$$

Then, Eq. (4.64) can be simplified as

$$\tilde{x}_j = \arg\max_{x_j} \sum_n \left[\sum_{q\in Q_{on}^{(j)}} z_{nq} \ln\left(\frac{n_s}{M} \sum_{n=1}^{N}\sum_{m=1}^{M} h_{nm}^2\right)\right] \tag{4.66}$$

When there is no fading in the system, i.e., $h_{nm} = 1$ ($n = 1, 2 \ldots, N$ and $m = 1, 2, \ldots, M$), then $\ln\left(\frac{n_s}{M} \sum_{n=1}^{N} \sum_{m=1}^{M} h_{nm}^2\right)$ is a constant for all $R_I(x)$, and Eq. (4.66) can be reduced to

$$\begin{aligned}\tilde{x}_j &= \arg\max_{x_j} \sum_n \left[\sum_{q\in Q_{on}^{(j)}} z_{nq} \right] \\ &= \arg\max_{x_j} \sum_{q\in Q_{on}^{(j)}} \left[\sum_n z_{nq} \right]\end{aligned} \tag{4.67}$$

Similarly, the likelihood decision criterion is $x_1 = x_2 or x_1 = -x_2$ in Eq. (4.67). As can be seen from Eq. (4.67), the likelihood decision criterion can be transformed into finding the maximum column sum on the symbol.

4.4.4 *Fading, Except for Background Radiation*

When only the background radiation in the system does not fade, i.e., $n_b \neq 0$ and $h_{nm} = 1$ $(n = 1, 2 \cdots, N$ and $m = 1, 2 \cdots, M)$, then $\sum_{n=1}^{N} \sum_{m=1}^{M} h_{nm}^2 = NM$. Taking $x_1 \neq x_2 and x_1 \neq -x_2$ as examples, Eq. (4.64) can be simplified as

$$\tilde{x}_j = \arg\max_{x_j} \sum_n \left[\sum_{q \in Q_{on}^{(j)}} z_{nq} \ln(Nn_s + NMn_b) + \sum_{q \in Q_{off}^{(j)}} z_{nq} \ln(NMn_b) \right] \tag{4.68}$$

Because $\ln(Nn_s + NMn_b) \geq \ln(NMn_b)$ also defines $\tau = \ln(NMn_b)$, $\tau + \Delta = \ln(Nn_s + NMn_b)$; then, Eq. (4.68) can be converted to

$$\begin{aligned}\tilde{x}_j &= \arg\max_{x_j} \sum_n \left[(\tau + \Delta) \sum_{q \in Q_{on}^{(j)}} z_{nq} + \tau \sum_{q \in Q_{off}^{(j)}} z_{nq}) \right] \\ &= \arg\max_{x_j} \sum_n \left[\Delta \sum_{q \in Q_{on}^{(j)}} z_{nq} + \tau \sum_{all\ q} z_{nq}) \right]\end{aligned} \tag{4.69}$$

Of these, $\tau \sum_{all\ q} z_{nq}$ and Δ are independent of x_j and can therefore be eliminated. Then, the likelihood decision criterion can be simplified as

$$\begin{aligned}\tilde{x}_j &= \arg\max_{x_j} \sum_n \left[\sum_{q \in Q_{on}^{(j)}} z_{nq} \right] \\ &= \arg\max_{x_j} \sum_{q \in Q_{on}^{(j)}} \left[\sum_n z_{nq} \right]\end{aligned} \tag{4.70}$$

Similarly, the likelihood decision criterion is $x_1 = x_2 or x_1 = -x_2$ in Eq. (4.70); only $\tau + \Delta = \ln\left(\frac{N(M+1)}{M} n_s + NMn_b\right)$ at this time. According to Eqs. (4.67) and (4.70), the likelihood decision criterion can also be regarded as finding the maximum column sum on the symbol.

4.4.5 No Background Radiation or Fading

In this case, there is always one "*on*" time slot with a non-zero count and several "*off*" time slots with zero counts on each received symbol. The likelihood decision criterion is still Eq. (4.66) because we are dealing with a slow-fading channel, that is, the fading coefficient of the channel remains constant over two consecutive sign periods. Assuming that the channel-fading coefficient has been accurately estimated, $\ln\left(\frac{n_s}{M}\sum_{n=1}^{N}\sum_{m=1}^{M}h_{nm}^2\right)$ or $\ln\left(\frac{n_s}{M}\left(\sum_{n=1}^{N}\sum_{m=1}^{M}h_{nm}^2+\sum_{n=1}^{N}\prod_{m=1}^{M}h_{nm}\right)\right)$ is also a constant over a symbolic period. The likelihood criterion can also be regarded as finding the maximum column sum on the symbol.

4.4.6 Both Background Radiation and Fading

When there is background radiation and fading in the system, $h_{nm} \neq 1$ and $n_b \neq 0$. In Eq. (4.63), because

$$\ln\left(\frac{n_s}{M}\sum_{n=1}^{N}\sum_{m=1}^{M}h_{nm}^2+n_b\sum_{n=1}^{N}\sum_{m=1}^{M}h_{nm}\right)\geq\ln\left(n_b\sum_{n=1}^{N}\sum_{m=1}^{M}h_{nm}\right) \tag{4.71}$$

we define

$$\tau_1=\ln\left(n_b\sum_{n=1}^{N}\sum_{m=1}^{M}h_{nm}\right)$$

$$\tau_1+\Delta_1=\ln\left(\frac{n_s}{M}\sum_{n=1}^{N}\sum_{m=1}^{M}h_{nm}^2+n_b\sum_{n=1}^{N}\sum_{m=1}^{M}h_{nm}\right)$$

If $x_1 \neq x_2$ and $x_1 \neq -x_2$ are still taken as examples, then Eq. (4.63) can be converted to

$$\begin{aligned}\tilde{x}_j&=\arg\max_{x_j}\sum_{n}\left[(\tau_1+\Delta_1)\sum_{q\in Q_{on}^{(j)}}z_{nq}+\tau_1\sum_{q\in Q_{off}^{(j)}}z_{nq})\right]\\&=\arg\max_{x_j}\sum_{n}\left[\Delta_1\sum_{q\in Q_{on}^{(j)}}z_{nq}+\tau_1\sum_{all\ q}z_{nq})\right]\end{aligned} \tag{4.72}$$

Because $\tau_1\sum_{all\ q}z_{nq}$ is independent of x_j, it can be eliminated. Then, the likelihood decision criterion can be simplified as

$$\tilde{x}_j = \arg\max_{x_j} \sum_n \left[\Delta_1 \sum_{q \in Q_{on}^{(j)}} z_{nq} \right]$$
$$= \arg\max_{x_j} \sum_{q \in Q_{on}^{(j)}} \left[\sum_n \Delta_1 z_{nq} \right] \tag{4.73}$$

where $\Delta_1 = \ln\left(\frac{n_s}{M} \sum_{n=1}^{N} \sum_{m=1}^{M} h_{nm}^2 + n_b \sum_{n=1}^{N} \sum_{m=1}^{M} h_{nm} \right) - \ln\left(n_b \sum_{n=1}^{N} \sum_{m=1}^{M} h_{nm} \right)$. Similarly, the likelihood decision criterion in $x_1 = x_2 or x_1 = -x_2$ is also Eq. (4.73); however, $H = \begin{bmatrix} \alpha^{n-1} & \alpha^{n-1} & \cdots \alpha & 1 \\ (\alpha^2)^{n-1} & (\alpha^2)^{n-1} & \cdots \alpha^2 & 1 \\ \cdots & \cdots & \cdots\cdots\cdots & \\ (\alpha^{2t})^{n-1} & (\alpha^{2t})^{n-2} & \cdots \alpha^{2t} & 1 \end{bmatrix}$ is different:

$$\Delta_1 = \ln\left(\frac{n_s}{M} \left(\sum_{n=1}^{N} \sum_{m=1}^{M} h_{nm}^2 + \sum_{n=1}^{N} \prod_{m=1}^{M} h_{nm} \right) + n_b \sum_{n=1}^{N} \sum_{m=1}^{M} h_{nm} \right) - \ln\left(n_b \sum_{n=1}^{N} \sum_{m=1}^{M} h_{nm} \right) \tag{4.74}$$

Similarly, the likelihood criterion can also be regarded as finding the maximum column sum on the symbol.

4.5 Error Performance of Pulse-Position Orthogonal Space–Time Coding

To better describe the system performance, we will discuss the following cases.

4.5.1 *Bit-Error Rate Without Background Radiation and Without Fading*

When there is no fading or background radiation in the system, $h_{nm} = 1$ ($n = 1, 2\cdots, N$ and $m = 1, 2\cdots, M$) and $n_b = 0$. The bit-error rate of the system is only caused by a wrong decision, and is defined as

$$P_s = \frac{Q-1}{Q} \exp(-N n_s) \tag{4.75}$$

Under constant total power, because the number of photons reaching the receiving end is related to N and unrelated to M, it can be seen from Eq. (4.75) that the bit-error rate of the system is only related to N. When the number of detectors N increases, it can be considered that the receiving aperture increases and the number of photons reaching the receiving end increases; thus, the bit-error rate of the system will decrease with an increase of N.

4.5.1.1 Bit-Error Rate with Background Radiation and no Fading

When there is no fading and only background radiation in the system, i.e., $h_{nm} = 1(n = 1, 2 \cdots, N, m = 1, 2 \cdots, M), n_b \neq 0$, it is very difficult to obtain the closed bit-error performance. Here, we only analyze the upper bound of its existence. To simplify the calculation, only the error codes caused by the number of photons in any "*off*" timeslot greater than those in the "*on*" timeslot are considered; then, the bit-error rate is

$$P_s = pP_1 + (1-p)P_2 \tag{4.76}$$

In Eq. (4.76),

$$P_1 \leq 1 - \left[\sum_{i=1}^{\infty}\sum_{j=0}^{i-1} \frac{(N(n_s + Mn_b))^i \exp(-N(n_s + Mn_b))}{i!} \times \frac{(NMn_b)^j \exp(-NMn_b)}{j!}\right]^{Q-1}$$

$$P_2 \leq 1 - \left[\sum_{i=1}^{\infty}\sum_{j=0}^{i-1} \frac{\left(N\left(\left(1+\frac{1}{M}\right)n_s + Mn_b\right)\right)^i \exp\left(-N\left(\left(1+\frac{1}{M}\right)n_s + Mn_b\right)\right)}{i!} \times \frac{(NMn_b)^j \exp(-NMn_b)}{j!}\right]^{Q-1}.$$

In this case, the number of photons from the origin to the "*on*" time slot is related to N and increases with an increase in N, and the system's bit-error performance is significantly improved with the increase in N. The number of photons in the "*on*" time slot has almost no relation to M. However, the number of photons generated by background radiation increases with the increase in M, indicating that increasing the number of lasers M has no obvious benefit to improving the system's bit-error performance. However, when the transmitting power of each laser is constant, increasing the number of lasers means that the total power of the system increases; thus, the system's bit-error performance will be significantly improved.

4.5.1.2 Bit-Error Rate with Fading and no Background Radiation

When only fading exists without background radiation in the system, $h_{nm} \neq 1(n = 1, 2\cdots, N, m = 1, 2\cdots, M)$ and $n_b = 0$. The atmospheric channel is assumed to be symmetric. The intensity-fading coefficient is $h_{nm} = a_{nm}^2$, where a_{nm} represents the signal-fading coefficient from the nth laser to the mth detector, which follows the lognormal distribution. Because the distance between the detectors is only a few tens of cm apart, the attenuation coefficients are considered to be approximately equal; thus, its bit-error rate is

$$\begin{aligned} P_s = \frac{Q-1}{Q}\Bigg\{ p\Big[\int \exp\Big(-\frac{\alpha^4 n_s}{M}\Big) f_A(\alpha)d\alpha\Big]^{NM} \\ + (1-p)\Big[\int \exp\Big(-\frac{\alpha^4 n_s}{M}\Big) f_A(\alpha)d\alpha\Big]^{NM}\Big[\int \exp\Big(-\frac{\alpha^4 n_s}{M}\Big) f_A(\alpha)d\alpha\Big]^{N}\Bigg\} \end{aligned} \tag{4.77}$$

As can be seen from Eq. (4.77), the number of photons reaching the "*on*" time slot at this time is related to both M and N, and increases with the increase of M and N, while the BER decreases with the increase of M and N.

4.5.1.3 Bit-Error Rate with Background Radiation and Fading

When both fading and background radiation exist in the system, i.e., $h_{nm} \neq 1$ and $n_b \neq 0$, both the fading and background radiation will affect the system performance. The assumption is the same as that in the case of fading without background radiation; that is, its fading coefficients are approximately equal. Then, the upper bound of its bit-error rate is still Eq. (4.76), where P_1 and P_2 are

$$\begin{aligned} P_1 \leq \int \Bigg\{ 1 - [\sum_{i=1}^{\infty}\sum_{j=0}^{i-1} \frac{(n_s N a^4 + n_b NMa^2)^i \exp(-(n_s N a^4 + n_b NMa^2))}{i!} \\ \times \frac{(n_b NMa^2)^j \exp(-n_b NMa^2)}{j!}]^{Q-1}\Bigg\} \cdot f_A(a)da \end{aligned}$$

$$\begin{aligned} P_2 \leq \int \Bigg\{ 1 - [\sum_{i=1}^{\infty}\sum_{j=0}^{i-1} \frac{(\frac{n_s N(M+1)}{M} a^4 + n_b NMa^2)^i \exp(-(\frac{n_s N(M+1)}{M} a^4 + n_b NMa^2))}{i!} \\ \times \frac{(n_b NMa^2)^j \exp(-n_b NMa^2)}{j!}]^{Q-1}\Bigg\} \cdot f_A(a)da \end{aligned}$$

In this case, the error performance of the system is related to the amount of background radiation, the number of detectors N, and the number of lasers M. The specific changes are analyzed using the Monte Carlo simulation method.

4.5.2 Simulation Analysis

The bit-error performance of 1×1, 1×2, 2×1, and 2×2 systems was simulated using the Monte Carlo method. If the total power of the system E_s remains unchanged, the simulation parameters are as follows: $S.I. = 0.6$, wavelength $\lambda = 1550nm$, $\eta = 0.5$, and $Q = 4$. Figures 4.7, 4.8, 4.9, 4.10 and 4.11 show the simulation results.

Figure 4.7 shows the bit-error performance without background radiation or fading. Figure 4.7 shows the following:

(1) The bit-error performance of the 2×2 and 1×2 systems is better than that of the 1×1 system. When $P_s = 10^{-6}$, the bit-error performance of the 2×2 and 1×2 systems is approximately $P_s = 2 \times 10^{-3}$ better than that of the 1×1 system. This shows that increasing the number of detectors is equivalent to increasing the aperture of the receiver, which helps improve the system performance.

(2) The error-code curves of the 1×1 and 2×1 systems coincide, and the error-code curves of the 1×2 and 2×2 systems coincide. This is caused by the constant total power of the system E_s. It shows that when the total power is constant, the bit-error performance of the system is not improved by increasing the number of lasers. The simulation results are consistent with the rule reflected in Eq. (4.75).

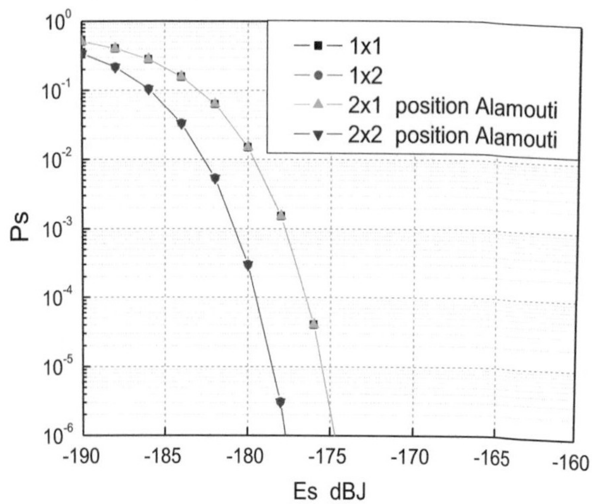

Fig. 4.7 BER without background radiation and without fading

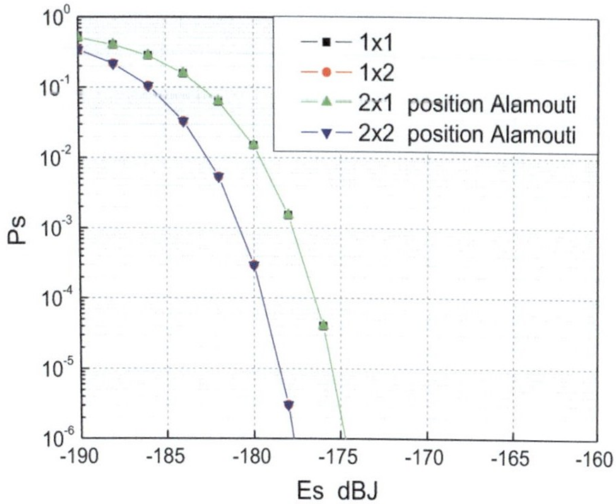

Fig. 4.8 BER without background radiation and with fading

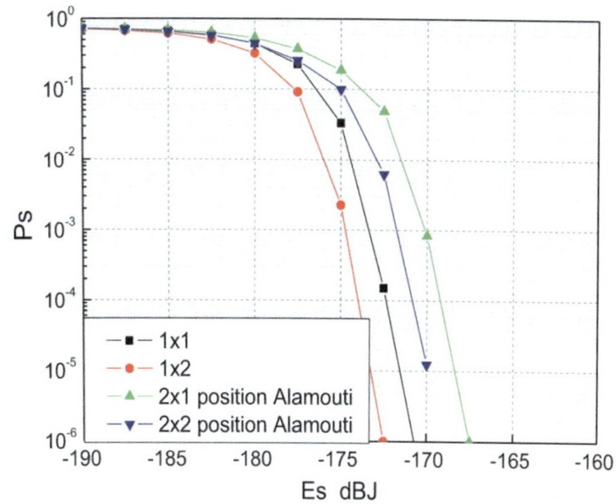

Fig. 4.9 BER without fading and with background radiation at constant total power

Figure 4.8 shows the bit-error performance without background radiation and with fading, assuming the scintillation index $S.I. = 0.6$. The following can be observed from Fig. 4.8:

(1) With the increase in the number of lasers and detectors, the bit-error performance of the system is significantly improved. When $P_s = 10^{-4}$, compared with the 1×1 system, the bit-error performance of the 2×1 system improves by approximately 6 dBJ, that of the 1×2 system improves by approximately 8 dBJ, and that of the 2×2 system improves by approximately 12 dBJ.

(2) The error-code–curve slopes of the 1×2 and 2×1 systems are almost the same, indicating that they have the same diversity gain. However, for the same

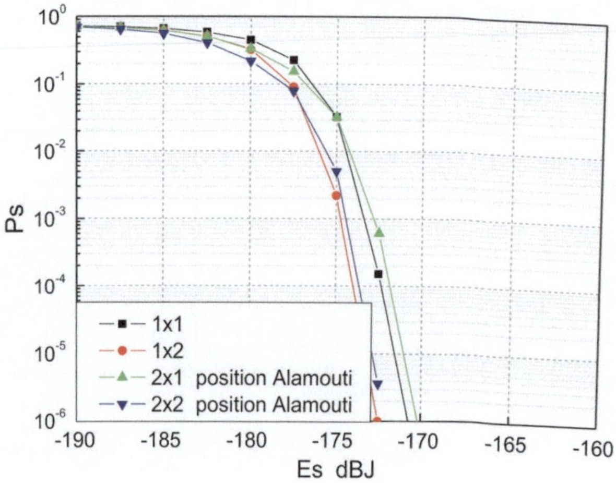

Fig. 4.10 BER of background radiation without fading when the transmitting power of each laser is equal

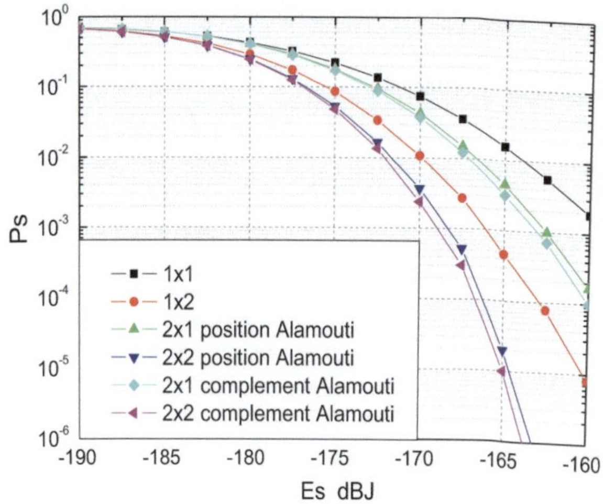

Fig. 4.11 BER with fading and background radiation

bit-error rate, the transmitting power required for receiving diversity is approximately 3 dBJ less than that for pulse-position-modulated Alamouti codes. This is because the total transmitting power of the two transmitting antennas is equal to that of the single antenna. However, if the transmitting power of two transmitting antennas is the same as that of a single antenna, the error-code performance curve will shift 3 dBJ to the left. The simulation results are consistent with the rule reflected in Eq. (4.76).

Figure 4.9 shows the bit-error performance without fading when the background radiation is $P_b T = -180$ dBJ. The following can be observed from Fig. 4.9:

(1) When the number of lasers is constant, increasing the number of detectors is beneficial to improving the system's bit-error performance.
(2) When the number of lasers changes, the bit-error performance of the system does not improve when the number of lasers increases. This is because increasing the number of lasers by M with the same total power means that the power on each laser will be reduced by $1/M$ times, and the background radiation introduced on each character will be M times that of a 1×1 system. Therefore, in this case, increasing the number of lasers does not improve the system's bit-error performance.

However, when the transmitting power of the two transmitting antennas is the same as that of a single antenna, the bit-error performance of the system is significantly improved, as shown in Fig. 4.10. The bit-error performance of the 2×1 system is close to that of the 1×1 system, and that of the 2×2 system is close to that of the 1×2 system. Compared with Fig. 4.9, the BER curves of the 2×2 and 2×1 systems in Fig. 4.10 are shifted to the left by approximately 3 dBJ.

(1) Compared with Fig. 4.7, the curves in Figs. 4.9 and 4.10 are steeper, which means that the power threshold for achieving reliable communication is determined by the background radiation. The rule reflected above is the same as that in Eq. (4.76).

Figure 4.11 shows the bit-error performance with both fading and background radiation, when $S.I. = 0.6$ and $P_b T = -180dBJ$. The curve in Fig. 4.11 becomes steeper relative to that in Fig. 4.8. The curve in Fig. 4.9 becomes smoother than that in Fig. 4.9. The curves in Fig. 4.11 reflect the simultaneous effects of fading and background radiation on the system performance. The following can be observed from Fig. 4.11:

(2) When the total power of the system remains unchanged, the bit-error performance of the system is significantly improved with the increase of the number of lasers and detectors. When $P_s = 2 \times 10^{-3}$, compared with the 1×1 system, the bit-error performance of the 2×1 system is approximately improved, and the bit-error performance of the 2×2 system is improved by 3.5 dBJ.
(3) Compared with the improved Alamouti code [2], the error performance of the proposed method is slightly reduced. When $P_s = 2 \times 10^{-4}$, the bit-error performance of the 2×1 system decreases by approximately 0.5 dBJ. When $P_s = 3 \times 10^{-6}$, the bit-error performance of the 2×2 system decreases by approximately 0.5 dBJ. The reduction of $Q - 2$ optical pulses in the representation of the complex conjugation of characters means that energy is saved.

4.6 Cascaded Orthogonal Space–Time Coding

Various researchers [2, 5–11] have proposed different space–time block-code schemes for intensity-modulation atmospheric laser communication. Although these coding schemes are different, they all have the same characteristic; that is, they can obtain a full transmit-diversity gain but no coding gain.

We know that cascaded code can bring the system performance close to the Shannon limit without increasing the complexity of the system hardware implementation. Cascaded codes in atmospheric laser communication have been studied [12–16] and some progress has been made. Therefore, the study of cascaded space–time coding in atmospheric laser communication is beneficial to obtaining coding gain and further improving the system's bit-error performance.

Compared with other channel codes, the Reed–Solomon (RS) code is a type of multi-base Bose–Chaudhuri–Hocquenghem (BCH) code with strong error-correction ability, which can correct not only burst errors but also random errors. Thus, it is especially suitable for burst-error correction, and can also easily match PPM modulation signals. A cascaded space–time block-coding scheme with the RS code as the outer code and orthogonal space–time code as the inner code is proposed. It can greatly improve the performance of an atmospheric laser-communication system and provide an effective means to suppress the effects of strong turbulence.

4.6.1 RS Codes

A Reed–Solomon code is a type of BCH code with a strong error-correction ability. It is also a typical algebraic–geometric code. It was first constructed by Reed and Solomon in 1960 using MS polynomials. It is a type of cyclic code with multiple random error corrections and a strict algebraic structure, which is easy to construct and study theoretically. Despite the complexity of the decoding algorithm, its error-correcting ability and decoding speed are unmatched by other code classes. Especially in the case of short and medium code lengths, its performance is close to the theoretical value [17].

RS-code coding and decoding methods are time-domain and frequency-domain algorithms. The time-domain coding and decoding algorithm appeared earlier and was widely used because of its maturity. The frequency-domain coding and decoding algorithm appeared later; however, because of its use of the fast Fourier inverse transform (FFT/IFFT) to improve the coding and decoding speed, it has great development potential.

In RS coding, these two algorithms are relatively simple; the time-domain coding only needs to divide a polynomial once, and the frequency-domain coding only needs to calculate an IFFT once. The difference between them is that the codeword obtained by the time-domain algorithm is a system code, which can be used to truncate the code. The codeword obtained by the frequency-domain algorithm is a non-system code and cannot be used for truncating coding. Therefore, only the time-domain coding method can be used in a cascaded space–time coding system.

4.6.1.1 RS-Code Encoding Method in the Time Domain

As the RS code is a type of cyclic code, its encoding method is the same as that of a general cyclic code. The coding method is mainly divided into two elements: Generation polynomial $g(x)$ and code polynomial $c(x)$.

(1) Generating the polynomial $g(x)$

Let α be a primitive element of the $GF(2^m)$ finite field; then, the generation polynomial of a primitive RS code with code length $n = 2^m - 1$ correcting t errors is

$$g(x) = (x+\alpha)(x+\alpha^2)\cdots(x+\alpha^{2t}) = g_0 + g_1x + g_2x^2 + \cdots g_{2t-1}x^{2t-1} + x^{2t} \tag{4.78}$$

It can also be

$$g(x) = \prod_{i=1}^{2t}(x+a^i) = \sum_{i=0}^{2t} g_ix^i \tag{4.79}$$

(2) Finding the code polynomial $c(x)$

Firstly, the information-element polynomial $u(x)$ is multiplied by x^{n-k} to become $x^{n-k}u(x)$; then, $x^{n-k}u(x)$ is divided by the generation polynomial $g(x)$ to obtain the remainder $r(x)$, which is the check-element polynomial. After the check-element polynomial is placed after the information-element polynomial of x^{n-k} bits, the codeword polynomial $c(x) = x^{n-k}u(x) + r(x)$ can be obtained. Thus, RS coding becomes a process of remainder by division.

a is a member of the $GF(2^m)$ finite field. We set the input information code as $u(x)$ and the code group after encoding as $c(x)$; then,

$$c(x) = x^{n-k}u(x) + x^{n-k}u(x) \bmod g(x) \tag{4.80}$$

Suppose the chosen RS code is of code length $n = 2^4 - 1 = 15$, defined in a $GF(2^4)$ finite field. For RS (15, 7) codes that can correct four code-element errors, the minimum distance between codes is $d = 2t + 1 = 9$, and their generating polynomial is

$$g(x) = (x-\alpha)(x-\alpha^2)\cdots(x-\alpha^{d-1}) = (x-\alpha)(x-\alpha^2)\cdots(x-\alpha^8) \tag{4.81}$$

where α is an element of the original field. For each RS code,

$$c = (c_{14}, c_{13}, \cdots, c_1, c_0) \tag{4.82}$$

which can be expressed as the following codeword:

$$c(x) = c_{14}x^{14} + c_{13}x^{13} + \cdots + c_1x + c_0 \tag{4.83}$$

Each codeword polynomial $c(x)$ is a multiple of $g(x)$:

$$c(x) = u(x) \cdot g(x) \tag{4.84}$$

where $u(x)$ is a polynomial with the highest degree of 6. To generate RS (15, 7) codes, we start with the following formula:

$$x^8u(x) = q(x)g(x) + r(x) \tag{4.85}$$

where $q(x)$ is the quotient of $\tilde{x}_2$ divided by $x^8u(x)$, and $r(x)$ is the remainder, whose degree is not greater than seven. The above equation can be written as

$$x^8u(x) + r(x) = q(x)g(x) \tag{4.86}$$

The left-hand side of the above equation is a multiple of $g(x)$, which can be used as a codeword polynomial:

$$c(x) = x^8u(x) + r(x) \tag{4.87}$$

If $u(x)$ is taken as an information polynomial composed of seven information symbols, and $r(x)$ is taken as the check polynomial composed of eight check symbols, it can be seen from the above formula that the information symbol and the check symbol are separated in an RS (15, 7) code without being confused, forming the system RS code.

4.6.1.2 RS-Code Decoding Method in the Time Domain

In channel coding, the decoding algorithm is usually the key. By contrast, the decoding algorithm of RS code is much more complex than the decoding algorithm of other codes, because RS code is a non-binary cyclic code, and it no longer has the properties to operate in the domain of feature 2.

The RS-code time-domain decoding algorithm also starts by calculating the adjoint of the received codeword. In the decoding process, not only the error position should be determined, but also the error size corresponding to the error position, which increases its decoding difficulty. The decoding mainly includes three steps: calculating the adjoint of the receiving polynomial, finding the corresponding error pattern of the adjoint, and decoding with the error pattern.

(1) Calculate the value of the adjoint S_k

Suppose the receiving polynomial is $R(x)$; its information part is $R_I(x)$ and its supervision part is $R_P(x)$; so,

$$S(x) = R\prime(x) + R_P(x) \quad (4.88)$$

In the formula, $R\prime(x)$ is the supervisory element polynomial for recoding $R_I(x)$. If the code-generation polynomial is $g(x)$, then

$$S(x) \equiv R_I(x) + R_P(x) \equiv R(x)(\bmod g(x)) \quad (4.89)$$

Equation (4.89) shows that the adjoint of the receiving polynomial of the cyclic code is the remainder of the receiving polynomial $R(x)$ divided by $g(x)$.

If $E(x)$ is set to be $R(x)$, then,

$$R(x) = C(x) + E(x) \quad (4.90)$$

However, $C(x)$ is a multiple of $g(x)$, so

$$S(x) \equiv [C(x) + E(x)] \equiv E(x)(\bmod g(x)) \quad (4.91)$$

Equation (4.91) also shows that the adjoint is determined by the error pattern and is unrelated to the specific codeword. The error pattern caused by the channel is

$$E(x) = R(x) - C(x) = e_0 + e_1X + ... + e_{n-1}X^{n-1} \quad (4.92)$$

Here, $e_i = r_i - c_i$ is a member of $GF(2^m)$. If the error pattern $e(x)$ contains t errors (non-zero elements) located at $X^1, X^2, \cdots, X^t$, then,

$$e(x) = e_1X + e_2X^2 + \cdots + e_tX^t \quad (4.93)$$

The most critical step in decoding is to determine $e(x)$ from $R(x)$; that is, the number of errors, the error position X^i, and the error value e_i. The $2t$ adjoint elements are obtained by substituting α^i into the receiving polynomial $R(x)$. If $S_k = 0$, the receiving is considered correct. If $S_k \neq 0$, we find an error pattern S_k. The RS (n, k) code on the finite field $\bar{s}_i(t) = \begin{cases} 0 & \frac{iT}{Q} \leq t \leq \frac{(i+1)T}{Q} \\ A & other \end{cases}$ takes $\alpha^1, \alpha^2, \ldots, \alpha^{2t}$ as the root, and the generated polynomial is

$$g(x) = (x - \alpha)(x - \alpha^2)\cdots(x - \alpha^{2t}) \quad (4.94)$$

where α is a primitive field element on $GF(q)$, and v is the number of error corrections.

The check matrix of the code is

$$H = \begin{bmatrix} \alpha^{n-1} & \alpha^{n-1} & \cdots & \alpha & 1 \\ (\alpha^2)^{n-1} & (\alpha^2)^{n-1} & \cdots & \alpha^2 & 1 \\ \cdots & \cdots & \cdots & \cdots & \cdots \\ (\alpha^{2t})^{n-1} & (\alpha^{2t})^{n-2} & \cdots & \alpha^{2t} & 1 \end{bmatrix} \tag{4.95}$$

Hence, the adjoint

$$S^T = HR^T = HE^T \tag{4.96}$$

We substitute into Eq. (4.92) to obtain

$$\begin{cases} S_1 = R(\alpha) = e_1 X^1 + \cdots + e_t X^t \\ S_2 = R(\alpha^2) = e_1 (X^1)^2 + \cdots + e_t (X^t)^2 \\ \cdots \\ S_{2t} = R(\alpha^{2t}) = e_1 (X^1)^{2t} + \cdots + e_t (X^t)^{2t} \end{cases} \tag{4.97}$$

(2) The error-location polynomial is determined by the adjoint value

It is very difficult to directly determine the error position X^i and error value e_i from Eq. (4.97), because it is a nonlinear simultaneous system of equations and generally not directly solved. Therefore, intermediate variables are introduced and defined.

$$\delta(x) = \delta_t x^t + \delta_{t-1} x^{t-1} + \cdots + \delta_1 x + 1 \tag{4.98}$$

is the error polynomial whose root is the inverse $(X^i)^{-1}$, $i = 1, 2, ..., v$ of the error position X^i; namely,

$$\delta(x) = (1 - xX_1)(1 - xX_2)\cdots(1 - xX_t) \tag{4.99}$$

If we know the coefficient $\delta_t, \delta_{t-1}, \cdots, \delta_1$ of $\delta(x)$ and solve for $\delta(x)$ to obtain its root $(X^i)^{-1}$, we can find the number of error positions X^i. Therefore, we should first find the coefficient of $\delta(x)$ from the known adjoint component $s_1, s_2, \cdots, s_{2t}$. If the number of errors actually generated is $v < t$, then, when $\delta_{v+1} = \delta_{v+2} = \cdots = \delta_t = 0$, Eq. (4.99) becomes

$$\delta(x) = \delta_v x^v + \delta_{v-1} x^{v-1} + \cdots + \delta_1 x + 1 \tag{4.100}$$

We multiply both sides of this equation by $e_i X_i^{j+v}$:

$$e_i X_i^{j+v} \delta(x) = e_i X_i^{j+v} (\delta_v x^v + \delta_{v-1} x^{v-1} + \cdots + \delta_1 x + 1) \quad j = 1, 2, \cdots, 2t$$
$$i = 1, 2, \cdots, v \tag{4.101}$$

Let $X_i^{-1} = x$ and X_i^{-1} be the root of $\delta(x)$, so

$$0 = e_i X_i^{j+v}(\delta_v X^{-v} + \delta_{v-1} X^{-(v-1)} + \cdots + \delta_1 X^{-1} + 1) \quad i = 1, 2, \cdots, v \quad (4.102)$$

or

$$e_i(\delta_v X_i^j + \delta_{v-1} X_i^{j+1} + \cdots + \delta_1 X_i^{j+v-1} + X_i^{j+v}) = 0$$

$$j = 1, 2, \ldots, 2t \quad i = 1, 2, \ldots, v \quad (4.103)$$

For every j, i from 1 to e, e equations of the above form can be obtained by adding them:

$$\sum_{i=1}^{v} e_i(\delta_v X_i^j + \delta_{v-1} X_i^{j+1} + \cdots + \delta_1 X_i^{j+v-1} + X_i^{j+v}) = 0 \quad j = 1, 2, \ldots, 2t \quad (4.104)$$

Compared with Eq. (4.97), it can be seen that each sum of the equation is an adjoint component; therefore, the above equation can be written as

$$\delta_v s_j + \delta_{v-1} s_{j+1} + \cdots + \delta_1 s_{j+v-1} + s_{j+v} = 0 \quad j = 1, 2, \cdots, 2t \quad (4.105)$$

or

$$\delta_v s_j + \delta_{v-1} s_{j+1} + \cdots + \delta_1 s_{j+v-1} = s_{j+v} \quad j = 1, 2, \cdots, 2t \quad (4.106)$$

We have $2t$ linear equations. Because $0 \le v \le t$, this is written in matrix form as

$$\begin{bmatrix} s_1 & s_2 & \cdots & s_{v-1} & s_v \\ s_2 & s_3 & \cdots & s_v & s_{v+1} \\ \cdots & \cdots & \cdots & \cdots & \cdots \\ s_v & s_{v+1} & \cdots & s_{2v-2} & s_{2v-1} \end{bmatrix} \begin{bmatrix} \delta_v \\ \delta_{v-1} \\ \cdots \\ \delta_1 \end{bmatrix} = \begin{bmatrix} s_{v+1} \\ s_{v+2} \\ \cdots \\ s_{2v} \end{bmatrix} \quad (4.107)$$

Let

$$[S] = \begin{bmatrix} s_1 & s_2 & \cdots & s_{v-1} & s_v \\ s_2 & s_3 & \cdots & s_v & s_{v+1} \\ \cdots & \cdots & \cdots & \cdots & \cdots \\ s_v & s_{v+1} & \cdots & s_{2v-2} & s_{2v-1} \end{bmatrix} \quad (4.108)$$

$$[\delta] = \begin{bmatrix} \delta_v \\ \delta_{v-1} \\ \cdots \\ \delta_1 \end{bmatrix} \quad [S\prime] = \begin{bmatrix} s_{v+1} \\ s_{v+2} \\ \cdots \\ s_{2v} \end{bmatrix} \quad (4.109)$$

then,

$$[S][\delta] = [S\prime], [\delta] = [S]^{-1}[S\prime] \tag{4.110}$$

If matrix $[S]$ is full rank, then $[S]^{-1}$ exists and the coefficient $\delta_v, \delta_{v-1}, \cdots, \delta_1$ of $\delta(x)$ can be obtained by the above equation. If the number of errors is $I_{on,i}$, $[S]$ can be proved to be a full-rank matrix. If the number of errors is less than v, $[S]$ is not of rank, and $[S]$ must be decanted to check if it is full rank. This process continues until the rank of $[S]$ is full. Then, the coefficient of $\delta(x)$ is found by Eq. (4.110). This method requires extensive computation and can be realized by computer programming. However, the computation is not complex and is simple to understand.

(3) Decoding with the error pattern

After finding the coefficient of $\delta(x)$ and substituting the error-position polynomial into Eq. (4.98), the root and error position can be solved. By substituting the error-position value into Eq. (4.97), the error value can be obtained by solving simultaneous equations. In the corresponding error position, the corresponding error value is added; that is, the error correction is completed.

4.6.2 *Cascaded Orthogonal Space–Time Coding*

Cascaded coding technology was proposed by Forney in 1966 [18]. Two or more codes are cascaded to increase the total code length, to improve the system performance without increasing the complexity of the system hardware. It is especially suitable for improving the performance of atmospheric laser-communication systems. A non-binary RS code is the most natural choice for a foreign code because it is a maximum-distance-separable code and easy to implement. The outer code adopts an RS code and the inner code adopts orthogonal space–time coding. The principle model of cascaded space–time block coding is shown in Fig. 4.12.

Its encoding principle is as follows [17]: After the serial and parallel transformation of the information sequence sent by the source, the sequence first enters the external-code encoder (RS code) for encoding. The external-code encoder outputs non-binary RS-code symbols. Because the used PPM modulation symbol is matched with the RS code, there is no need to convert the RS code into binary code, and the

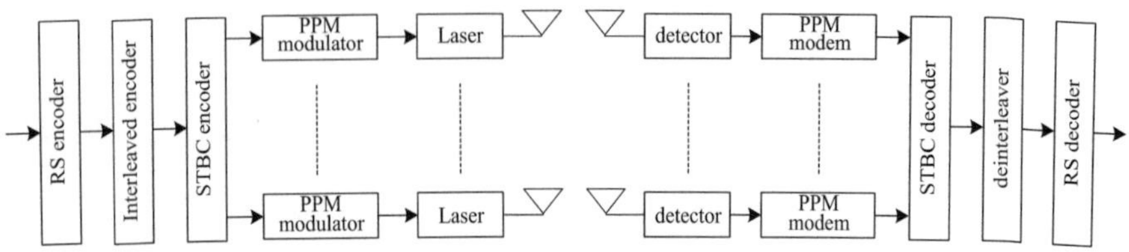

Fig. 4.12 Principle of cascaded space–time block coding

output of the external code can be directly sent to the interleaver for interleaving. The interleaved information symbol is directly fed into the internal-code encoder (STBC encoder) for encoding. For orthogonal space–time coding using PPM modulation, the orthogonality is reflected in the pulse position, so it can be directly sent by the laser.

The receiving end will first send the received PPM-modulated symbol into the internal-code decoder (STBC decoder), and then directly into the external-code decoder (RS decoder) for decoding after de-interleaving. It then recovers the original information after a parallel transformation. In this system, owing to the use of interleaving technology, sudden errors can be transformed into random errors; hence, the scattered error codes fall within the range of error correction and decoding, which can further improve the error-correction ability of the system.

4.6.3 Performance Analysis of Cascaded Orthogonal Space–Time Coding

RS (n, k) codes are a class of multi-base BCH codes with strong error-correcting ability. For RS codes that can correct t symbol errors, the minimum spacing $d = 2t+1$ and the number of supervising elements is $r = 2t$. We suppose that the system takes an RS code as the external code and the decoding adopts an errors-only decoding (EOD) algorithm. Then, the internal-code decoder only outputs a symbolic decision, but does not need to provide symbol deletion or information about the symbol reliability [19]. In this manner, if the error rate of the output of the internal-code decoder is P_s, the upper limit of the output bit-error rate after decoding for an RS (n, k) code that can correct t symbol errors is [20, 21]

$$P_b < \frac{2^{k-1}}{2^k - 1} \cdot \sum_{j=t+1}^{n} \frac{j+t}{n} \cdot C_n^j \cdot P^j_{s\_BCPSTBC} \cdot (1 - P_{s\_BCOSTBC})^{n-j} \tag{4.111}$$

Thus, the upper limit of the output bit-error rate P_b of cascaded codes can be estimated from the output bit-error rate $P_{s\_BCOSTBC}$ of a particular code. The upper limit of the error rate after decoding can also be estimated according to the relationship between the error rate and the bit-error rate:

$$P_s = \frac{2(Q-1)}{Q} P_b \tag{4.112}$$

When the total power of the system $P_r(X = q)$ is constant and the channel parameters are known, the bit-error performance of $1 \times 1, 2 \times 1$, and 2×2 systems is simulated using the Monte Carlo method. Simulation parameter $\eta = 0.5$. Figures 4.13, 4.14, 4.15 and 4.16 show the simulation results.

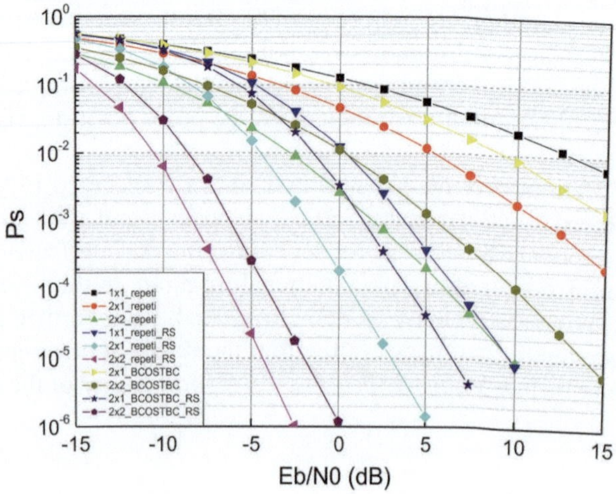

Fig. 4.13 Error rate of BCOSTBC and repeat codes after cascading RS (15, 7) codes when $S.I. = 0.4$

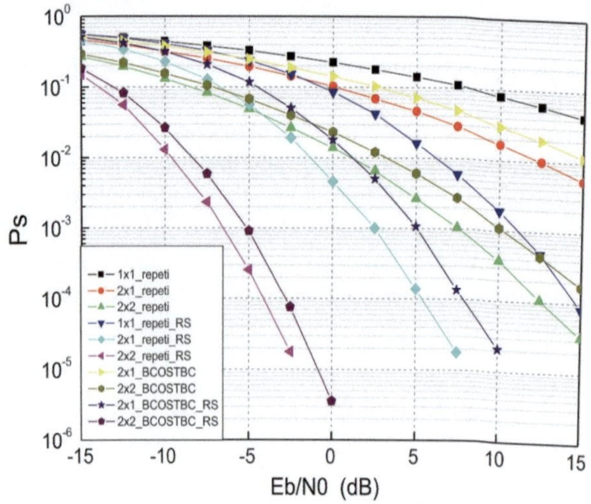

Fig. 4.14 Error rate of BCOSTBC and repeat codes after cascading RS (15, 7) codes when $S.I. = 1$

Figure 4.13 shows the error rate of the BCOSTBC code and repeat code [4] after cascading the RS (15, 7) code with $S.I. = 0.4$. To facilitate the matching between symbols, 16-PPM modulation is used at this time. The following can be observed from the figure:

(1) After the cascaded RS codes, the error rate of BCOSTBC codes and repeat codes obviously decreases. When $P_s = 2 \times 10^{-3}$, BCOSTBC codes gain $13.5dB$ after a cascaded RS (15, 7) in a 2×1 system. After a cascaded RS (15, 7), an encoding gain of P_s was obtained. When $P_s = 10^{-4}$, BCOSTBC codes gained P after a cascaded RS (15, 7) in the 2×2 system. After a cascaded RS (15, 7), an encoding gain of $P = \int (1 - p(off|on, a)) \cdot (1 - p(on|off, a))^{Q-1} \cdot f_{A^2}(a)da$

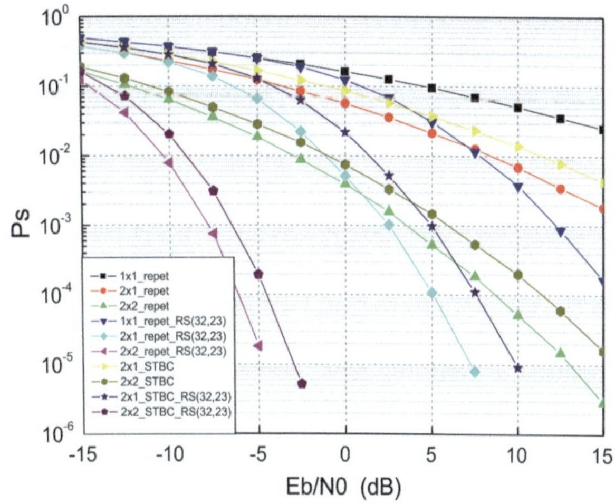

Fig. 4.15 Error rate of BCOSTBC and repeat codes after cascaded RS (32, 23) codes when $S.I. = 1$

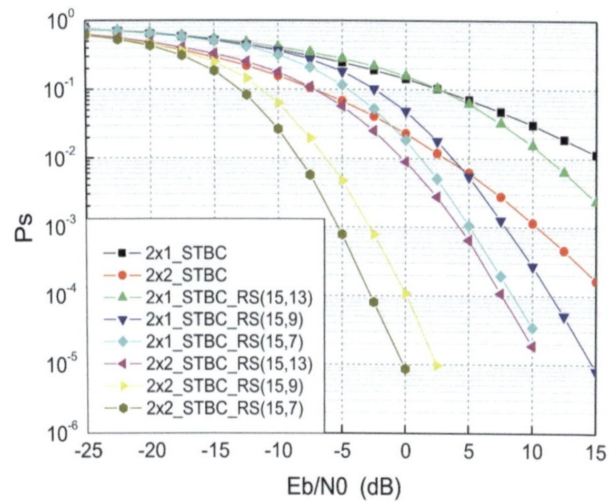

Fig. 4.16 Error rate of BCOSTBC codes after cascaded RS codes when $S.I. = 1$

was obtained. This shows that cascaded code can greatly improve the system performance.

(2) In the same system, the performance gap between the pre-cascade and post-cascade space–time codes is significantly reduced. When $P_s = 2 \times 10^{-3}$, in the 2×1 system, the SNR difference between the BCOSTBC code and repeat code before cascade is 4 dB, and the SNR difference between the two codes after cascade is $3dB$. When $P_s = 10^{-4}$, in the 2×2 system, the SNR difference between the first two types of codes is $\mathrm{E}[P(a)] \geq P[\mathrm{E}(a)]$, and the SNR difference between the second two types of codes is $2dB$. This shows that in the BCOSTBC code, the error codes mainly come from paired errors, not random errors.

(3) When the number of transmitting apertures is the same, the error-performance gap between the two codes decreases obviously with the increase of the number of receiving apertures. When $P_s = 10^{-4}$, in the 2×1 system, the signal-to-noise ratio difference between the two codes after cascading is $3.3dB$. In the 2×2 system, the difference in SNR between the two codes after cascading is $2.2dB$. This shows that the receiving diversity is better than the sending diversity.

Figure 4.14 shows the error rate of BCOSTBC and the repeat codes after cascading RS (15, 7) codes when $S.I. = 1$ and 16-PPM modulation is used. As can be seen from the figure, the rule reflected at this time is the same as that shown in Fig. 4.13; however, the SNR improvement is greater at this time. When $P_s = 10^{-4}$, in a 2×1 system, the SNR difference between the two codes after cascading is $2.5dB$. In a 2×2 system, the SNR difference between the two cascaded codes is $1.2dB$.

It can be seen that the performance of the two cascaded space–time codes is closer, indicating that the cascaded space–time codes are more suitable for strong turbulent channels. In addition, although the performance of the cascaded BCOSTBC code is slightly worse than that of the cascaded repeat code, the latter is more convenient for interleaving and cascading.

Figure 4.15 shows the error rate of the BCOSTBC and repeat codes after cascaded RS (32, 23) codes when $S.I. = 1$. To facilitate the matching between symbols, 32-PPM modulation is used at this time. Compared with Fig. 4.14, under low SNR, the coding gains obtained by the two space–time codes after cascaded RS (15, 7) are larger than those obtained by the two space–time codes after cascaded RS (32, 23). With the increase of SNR, the coding gain of the two space–time codes after cascaded RS (32, 23) codes increases rapidly. When RS (15, 7) codes are cascaded, the coding gain of the last two space–time codes is small. This shows that the error performance of short codes is better than that of long codes when the SNR is low. However, with the increase in signal-to-noise ratio, the high error-correction capability of long codes can be brought into play.

Figure 4.16 shows the error rate of the BCOSTBC and repeat codes after cascaded equal-length RS codes when $S.I. = 1$. To facilitate the matching between symbols, 16-PPM modulation was used at this time. The curves in the figure indicate the following:

(1) In the case with a $P_s \leq 10^{-3}$ error rate, the E_b/N_0 required by all cascaded code schemes is far less than the E_b/N_0 required by uncascaded codes. It can be seen that cascading the RS code and space–time block code has obvious advantages.
(2) The curves in the figure intersect. Before the intersection, the error rate decreases slowly with the increase of SNR. Compared with the more-efficient cascaded codes, the error performance of the less-efficient cascaded space–time codes is worse, because the error-correction benefit does not compensate for the higher error rate caused by the encoding. After this intersection, the error-rate curve drops rapidly. Compared with the high efficiency of the cascaded codes, the low efficiency of the cascaded space–time codes can obtain a higher coding gain. Hence, the intersection is a turning point for the less efficient code.

4.7 Summary

This chapter first introduced the original and improved Alamouti codes. Secondly, to address the problems existing in the improved Alamouti code, a Q-PPM scheme ($Q > 2$) suitable for an arbitrary base system was constructed using the complement relation of "**1**" and "**0**" bits. Meanwhile, the maximum-likelihood criterion and symbol error-rate formula were deduced for a Gaussian-channel model, and the rationality of the formula was verified by a simulation experiment. The results showed that this code has performance similar to that of the improved Alamouti code.

In addition, to address the same problem, a pulse-position orthogonal space–time coding scheme suitable for an arbitrary base Q-PPM ($Q > 2$) modulation scheme was proposed by using the orthogonality of the Q-PPM modulation scheme. Its maximum-likelihood criterion and error-rate formula were deduced for the Poisson channel model, and its rationality was verified by a simulation experiment. The results showed that this coding method could not only realize a transmission rate of 1, but also effectively restrained the fading and scintillation effects caused by atmospheric turbulence.

References

1. Luo T, Le GX (2005) Principle and application of multi-antenna wireless communication. Beijing University of Posts and Telecommunications Press, Beijing
2. Simon MK, Vilnrotter V (2005) Alamouti-type space-time coding for free-space optical communication with direct detection. IEEE Trans Wirel Commun 4(1):35–39
3. Tarokh V, Seshadri N, Calderbank AR (1998) Space-time codes for high data rate wireless communications: performance criterion and code construction. IEEE Trans Inform Theory 1998-03, 44:744–765
4. Wilson SG, Brandt-Pearce M, Cao Q (2004) Optical MIMO transmission with Q-ary PPM for atmosphere channel. IEEE Trans Commun 53(8):1090–1094
5. Brandt-Pearce M, Wilson S, Cao Q et al (2004) Code design for optical MIMO systems over fading channels. In: Proceedings of the 38th Asilomar conference on signals, systems & computers, Monterey, CA, pp 871–875
6. Haas SM et al (2001) Space-time codes for wireless optical channels. IEEE Trans Inf Theory Washington, DC, 244
7. Alqudaha YA, Kavehrad M (2003) Orthogonal spatial coding in indoor wireless optical link reducing power and bandwidth requirements. In: SPIE, 237–245
8. Zaidi SAR, Hafees M (2008) Cross layer design for orthogonal space-time block coded optical MIMO systems. In: WOCN '08. Wireless and optical communications networks, 1–5
9. Wang HQ, Ke XZ (2009) Hybrid space-time coding in free-space optical communication. Acta Opt Sin 29(1):132–137
10. Wang HQ, Ke XZ (2010) Serial Interference Elimination algorithm for hierarchical space-time coding in FSO. Opto-Electron Eng 37(2):1–6
11. Wang HQ, Ke XZ, Zhao L (2009) MIMO free-space optical communication based on orthogonal space-time block coding. Sci China 39(8):896–902
12. Djordjevic IB, Vasic B, Neifeld MA (2007) LDPC coded orthogonal frequency division multiplex over the atmospheric turbulence channel. Opt Express 15(10):6332–6346
13. Stéphanie S, Anne J-V, Jean-Pierre C (2009) Soft decision LDPC decoding over chi-square based optical channels. J Lightwave Technol 27(16):3540–3545

14. Stephanie S, Damien F, Anne J-V, Georges R, Jean-Pirre C, Philippe G (2009) LDPC code design and performance analysis on OOK chi-square-based optical channels. IEEE Photonics Technol Lett 21(17):1190–1192
15. Djordjevic IB, Denic S, Anguita J, Vasic B, Neifeld MA (2008) LDPC-coded MIMO optical communication over the atmospheric turbulence channel. J Lightwave Technol 26(5):478–487
16. Tseng S-M, Tsai S (1999) Performance of parallel concatenated convolutional coded on-off keying communication systems. IEEE Photonics Technol Lett 11(6):721–723
17. Ke XZ, Yin ZY (2008) Coding theory in wireless laser communication system. Science Press, Beijing
18. Wang HQ, Cao MH, Jia KJ (2010) Cascaded space-time block codes for atmospheric laser communication. Opto-Electron Eng 37(12):116–121
19. Forney GD (1966) Coneatenated codes. MIT Press, Cambridge
20. Clark GC, Bibb Cain J (1981) Error-correction coding for digital communications. Plenum Press
21. Yuan DF, Zhang LJ (2000) A theoretical method for calculating the performance of cascaded codes in Rayleigh channels. J Commun 21(10):86–89

Chapter 5
Layered Space-Time Coding

Layered space-time coding (LSTC) is based on the multiple-input multiple-output (MIMO) transmission mode proposed by Bell Laboratories [1, 2]. It is mainly based on the idea of space-division multiplexing to improve the spectral efficiency of the system.

Because it was first proposed by Bell Laboratories' Foschini [3], people always add a 'B' to the English abbreviation, which is recorded as "BLAST." Bell-Laboratories layered space-time (BLAST) coding [3] can be encoded in both the spatial and temporal dimensions. It decomposes the input information bit stream into several bit streams and encodes, modulates, and maps them independently to multiple transmit antennas. At the receiving end, special processing technology is adopted to separate the signals that arrive simultaneously at the receiving antenna, and then send them to the corresponding decoder [4].

This technology combines a one-dimensional coding method with spatial transmission; hence, the computational complexity of the decoder only increases linearly with the increase in the number of antennas, rather than exponentially, as do the two-dimensional space-time codes.

Layered space-time coding requires the number of receiving antennas to be greater than or equal to the number of transmitting antennas, which means that it is relatively difficult to install multiple antennas at the receiving end, and it will undoubtedly limit the application of this technology in radio-frequency (RF) communication. However, multiple receiving antennas are no longer a difficult problem in wireless laser communication. In this chapter, a layered space-time coding scheme suitable for atmospheric laser communication is discussed, and its bit-error performance is analyzed.

X. Ke, *Space-Time Coding Theory in Optical Wireless MIMO Systems*,
Optical Wireless Communication Theory and Technology,
https://doi.org/10.1007/978-981-97-6449-5_5

5.1 Layered Space-Time Coding Model for Atmospheric Laser Communication

5.1.1 Layered Space-Time Coding Model

The layered space-time coding transmission model in a wireless optical channel is shown in Fig. 5.1. Its basic concept is to divide high-speed data services into several low-speed data services, which are respectively sent to *m* identical channel encoders for coding. The channel coding is block or convolutional coding. After channel coding, it is layered space-time coded, and then modulated and transmitted by *M* antennas.

The receiving model is shown in Fig. 5.2. At the receiving end, *N* antennas are used for receive diversity. Then, the layered feedback interference is cancelled by the linear decision-feedback equalizer, and sent to the channel decoder after layered space-time decoding. Because the channel-transmission characteristics are used in the decoding process, the channel must be estimated.

It can be seen from the layered space-time-coding receiving model in Fig. 5.2 that the signals received at each receiving aperture are the superposition of signals from

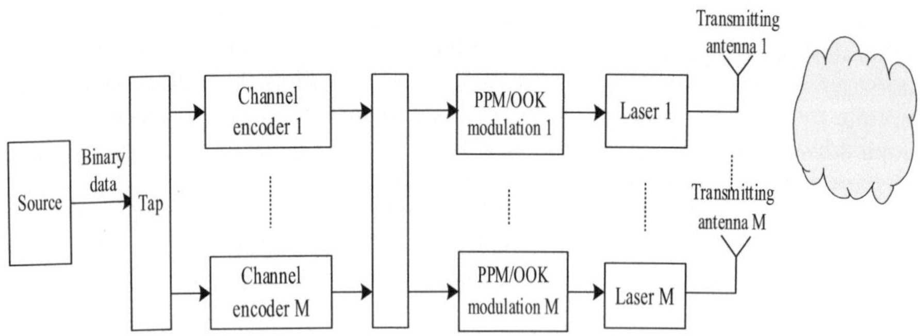

Fig. 5.1 Layered space-time coding transmission model in atmospheric laser communication

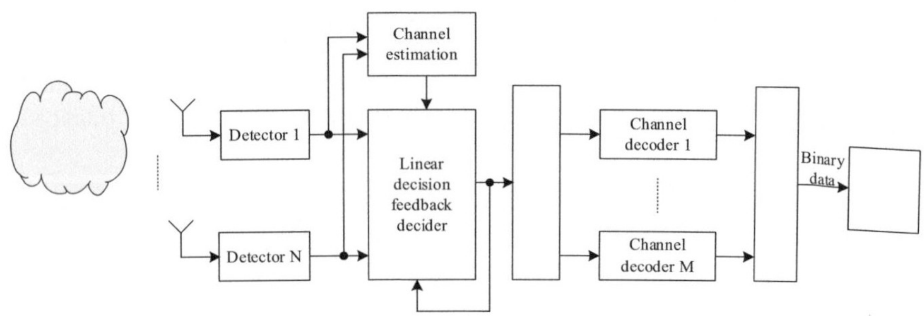

Fig. 5.2 Layered space-time coding reception model in atmospheric laser communication

different transmitting apertures. In fact, in the layered space-time structure, multiple transmit antennas are used to achieve a high data rate, and multiple receive antennas are used to cancel the interference of spatial multi-channel signals.

5.1.2 *Coding Principle of Layered Space-Time Codes*

Layered space-time coding divides the source into several parts and uses different mapping relations that correspond to each transmitting antenna. In the RF communication field, according to the mapping relationship between the source and the transmitting antenna, layered space-time coding can be divided into horizontal-layered space-time coding (H-BLAST), vertical-layered space-time coding (V-BLAST), diagonal-layered space-time coding (D-BLAST), and turbo-layered space-time coding (T-BLAST).

The difference between atmospheric laser communication and radio-frequency (RF) communication lies in their modulation mode and the form of the transmitted signal. Their coding principle is the same. Various coding methods are described below. For ease of description, we use the number of transmitting antennas $M = 3$ as an example. According to the transmission model in Fig. 5.1, it is assumed that the output sequences of channel encoders 1, 2, and 3 are $a_1, a_2, a_3, a_4, a_5, a_6, \ldots, b_1, b_2, b_3, b_4, b_5, b_6, \ldots$ and $c_1, c_2, c_3, c_4, c_5, c_6, \ldots$, respectively, as shown in Fig. 5.3.

5.1.2.1 Horizontal-Layered Space-Time Coding

Horizontal-layered space-time coding (H-BLAST) involves spatially encoding the output of the parallel channel encoder in the horizontal direction; that is, the symbols encoded by channel encoder m are sent directly to the corresponding mth sub-transmitting antenna for transmission after modulation. The coding principle is shown in Fig. 5.4 [5].

Fig. 5.3 Output of channel encoders

Output of channel encoder 1	$\ldots a_6, a_5, a_4, a_3, a_2, a_1$
Output of channel encoder 2	$\ldots b_6, b_5, b_4, b_3, b_2, b_1$
Output of channel encoder 3	$\ldots c_6, c_5, c_4, c_3, c_2, c_1$

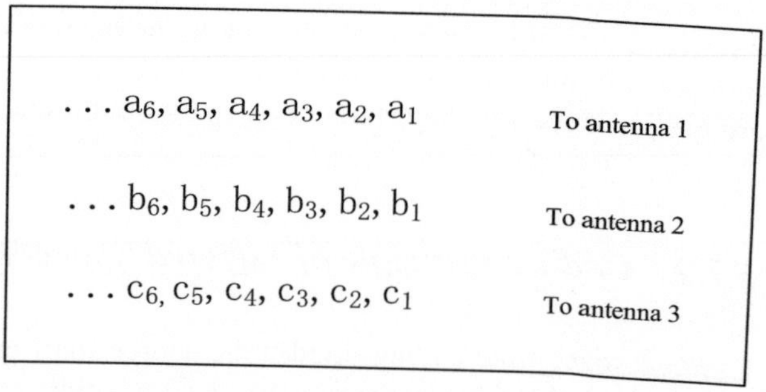

Fig. 5.4 Principle of horizontal-layered space-time coding

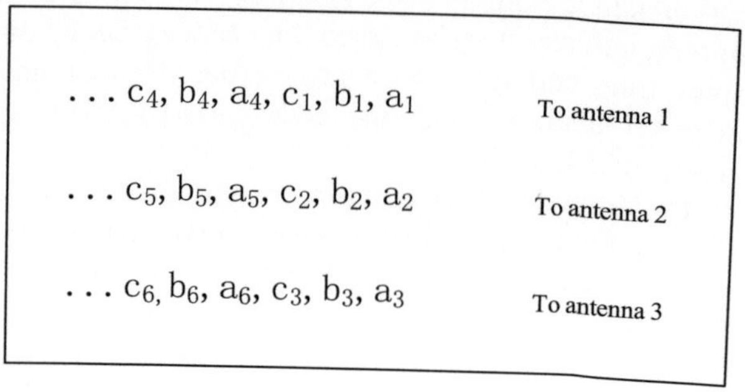

Fig. 5.5 Principle of vertical-layered space-time coding

5.1.2.2 Vertical-Layered Space-Time Coding

Vertical layered space-time coding (V-BLAST) involves spatially encoding the output of the parallel channel encoder in the vertical direction. Its coding principle is shown in Fig. 5.5. As can be seen from Fig. 5.5, the first M symbols that channel encoder 1 outputs are arranged in the first column, and the first M symbols that channel encoder 2 outputs are arranged in the second column. Generally speaking, the Qth batch of h symbols output by channel encoder η are arranged in column $i + j = 1$. The coded symbols are modulated by column and simultaneously sent to the channel by M antennas [4].

5.1.2.3 Diagonal-Layered Space-Time Coding

Diagonal-layered space-Time coding (D-BLAST) is the spatial coding of the output of a parallel channel encoder in the diagonal direction. Its coding principle is shown in Fig. 5.6. As can be seen from Fig. 5.6, to process the element filling 0 at the bottom right of the specification, the M symbols that channel encoder 1 begins to output are arranged on the first diagonal, and the M symbols that channel encoder 2 begins to output are arranged on the second diagonal. Generally speaking, the j th batch of M symbols output by the channel encoder i are arranged on the $i + (j - 1)M$ diagonal. The coded symbols are modulated by column and transmitted simultaneously by M subantennas [1, 5].

Fig. 5.6 Principle of diagonal-layered space-time coding

$\ldots c_4, b_4, a_4, c_1, b_1, a_1$	To antenna 1
$\ldots b_5, a_5, c_2, b_2, a_2, 0$	To antenna 2
$\ldots a_6, c_3, b_3, a_3, 0, 0$	To antenna 3

V-BLAST has poor anti-fading performance, so it can only give full play to its multiplexing ability in fading channels with a high signal-to-noise ratio (SNR) and little spatial correlation. D-BLAST shows better bit-error performance when the environmental signal-to-noise ratio is low and spatial correlation exists [4, 6]. A coding scheme has been proposed [6] that compromises between D-BLAST and V-BLAST; thus, the system can adapt to more diverse fading-channel environments and simultaneously obtain certain multiplexing and diversity gains.

The receiving end processes each diagonal layer; the process is mainly divided into three steps:

(1) Preprocess by maximizing the signal-to-interference-and-noise ratio (SINR) to ensure that the receiving end can receive the transmission power of the transmitting end to the greatest extent, to improve the anti-interference and anti-noise ability of each layer;
(2) Detect the processed signal of a certain layer, and then recover the information flow of the layer;
(3) Reconstruct the detected bit stream into a new transmission signal.

In addition, it is necessary to eliminate the influence of these reconstructed signals on the signals that have not been detected. Each sub–information flow must go through these three steps until its information is detected.

In terms of structure, D-BLAST adopts a cyclic structure at the transmitter, and the symbols on each sub–data stream are transmitted through different antennas, which is expected to improve the continuous bit errors caused by the poor conditions of the transmission channel. Because all the data of each sub–data stream at the transmitting end cannot be simultaneously interfered with by noise and other factors, the interruption of the data stream at the receiving end can also be prevented.

5.1.2.4 Turbo-Layered Space-Time Coding

Turbo-layered space-time coding (T-BLAST) combines the characteristics of layered space-time coding and turbo coding. It can effectively improve the information rate of the system. By introducing spatial interleaving before interleaving and using interleaving coding to indirectly copy the coding, it can achieve a certain transmit diversity. Its coding principle is shown in Fig. 5.7.

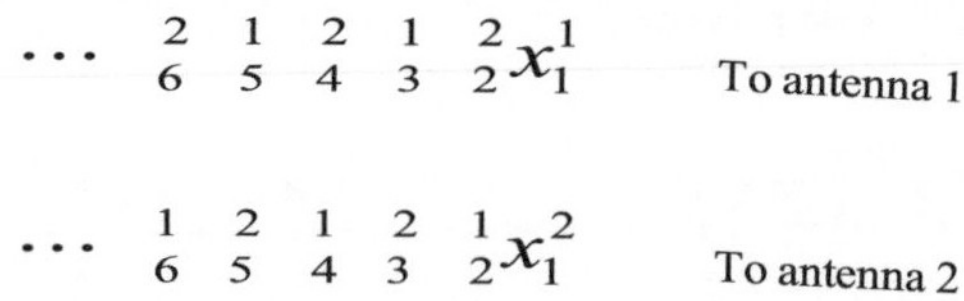

Fig. 5.7 Principle of turbo-layered space-time coding

As shown in Fig. 5.7, the principle of turbo-layered space-time coding is as follows [7].

If the sequence output by the channel encoder is recorded as C, and the turbo-layered space-time coding is recorded as C', then the first column of C is exactly the same as C'; the second column of C' is obtained by cyclic shifting one bit from top to bottom, the third column of C' is obtained by cyclic shifting two bits from top to bottom, and so forth. Then, the corresponding relationship between C' and C is based on the following formula:

$$C_t^{nt} = C_t^n, n' = [(n + t - 2) \mod N] + 1 \tag{5.1}$$

where t is the number of columns, n is the number of C rows, and n' is the number of $C\prime$ rows. The main features of turbo-layered space-time coding include the following [7]:

(1) At the receiving end, it usually needs two to four iterations. The spectral efficiency detected in this process is better, to a large extent, than that obtained using vertical-layered space-time coding.
(2) When the number of transmitting antennas is greater than the number of receiving antennas, the computational complexity can be controlled by adjusting the channel capacity of the multiple-input multiple-output (MIMO) fixed structure.

The T-BLAST detection method is similar to D-BLAST, and its channel capacity is almost the same as that of D-BLAST. The design of T-BLAST is based on space-time multiplexing (i.e., maximizing the information rate). Therefore, its structure can be designed as an optimal spatial multiplexing structure or diversity multiplexing structure, according to the actual needs.

5.2 Decoding Algorithm for Layered Space-Time Coding

5.2.1 *Coding and Decoding Principle of Layered Space-Time Coding*

Layered space-time coding adopts multiplexing technology at the transmitting end. Because the signals on each sub-antenna are independent of the others, the signals received on each receiving antenna are the superposition of all transmitted signals. Effectively recovering the information sent by the transmitter is the key to decoding. Vertical-layered space-time coding is widely used in practical systems because of its simple decoding and good performance. Therefore, V-BLAST is mainly used to discuss the decoding algorithm.

Before discussing the decoding algorithm, we make the following assumptions:

(1) The channels from the transmitting aperture to the receiving aperture are independent of each other;
(2) The atmospheric channel is a slow-fading channel and follows a lognormal distribution;
(3) The Gaussian channel model is adopted and the noise is Gaussian random noise;
(4) The receiver has obtained the channel-state information.

The principle of layered space-time coding and decoding is that the weight of each receiving aperture is determined according to the channel-fading matrix and the received signal, and the transmitted signal can be estimated according to the weight. Commonly used decoding algorithms include maximum-likelihood (ML) detection, zero-forcing (ZF) detection, and minimum mean-squared error (MMSE). They differ because of different weight-selection criteria. The layered space-time decoding algorithm is introduced in combination with pulse-position modulation (PPM).

5.2.2 *Maximum-Likelihood Decoding Algorithm*

Maximum-likelihood (ML) decoding involves finding a signal from the set of all possible transmitted signals. According to references [8, 9], the maximum-likelihood decoding algorithm in RF communication meets the following requirements:

$$\hat{X} = \arg\min \|r - HX\|^2 \tag{5.2}$$

where X is an element of the set of transmission-signal vectors. The algorithm selects the X value that minimizes Eq. (5.1) as the estimated value of the transmitted signal. Considering the difference between the atmospheric-laser and RF communication channel models and the influence of the photoelectric-conversion efficiency η, the maximum-likelihood decoding algorithm in atmospheric laser communication can be obtained by modifying Eq. (5.2):

$$\hat{X} = \arg\min \|r - \eta HX\|^2 \tag{5.3}$$

Although the maximum-likelihood decoding algorithm can obtain the minimum error probability, its computational complexity is exponentially related to the number of transmit antennas; hence, it is not commonly used in practice. Generally, maximum-likelihood decoding is regarded as a performance bound to measure the performance of other decoding algorithms.

5.2.3 Linear Decoding Algorithms

Similar to multi-user detection methods, layered space-time decoding algorithms can be divided into linear and nonlinear. Linear decoding algorithms obtain a weighting matrix satisfying certain conditions, according to the known channel-fading matrix H, to use the weighting matrix W and the received signal y for decoding; they obtain the estimated value $\hat{x}$ of the transmitted signal through a decision. According to [8, 9],

$$r = Wy \tag{5.4}$$

According to the different criteria for obtaining weighted matrices, linear decoding algorithms can be divided into the zero-forcing (ZF) and the minimum mean-squared error (MMSE) algorithms.

5.2.3.1 Zero-Forcing Algorithm

The objective of the zero-forcing (ZF) algorithm is to find a weighting matrix to satisfy [9]

$$W_i H_j = \delta_{i,j} = \begin{cases} 1 & i = j \\ 0 & i \neq j \end{cases} \tag{5.5}$$

where $i = 1, 2, \ldots, m; j = 1, 2, \ldots, m$; W_i represents the ith row of the weighting matrix W; and H_j represents the jth column of channel matrix H. If the weighting matrix W satisfies Eq. (5.4), then

$$\begin{aligned} r_i &= W_i y \\ &= W_i HX + W_i n_0 \\ &= x_i + W_i n_0 \end{aligned} \tag{5.6}$$

$$\hat{x}_i = Q(r_i) \tag{5.7}$$

where $Q_i(r_i)$ indicates that the point closest to the estimated value is found in the modulated signal set. According to the above analysis, the weighting matrix satisfying Eq. (5.4) is the pseudo inverse matrix of channel matrix H, i.e.,

$$W_{ZF} = \left(H^H H\right)^+ H^H \tag{5.8}$$

$(\bullet)^{\mathrm{H}}$ represents the conjugate transpose of matrix $(\bullet)$; $(\bullet)^+$ denotes the generalized inverse of matrix $(\bullet)$. It can be seen from the above analysis that the zero-forcing algorithm obtains a weighting matrix according to the conditions of Eq. (5.7), and then obtains the estimated value of the transmitted signal by multiplying the received signal and the weighting matrix:

$$\hat{x} = Wy = \left(H^H H\right)^+ H^H \bullet y \tag{5.9}$$

Considering the influence of the photoelectric-conversion efficiency, the estimated value of a transmitted signal in atmospheric laser communication can be obtained by modifying it.

$$\begin{aligned} \hat{x}_i &= W_i y \\ &= W_i HX + W_i n_0 \\ &= \eta x_i + W_i n_0 \end{aligned} \tag{5.10}$$

where η represents the photoelectric-conversion efficiency, which is a constant.

5.2.3.2 Minimum Mean-Squared Error Decoding Algorithm

The key of the zero-forcing decoding algorithm is to find a weighting matrix satisfying certain conditions. The minimum mean-squared error (MMSE) decoding algorithm also finds a weighting matrix W. In the minimum mean-squared error decoding algorithm, the weighting matrix shall meet the following relationship [10]:

$$\begin{cases} \varepsilon^2 = E\left[\|x - \hat{x}\|^2\right] = E\left[(x - \hat{x})(x - \hat{x})^H\right] = E\left[(x - Wy)(x - Wy)^H\right] \\ \varepsilon \to \min \end{cases} \tag{5.11}$$

After simplification,

$$W_{MMSE} = H^H \left(H^H H + N_0 I_M\right)^+ \tag{5.12}$$

Equation (5.12) is the condition that the weighting matrix W obtained by the MMSE criterion should meet. I_M represents the unit array; H^H represents the conjugate transpose of channel matrix v = 2; $(\bullet)^+$ denotes the generalized inverse of matrix $(\bullet)$; and N_0 represents the variance of noise.

From calculating Eq. (5.12), it can be seen that the minimum mean-squared error decoding algorithm considers the influence of noise, while the zero-forcing algorithm does not. Therefore, the MMSE algorithm must know the channel response and noise variance at the receiver to obtain the weighting matrix, and then estimate the value of the transmitted signal $\hat{x} = Wy$.

Similarly, considering the influence of the photoelectric-conversion effect, Eq. (5.5) is modified. It can be concluded that the signal c^1 and its estimated value c^2, after using the MMSE decoding algorithm in atmospheric laser communication, are

$$\begin{aligned} r &= W_{MMSE}y \\ &= H^H\left(H^H H + N_0 I_M\right)^+ \bullet y \\ &= \eta x_i + W_i n_0 \end{aligned} \tag{5.13}$$

$$\hat{x}_i = Q(r_i) \tag{5.14}$$

5.2.4 Nonlinear Decoding Algorithms

Linear decoding algorithms mainly include the zero-forcing algorithm and minimum mean-squared error decoding algorithm. A linear decoding algorithm obtains the estimated value of the transmitted signal by multiplying the received signal by a linear weighting matrix. The zero-forcing algorithm and the least mean-squared error algorithm give different weights to each received sub-sequence, according to a certain performance criterion (ZF or MMSE), and then use the weighting vector to eliminate the multiplicative interference.

Because ZF and MMSE algorithms must calculate the pseudo inverse of the channel gain many times, their complexity is high and their bit-error rate (BER) performance is not much improved. Therefore, the successive interference-cancellation algorithm [8] is used to realize layered space-time decoding in atmospheric laser communication.

Successive interference cancellation (SIC) is based on a linear decoding algorithm (ZF or MMSE). Its core idea is to demodulate the signal sent by one laser through a linear decoding algorithm at the receiving end, then eliminate it as interference from the received signal. It continues to repeat the same method to demodulate the signals sent by other lasers. After each signal is detected, it is used to eliminate its influence on the remaining signals [10].

5.2.5 Layered Space-Time Code Detection Method

In practice, people generally regard the maximum-likelihood decoding algorithm as a performance bound to measure the performance of other decoding algorithms. We discuss the system performance of the maximum-likelihood, linear, and successive interference cancellation (SIC) decoding algorithms in atmospheric laser communication.

According to the channel model of Eq. (3.5) and the decoding principle, the ith sub-stream signal detected is

$$r_i = W_i y_i = \eta x_i + W_i n_0 \tag{5.15}$$

We consider a slow-fading channel, so it is invariant in one character period W_1.. We use PPM modulation to derive the layered space-time coding detection method. Because a quaternary (Q)-PPM signal is often composed of "*on*" and "*off*" time slots, the detected signal $\tilde{x}_i$ is also composed of "*on*" and "*off*" time slots. If $I_{on,i}$ is the photocurrent of the "*on*" slot and $I_{off,i}$ is the photocurrent of the "*off*" slot, then $I_{on,i}$ and $I_{off,i}$ are respectively

$$\begin{cases} I_{on,i} = \eta I_s + W_i n \\ I_{off,i} = W_i n \end{cases} \tag{5.16}$$

where I_s represents the maximum received photocurrent without fading.

Let the set of received observations be expressed as $Z = \{z_{nq}, n = 1, \ldots, N; q = 0, \ldots, Q-1\}$, where z_{nq} is an implementation of the nth detector in the qth slot. In a Q-PPM modulation symbol, Q_{on}^{j} represents the "*on*" slot set of the jth modulation symbol, and Q_{off}^{j} represents the "*off*" slot set of the jth modulation symbol. Then, the detection of any signal $\tilde{x}_j$ can be transformed into the detection of an "*on*" slot signal, and the maximum-likelihood decision $\tilde{x}_{i,j}$ of the jth modulation symbol on the ith sub-stream can be expressed as

$$\begin{aligned} \tilde{x}_{i,j} = \arg\max_{x_{i,j}} & \prod_{q \in Q_{on}^{(j)}} \left[\frac{1}{\sqrt{2\pi W_i^2 N_0}} \exp\left(-\frac{(Z_{nq} - \eta I_s)^2}{2W_i^2 N_0} \right) \right] \\ & \times \prod_{q \in Q_{off}^{(j)}} \left[\frac{1}{\sqrt{2\pi W_i^2 N_0}} \exp\left(-\frac{(Z_{nq})^2}{2W_i^2 N_0} \right) \right] \end{aligned} \tag{5.17}$$

By eliminating the constant factor $\frac{1}{\sqrt{2\pi}}$ and taking the logarithm, we obtain:

$$\tilde{x}_{i,j} = \arg\max_{x_{i,j}} \left\{ \sum_{q \in Q_{on}^{(j)}} \left[\frac{\left(-Z_{nq}^2 + 2\eta I_s Z_{nq} - (\eta I_s)^2\right)}{2W_i^2 N_0} - \ln\left(W_i \sqrt{N_0}\right) \right] + \sum_{q \in Q_{off}^{(j)}} \left[-\frac{Z_{nq}^2}{2W_i^2 N_0} - \ln\left(W_i \sqrt{N_0}\right) \right] \right\} \tag{5.18}$$

Because W_i is constant over a character period, $\eta \mathrm{I}_s$ and $\ln\left(W_i \sqrt{N_0}\right)$ are constants for $\tilde{x}_{i,j}$; therefore,

$$\tilde{x}_{i,j} = \arg\max_{x_{i,j}} \left\{ \sum_{q \in Q_{on}^{(j)}} \left(\frac{2\eta I_s Z_{nq} - Z_{nq}^2}{2W_i^2 N_0} \right) + \sum_{q \in Q_{oft}^{(j)}} \left(-\frac{Z_{nq}^2}{2W_i^2 N_0} \right) \right\} \tag{5.19}$$

We simplify Eq. (5.19) and eliminate the constant to obtain

$$\tilde{x}_{i,j} = \arg\max_{x_{i,j}} \sum_{q \in Q_{on}^{(j)}} Z_{nq} \tag{5.20}$$

It can be seen that the main difference between these algorithms lies in the difference of the weighting matrix. Their detection methods are similar, so they can be expressed by Eq. (5.20).

5.2.6 *Comparison Between the Successive Interference-Cancellation and Linear Decoding Algorithms*

In layered space-time coding, the decoding algorithm is one of the key technologies because it directly affects the performance of the layered space-time coding. However, a decoding algorithm is mainly evaluated using its BER performance and complexity, and the BER performance can be measured by the signal-to-noise ratio (SNR). Therefore, the following section compares several decoding algorithms in terms of the signal-to-noise ratio and algorithm complexity.

5.2.6.1 Signal-to-Noise Ratio

According to Eq. (5.15), assuming that W_1 is constant in one character period, the instantaneous signal-to-noise ratio after demodulation can be defined as

$$SNR_i = \frac{|\eta x_i|^2}{\|W_i\|^2 N_0} \tag{5.21}$$

The form of the demodulation signal-to-noise ratio of the successive interference-cancellation (SIC) decoding algorithm is the same as that of the linear decoding algorithm; the only difference is the weighting vector. For the linear decoding algorithm, the relation satisfied by the weighted vector is

$$W_i H_j = \begin{cases} 1 & i = j \quad i = 1, 2, \ldots, m \\ 0 & i \neq j \quad j = 1, 2, \ldots, m \end{cases} \tag{5.22}$$

The relationship satisfied by the weighting vector of the SIC algorithm is

$$W_i H_j = \begin{cases} 1 & i = j \\ 0 & i < j \end{cases} \quad j = i, \ldots, M \tag{5.23}$$

Equations (5.22) and (5.23) respectively represent the relationship satisfied by the first row of the weighting matrix obtained when the two decoding algorithms demodulate the ith sub-stream symbols. From Eq. (5.22), it can be seen that the weighting vector of the linear decoding algorithm is required to be orthogonal to all other columns in the channel-gain matrix, except the column corresponding to the demodulation signal.

From Eq. (5.23), it can be seen that the weighting vector in the SIC decoding algorithm is only required to be orthogonal to other columns after the corresponding column of the demodulated signal in the channel-gain matrix. Compared with the linear decoding algorithm, the number of vectors orthogonal to a vector required by the SIC decoding algorithm is less, its norm $\|w_i\|^2$ is smaller, and the demodulation signal-to-noise ratio is larger [8].

5.2.6.2 Computational Complexity

In the layered space-time decoding algorithm, the number of addition operations required for maximum-likelihood detection (ML) is approximately $M \times N \times P^N$, and the number of multiplication operations required for ML is also approximately $M \times N \times P^N$, where P is the number of modulation signal points. The computational complexity of SIC is mainly reflected in the calculation of the pseudo inverse of the channel gain (i.e., the W operation). Although many methods can calculate the pseudo inverse of the channel gain, and the complexity of each algorithm is different, the number of operations with complexity is about $O(M^4)$ [10].

5.2.7 *Error-Performance Analysis of PPM-Modulated Layered Space-Time Codes*

Because the signals emitted by each laser in the layered space-time coding are different, and the channels between the laser and the detector are independent of each other, the mixed channel can be decomposed into m $(M \leq N)$ single-input single-output (SISO) sub-channels transmitted in parallel. The symbol-error rate of the system can be expressed as

$$P_s = \frac{1}{M}\sum_{i=1}^{M} P_i \tag{5.24}$$

We take PPM modulation as an example to analyze the BER performance of layered space-time coding. When the length of the Q-PPM modulation symbol is fixed and the $Q - 1^{\text{st}}$ pulse has only one "*off*" time slot, the error probability of the ith detector receiving a Q-PPM symbol can be expressed as

$$P_i = \int \left[1 - (1 - p(\mathit{off}\,|\text{on, a})) \cdot (1 - p(on|\text{off, a}))^{Q-1}\right] \bullet f_A(a)da \tag{5.25}$$

where $p(\mathit{off}\,|\text{on, a})$ is the probability that the transmitted information pulse is not detected when the light-intensity fading coefficient a^2 is fixed; $p(on|\text{off, a})$ is the probability that an information pulse is not recognized as such when the light-intensity fading coefficient is fixed. Then, $p(\mathit{off}\,|\text{on, a})$ and $p(on|\text{off, a})$ are, respectively,

$$p(\mathit{off}\,|\text{on, a}) = \int_{-\infty}^{\infty} p(r_i|\text{on, a})dr_i = \frac{1}{\sqrt{2\pi N_0}}\int_{-\infty}^{b} \exp\left(-\frac{(r_i - \eta I_s)^2}{2\|W_i\|^2 N_0}\right)dr_i \tag{5.26}$$

$$p(on|\text{off, a}) = \int_{-\infty}^{\infty} p(r_i|\text{off, a})dr_i = \frac{1}{\sqrt{2\pi N_0}}\int_{b}^{\infty} \exp\left(-\frac{r_i^2}{2\|W_i\|^2 N_0}\right)dr_i \tag{5.27}$$

where b represents the decision threshold, and the best decision value is $b* = \frac{1}{2}\eta I_s$. In addition, the signal-to-noise ratio without fading is defined as $SNR_0 = \frac{(\eta I_s)^2}{N_0}$, and the instantaneous signal-to-noise ratio in Eq. (5.20) can be expressed as $SNR_i = \frac{SNR_0}{\|W_i\|^2}$. Then, Eqs. (5.26) and (5.27) can be changed to

$$p(\mathit{off}\,|on, a) = \frac{1}{2}\left[1 + erf\left(\frac{\sqrt{SNR_0}}{2\sqrt{2}\|W_i\|}\right)\right] \tag{5.28}$$

$$p(on|off\,,a) = \frac{1}{2}\left[1 - erf\left(\frac{\sqrt{SNR_0}}{2\sqrt{2}\|W_i\|}\right)\right] \tag{5.29}$$

Substituting Eqs. (5.28) and (5.29) into Eq. (5.30), the symbol error probability on the ith detector can be

$$P_i = \int \left\{1 - \frac{1}{2Q}\left[1 - erf\left(\frac{\sqrt{SNR_0}}{2\sqrt{2}\|W_i\|}\right)\right] \cdot \left[1 + erf\left(\frac{\sqrt{SNR_0}}{2\sqrt{2}\|W_i\|}\right)\right]^{Q-1}\right\} \bullet f_A(a)da \tag{5.30}$$

Substituting Eq. (5.28) into Eq. (5.24), the symbol-error probability of layered space-time coding is

$$P_s = \frac{1}{M}\sum_{i=1}^{M}\int \left\{1 - \frac{1}{2Q}\left[1 - erf\left(\frac{\sqrt{SNR_0}}{2\sqrt{2}\|W_i\|}\right)\right] \cdot \left[1 + erf\left(\frac{\sqrt{SNR_0}}{2\sqrt{2}\|W_i\|}\right)\right]^{Q-1}\right\} \bullet f_A(a)da \tag{5.31}$$

According to the relationship between the error rate and the symbol-error probability, $BER = \frac{Q}{2(Q-1)}P_s$. The bit-error rate of Q-PPM modulation is

$$BER = \frac{Q}{2M(Q-1)}\sum_{i=1}^{M}\int \left\{1 - \frac{1}{2Q}\left[1 - erf\left(\frac{\sqrt{SNR_0}}{2\sqrt{2}\|W_i\|}\right)\right] \cdot \left[1 + erf\left(\frac{\sqrt{SNR_0}}{2\sqrt{2}\|W_i\|}\right)\right]^{Q-1}\right\} \bullet f_A(a)da \tag{5.32}$$

In Eqs. (5.31) and (5.32), the closed form of the bit-error rate cannot be obtained, owing to the complexity of integration. Except for the maximum-likelihood decoding algorithm, the basic principles of the other decoding algorithms used in this study are the same; hence, the analyses of their bit-error performances are also very similar. Therefore, their bit-error performance scan be expressed by Eqs. (5.31) or (5.32). The only difference is that the weighting matrix W is different. Hence, we can obtain the bit-error performance from the size of $\|W_i\|$ in the different decoding algorithms. For example, the value of the last $\|W_i\|$ of the SIC decoding algorithm is smaller than that of the linear decoding algorithm; hence, the BER performance of the former is significantly better than that of the latter.

5.3 Performance Analysis of Layered Space-Time Codes

The performance of horizontal-, turbo-, and diagonal-layered space-time codes is simulated and analyzed. The simulation conditions are as follows:

(1) Assuming that the total power E_S remains unchanged, the value is 1 (normalized);
(2) The channel characteristics and noise can be accurately estimated by the receiver;
(3) The photoelectric-conversion efficiency is 0.6;
(4) In weak turbulence, the atmospheric channel follows a lognormal distribution, and the scintillation index SI is 0.6. For medium and strong turbulence, the atmospheric channel follows a gamma–gamma distribution and the scintillation index SI is 1 and 3, respectively;
(5) The system adopts pulse-position modulation, taking 4-PPM as an example;
(6) The number of antennas is 2×2, 2×4, 4×4, and 4×6.

5.3.1 Horizontal-Layered Space-Time Coding

The zero-forcing detection, minimum mean-squared error detection, maximum-likelihood detection, successive interference-cancellation (SIC), and ordered successive interference-cancellation (OSIC) algorithms are used to analyze the performance of horizontal-layered space-time codes.

Figures 5.8 and 5.9 show the bit-error rate curves for the zero-forcing, minimum mean-squared error, and maximum-likelihood detection algorithms when SI = 0.6 and the atmospheric channel is subject to weak turbulence with a lognormal distribution. The following can be observed:

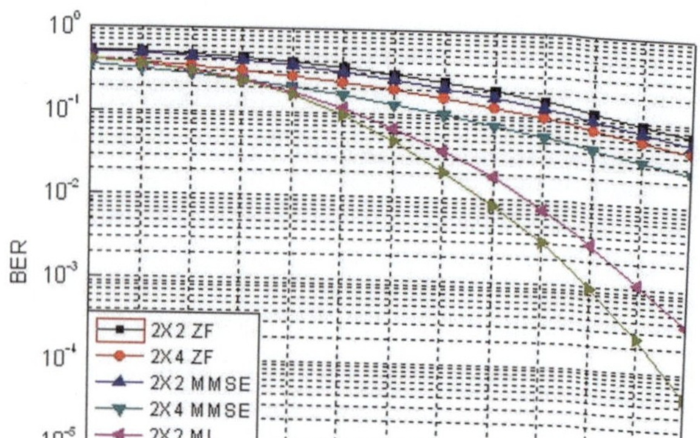

Fig. 5.8 2×2 and 2×4 simulation diagram (SI = 0.6) with weak turbulence

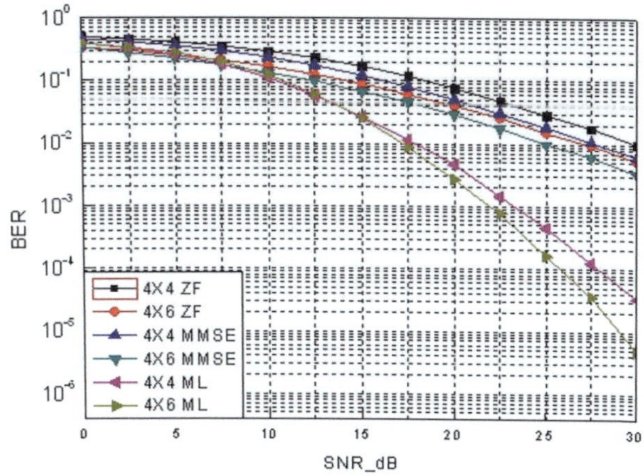

Fig. 5.9 4×4 and 4×6 simulation diagram (SI = 0.6) with weak turbulence

(1) The maximum-likelihood detection algorithm has the best performance, followed by the minimum mean-squared error detection algorithm, and finally, the zero-forcing detection algorithm. The simulation results are consistent with the theoretical analysis.
(2) The diversity gain is related to the number of detectors. For example, as shown in Fig. 5.9, when the bit-error rate is 10^{-4}, the signal-to-noise ratio of the maximum-likelihood algorithm is improved by 2.5 dB.

Figures 5.10 and 5.11 are the BER curves of the zero-forcing detection, minimum mean-squared error detection, and maximum-likelihood detection algorithms in the case of medium turbulence, in which the atmospheric channel follows a gamma–gamma distribution when SI = 1. Comparing Figs. 5.10 and 5.11, it can be seen that atmospheric turbulence affects the performance of the system to a certain extent. With the increase of the scintillation index, the anti-fading performance of the system decreases to some extent; however, the same bit-error rate characteristics can be obtained with an increase in the number of antennas.

Figures 5.12 and 5.13 show the bit-error rate characteristics of the system under strong turbulence, and the scintillation index SI = 3. It can be seen that the bit-error rate of the system is affected by strong turbulence in the case of two transmitters and multiple receivers and four transmitters and multiple receivers, and the bit-error rate increases. However, the bit-error rate of the system with two transmitters and multiple receivers increases greatly, from nearly 10^{-6} to 10^{-4} at 30 dB. The bit-error rate of the system with four transmitters and multiple receivers rises to 10^{-5}.

Horizontal-layered space-time coding cannot effectively suppress the impact of atmospheric turbulence on the system because of its poor coding characteristics. Figure 5.14 shows the comparison of the ZF, SIC-ZF, and OSIC-ZF detection algorithms in a horizontal-layered space-time coding system. It can be seen that SIC-ZF and OSIC-ZF have better algorithm performance than ZF, which improves the performance of the horizontal layered space-time coding system to a certain extent, consistent with the theoretical analysis results.

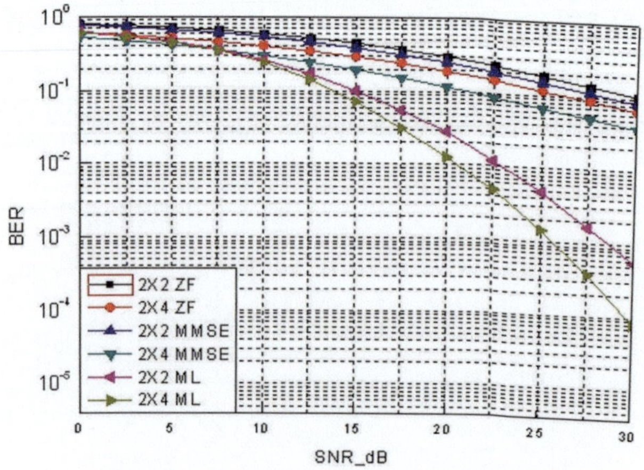

Fig. 5.10 2 × 2 and 2 × 4 simulation diagram (SI = 1) with medium turbulence

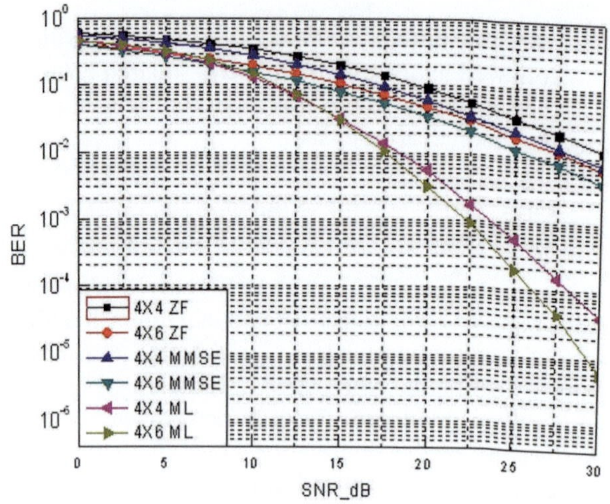

Fig. 5.11 4 × 4 and 4 × 6 simulation diagram (SI = 1) with medium turbulence

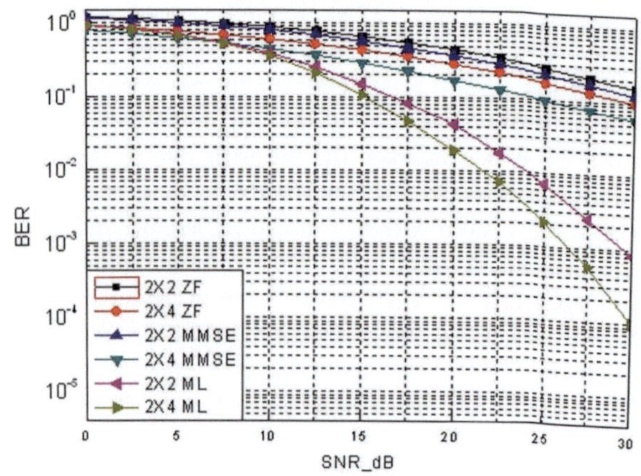

Fig. 5.12 2 × 2 and 2 × 4 simulation diagram (SI = 3) with strong turbulence

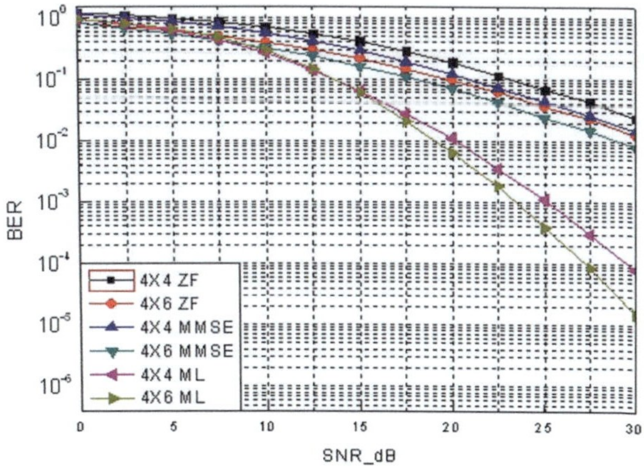

Fig. 5.13 4 × 4 and 4 × 6 simulation diagram (SI = 3) with strong turbulence

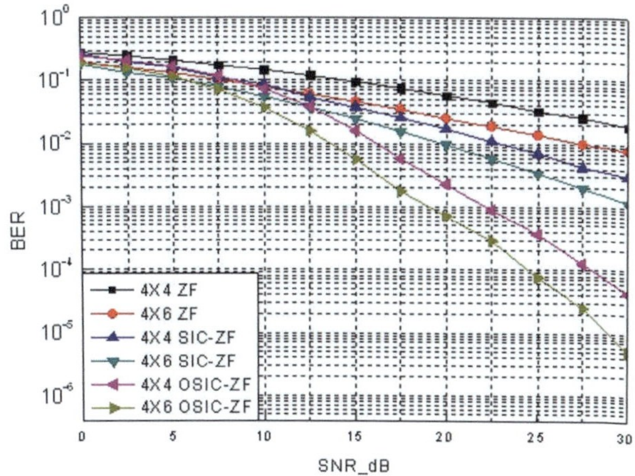

Fig. 5.14 Simulation of ZF, SIC-ZF, and OSIC-ZF

The list of reference procedures for horizontal-layered space-time coding is as follows:

```
function change = TBLAST_encode22(temp, n)  % Code for transmitting
and receiving multiple messages
sendin = reshape(temp, 2, n/2); % n is the length of the code after
channel coding
for i = 1: 2: n/2
   change(:, (i + 1) / 2) = 2 * sendin(:, i) + sendin(:, i + 1) + 1;
end
function change24 = TBLAST_encode44(temp, n) % Coding for multiple
transmissions and multiple receptions
sendin = reshape(temp, 4, n / 4) % n is the length of the code after
channel coding and T is the number of transmitting antennas
for i = 1: 2: n/4
   change24(:, (i + 1) / 2) = 2 * sendin(:, i) + sendin(:, i + 1) + 1;
end
```

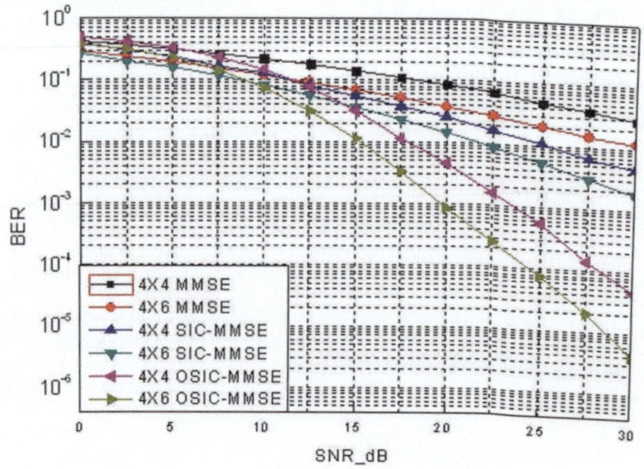

Fig. 5.15 Simulation of MMSE, SIC-MMSE, and OSIC-MMSE

Figure 5.15 shows a comparison of the results of the MMSE, SIC-MMSE, and OSIC-MMSE detection algorithms in the horizontal-layered space-time coding system. It can be seen that the ordered successive interference-cancellation (OSIC) algorithm improves the performance of the MMSE detection algorithm, and the improvement range is higher than that of the SIC detection algorithm. The simulation results are consistent with the theoretical analysis of the SIC and OSIC algorithms.

5.3.2 *Diagonal-Layered Space-Time Code*

Considering different transmit and receive antennas, different noise levels, and different atmospheric scintillation-index parameters, the impact of diagonal-layered space-time coding on the bit-error rate of the system is analyzed. The zero-forcing, minimum mean-squared error, and maximum-likelihood detection algorithms, and the successive interference-cancellation (SIC) and ordered successive interference-cancellation (OSIC) algorithms are selected for analysis and comparison.

Comparing Figs. 5.16 and 5.18, when the scintillation index SI is 0.6 and 1, respectively, in the case of two transmitters and multiple receivers, the bit-error rate of the system increases slightly in the case of medium turbulence with the increase of the scintillation index, indicating that the system cannot well suppress the impact of atmospheric turbulence in this case (Fig. 5.17).

Comparing Figs. 5.17 and 5.19, in the case of four transmitters and multiple receivers, the turbulence-intensity enhancement does not affect the bit-error rate of the system. In the case of medium and strong turbulence, increasing the number of antennas enhances the ability of the system to suppress atmospheric turbulence. It can be seen that the D-BLAST system has a certain ability to suppress the scintillation effect caused by atmospheric turbulence.

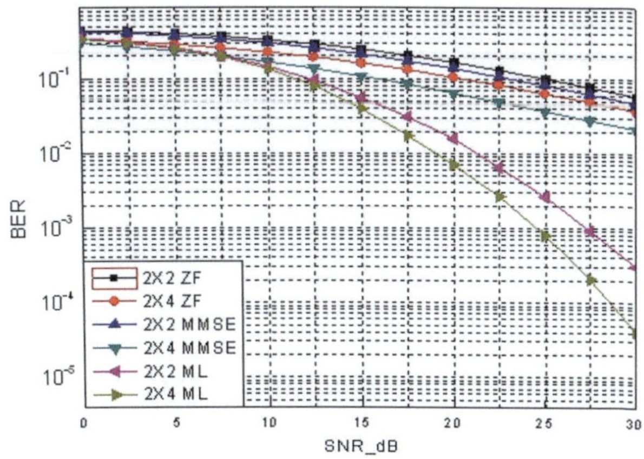

Fig. 5.16 2 × 2 and 2 × 4 simulation diagram (SI = 0.6)

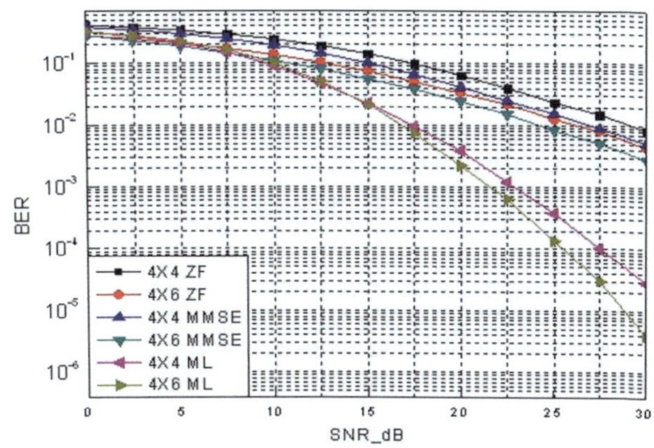

Fig. 5.17 4 × 4 and 4 × 6 simulation diagram (SI = 0.6)

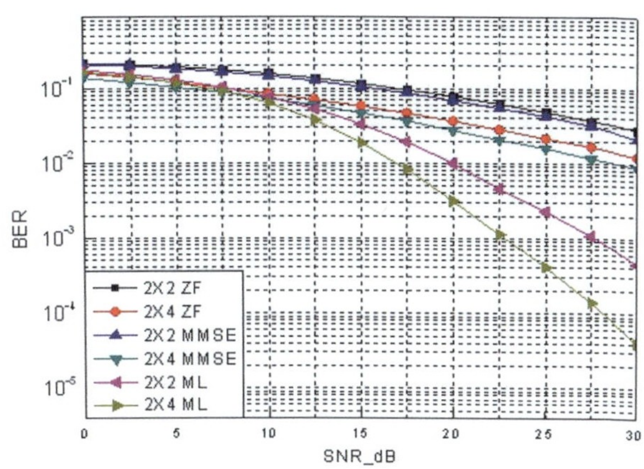

Fig. 5.18 2 × 2 and 2 × 4 simulation diagram (SI = 1)

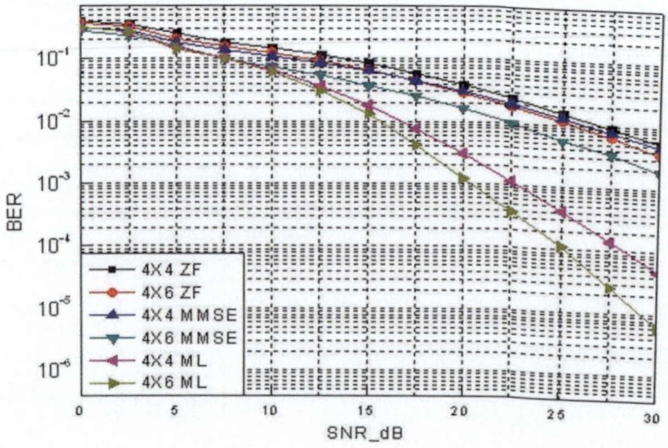

Fig. 5.19 4 × 4 and 4 × 6 simulation diagram (SI = 1)

In the D-BLAST coding system, the selection of the detection algorithm also has a great impact on the system performance. As shown in Fig. 5.19, when the scintillation index SI = 1, the signal-to-noise ratio is approximately 30 dB, and the number of antennas is respectively 4 × 4 and 4 × 6, the BER of ZF is 3×10^{-2} and 10^{-2}, the BER of MMSE is 8×10^{-2} and 5×10^{-3}, and the BER of ML is 10^{-4} and 4×10^{-6}. This shows that in the D-BLAST coding system, the performance of ML was verified to be much better than that of MMSE and ZF.

Figures 5.20 and 5.21 are the curves of the bit-error rate and signal-to-noise ratio of the system under strong turbulence. At this time, the scintillation index SI = 3. It can be seen that in the case of strong turbulence, the bit-error rate of the two-transmitter and multiple-receiver system increases from nearly 10^{-6} to 10^{-5} at 30 dB. However, in the case of four transmitters and multiple receivers, the bit-error rate of the system is not significantly improved, which indicates that increasing the number of antennas can suppress the impact of atmospheric turbulence to a certain extent.

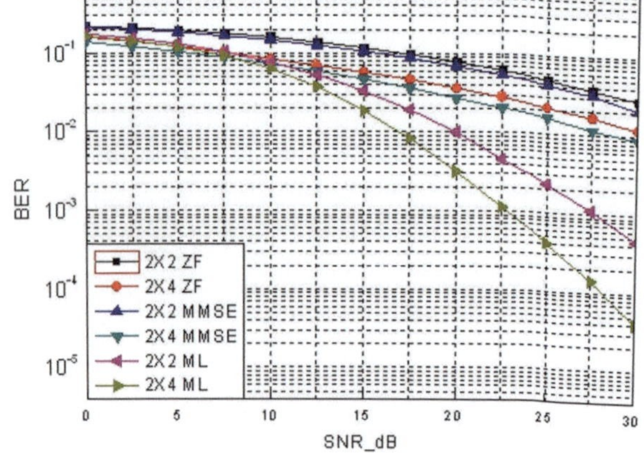

Fig. 5.20 2 × 2 and 2 × 4 simulation diagram (SI = 3) with strong turbulence

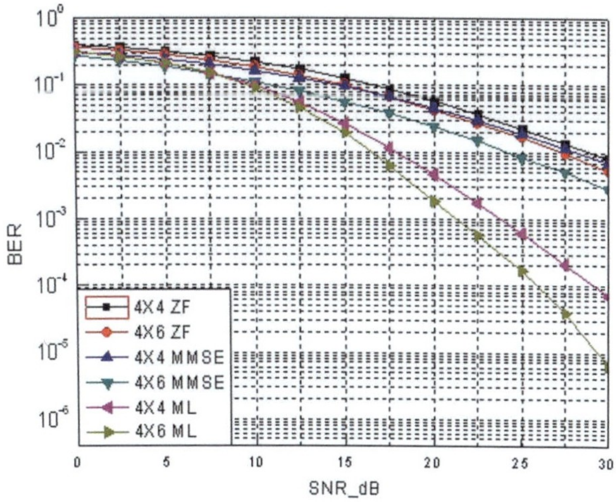

Fig. 5.21 4 × 4 and 4 × 6 simulation diagram (SI = 3) with strong turbulence

As shown in Figs. 5.22 and 5.23, the performance of SIC-ZF, OSIC-ZF, SIC-MMSE, and OSIC-MMSE is significantly better than that of ZF and MMSE, and the bit-error performance of the system can be significantly improved. OSIC is better than SIC in improving performance.

For example, when the bit-error rate (BER) is 10^{-2}, the signal-to-noise ratio of the system using the OSIC-MMSE algorithm is about 15 dB higher than that using the MMSE algorithm, and the signal-to-noise ratio of the system using the SIC-MMSE algorithm is about 10 dB higher. The slope of the system bit-error rate curve using the SIC-ZF, OSIC-ZF, SIC-MMSE, and OSIC-MMSE algorithms is very large, which shows that the detection algorithm can accurately eliminate the intersymbol interference after subtracting the detected-signal decision component

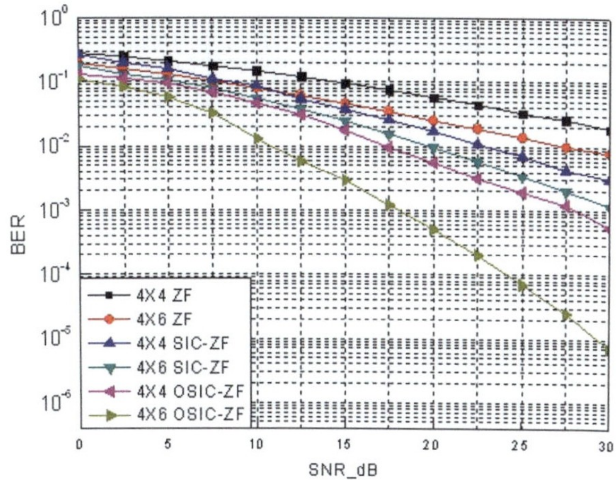

Fig. 5.22 Simulation diagram of ZF, SIC-ZF, and OSIC-ZF

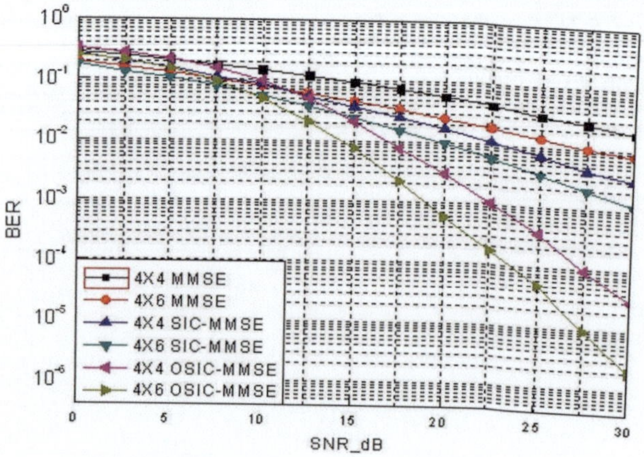

Fig. 5.23 Simulation diagram of MMSE, SIC-MMSE, and OSIC-MMSE

from the received signal. This not only improves the performance of the ZF and MMSE detection algorithms, but also retains the multiplexing gain of the system.

The list of diagonal layered space-time coding reference procedures is as follows:

```
function change24 = encode234_new(temp, n) % Code for transmitting
and receiving multiple messages
sendin = reshape(temp, 2, n/2);
senda0 = reshape(sendin(1, :), [2, n/4]);
sendb0 = reshape(sendin(2, :), [2, n/4]);
send = insert(senda0, sendb0, n/2);
extend = zeros(2, 2);
send1 = [send, extend];
n = n/2;
send2 = zeros(2, n+2);
for i = 1: n
   send2(1, i) = send1(1, i);
   send2(2, i+1) = send1(2, i);
end
n = n + 2;
for i = 1: 2: n
   change24(:, (i + 1)/2) = 2 * send2(:, i) + send2(:, i + 1) + 1;
% Because this is 4-PPM, it is necessary to form a pair of adja-
cent symbols, such as 01 10.... In this manner, it can be modulated
according to the 4-PPM principle.
end
function change24 = encode44_new(temp, n) % Coding for multiple
transmissions and multiple receptions
sendin = reshape(temp, 4, n/4);
senda0 = reshape(sendin(1, :), [4, n/16]);
sendb0 = reshape(sendin(2, :), [4, n/16]);
sendc0 = reshape(sendin(3, :), [4, n/16]);
sendd0 = reshape(sendin(4, :), [4, n/16]);
send = insert44(senda0, sendb0, sendc0, sendd0, n/4);
extend = zeros(4, 4);
send = [send, extend];
n = n/4;
```

```
send1 = zeros (4, n+4);
for i = 1: n
   send1(1, i) = send(1, i);
   send1(2, i + 1) = send(2, i);
   send1(3, i + 2) = send(3, i);
   send1(4, i + 3) = send(4, i);
end
n = n + 4;
for i = 1: 2: n
   change24(:, (i + 1)/2) = 2 * send1(:, i) + send1(:, i + 1) + 1;
end
```

5.3.3 *Turbo-Layered Space-Time Coding*

The T-BLAST coding system is simulated to analyze the bit-error rate and signal-to-noise ratio curves of the system, in terms of the number of transmitting and receiving antennas and noise. The zero-forcing detection, minimum mean-squared error detection, maximum-likelihood detection, successive interference-cancellation, and ordered successive interference-cancellation algorithm are used.

The conditional assumptions of the turbo-layered space-time code simulation are as follows:

(1) Assuming that the total power E_S is constant, the value is 1 (normalized);
(2) The channel characteristics and noise can be accurately estimated by the receiver;
(3) The photoelectric-conversion efficiency is 0.6;
(4) In weak turbulence, the atmospheric channel follows a lognormal distribution, and the scintillation index SI is 0.6; for medium and strong turbulence, the atmospheric channel obeys a gamma–gamma distribution, and the scintillation index SI is 1 and 3, respectively;
(5) The system adopts 4-PPM modulation.

The bit-error rate of the T-BLAST coding system decreases with the increase of the number of transmit and receive antennas. Whether the number of transmit or receive antennas is changed, the bit-error rate decreases greatly. For example, in Fig. 5.24, when the signal-to-noise ratio is 30 dB and the detection algorithm is ML, the bit-error rates for 2×2 and 2×4 are 10^{-3} and 10^{-4}, respectively, because the MMSE algorithm considers the impact of noise; hence, its bit-error rate performance is significantly better than that of the ZF algorithm.

When the signal-to-noise ratio is small, the performance difference between the ZF and MMSE systems is large, which indicates that the MMSE detection algorithm has better performance under a low signal-to-noise ratio. In the case of a high SNR, the bit-error rate difference between the ZF and MMSE systems is not very large, which indicates that the MMSE algorithm is more suitable for low SNRs.

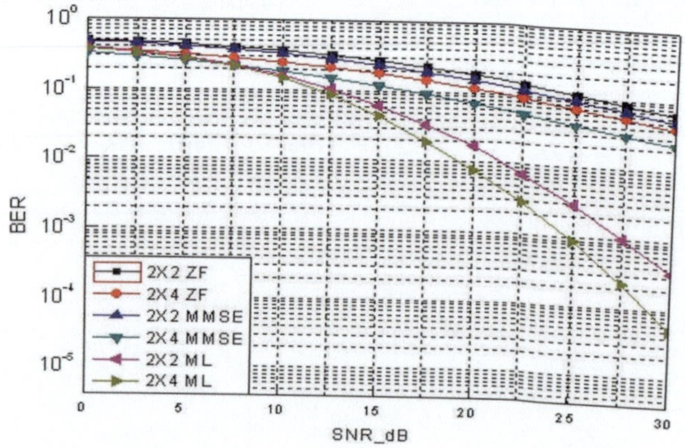

Fig. 5.24 2 × 2 and 2 × 4 simulation diagram (SI = 0.6)

As can be seen from Fig. 5.25, when the maximum-likelihood detection algorithm is used in the T-BLAST coded system, the bit-error rate of the system decreases from 10^{-4} to nearly 10^{-6} when the signal-to-noise ratio is 30 dB. The results show that the performance of the maximum-likelihood detection algorithm is better than that of the MMSE and ZF detection algorithms (Fig. 5.26).

Comparing Figs. 5.25 and 5.27, the scintillation index SI is 0.6 and 1, respectively; that is, in the case of weak and medium turbulence, the T-BLAST system must increase the number of transmit and receive antennas to reduce the bit-error rate. T-BLAST and D-BLAST coding in a free-space optical (FSO) MIMO system can overcome the impact of medium and strong turbulence on the system.

Figures 5.28 and 5.29 show the simulation curves of the bit-error rate and signal-to-noise ratio of the system under strong turbulence. The scintillation index SI = 3 shows that the system will be affected by strong atmospheric turbulence. The bit-error rate of the system with two transmitters and multiple receivers increases to a certain extent. At 30 dB, the bit-error rate increases from close to 10^{-6} to 10^{-4}. In the four-transmitter and multiple-receiver system, the improvement of the bit-error

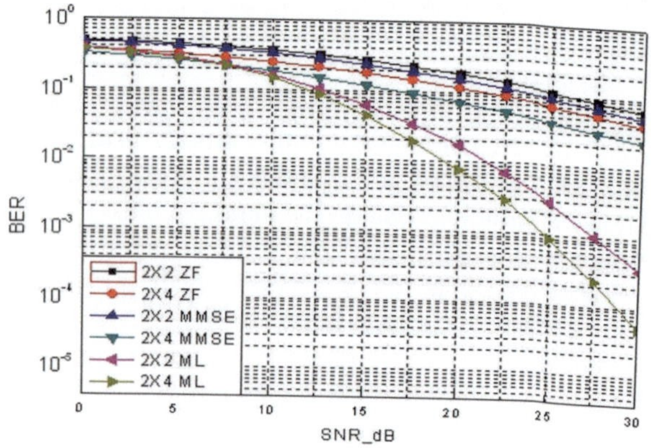

Fig. 5.25 4 × 4 and 4 × 6 simulation diagram (SI = 0.6)

Fig. 5.26 2 × 2 and 2 × 4 simulation diagram (SI = 1) with medium turbulence

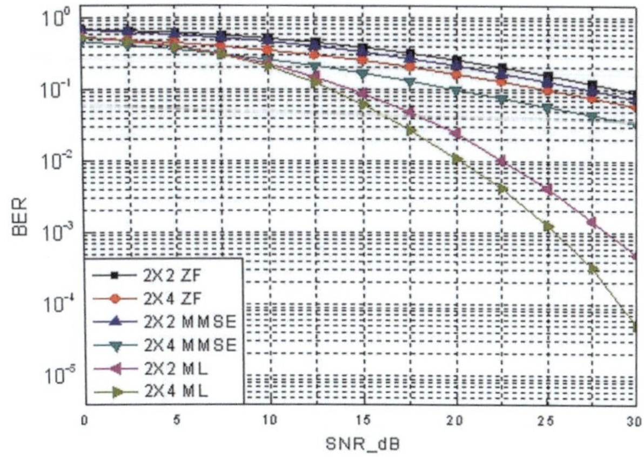

Fig. 5.27 4 × 4 and 4 × 6 simulation diagram (SI = 1) with medium turbulence

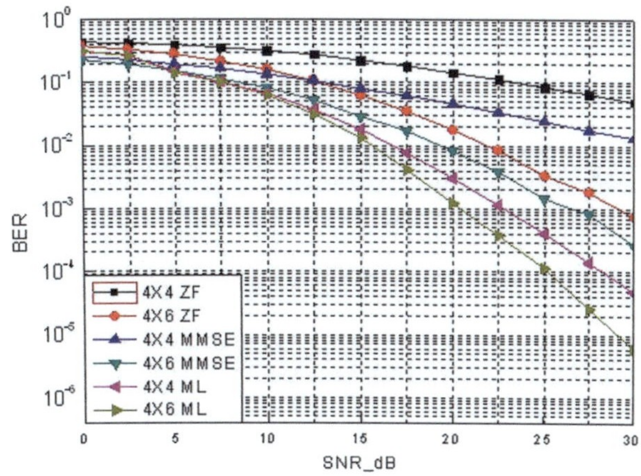

rate is not significant, indicating that increasing the number of transmit and receive antennas will suppress the impact of turbulence on the system performance to a certain extent.

Figures 5.30 and 5.31 show the simulation results of successive interference cancellation (SIC) and ordered successive interference cancelation (OSIC) using the T-BLAST coding system. It can be seen that the bit-error rate of the system using turbo-layered space-time coding is also close to 10^{-6}. It also shows that the performance of the improved SIC-ZF, OSIC-ZF, SIC-MMSE, and OSIC-MMSE detection algorithms is significantly better than that of the ZF and MMSE detection algorithms, and their bit-error rate is close to that of the maximum-likelihood detection algorithm.

The system using T-BLAST coding can obtain a lower bit-error rate using D-BLAST. Although the bit-error rate is not as low as D-BLAST, its coding complexity is much lower than that of D-BLAST.

The list of turbo-layered space-time coding reference procedures is as follows:

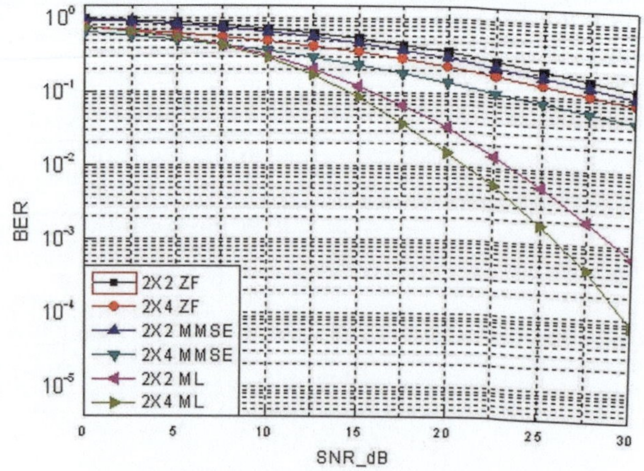

Fig. 5.28 2 × 2 and 2 × 4 simulation diagram (SI = 3) with strong turbulence

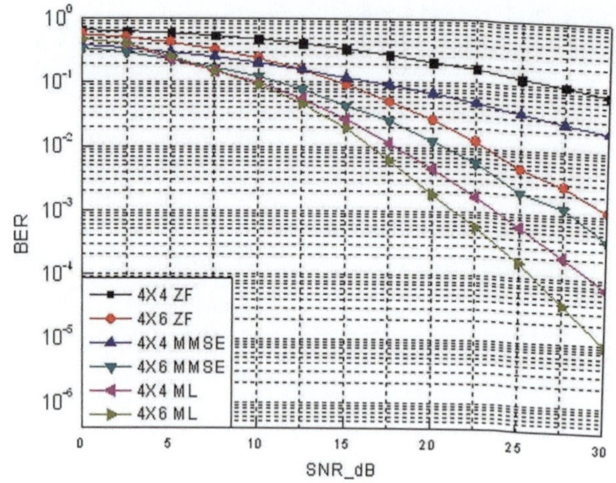

Fig. 5.29 4 × 4 and 4 × 6 simulation diagram (SI = 3) with strong turbulence

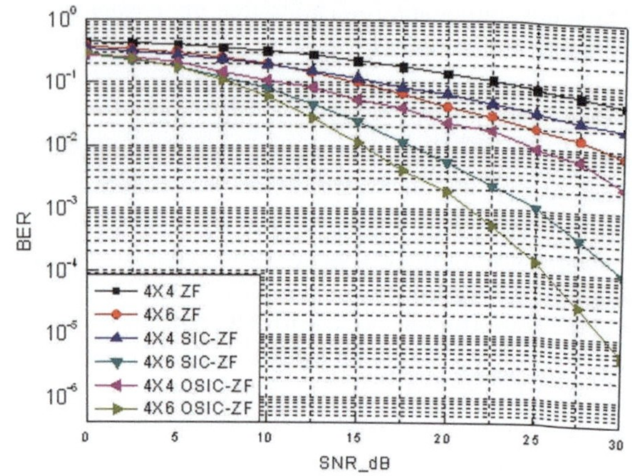

Fig. 5.30 Simulation diagram of ZF, SIC-ZF, and OSIC-ZF

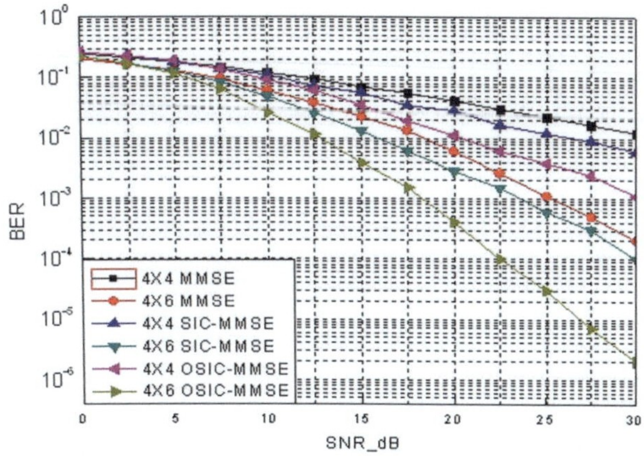

Fig. 5.31 Simulation diagram of MMSE, SIC-MMSE, and OSIC-MMSE

```
function change = TBLAST_encode22(temp, n) % Code for transmitting
and receiving multiple messages
sendin = reshape(temp, 2, n/2); % n is the length of the code after
channel coding
send = zeros(2, n/2);
for i = 1: n/2
   if mod(i, 2) == 0
      send(1, i) = sendin(2, i);
      send(2, i) = sendin(1, i);
   else
      send(1, i) = sendin(1, i);
      send(2, i) = sendin(2, i);
   end
end
for i = 1: 2: n/2
   change(:, (i + 1)/2) = 2 * send(:, i) + send(:, i + 1) + 1;
end
function change24 = TBLAST_encode44(temp, n) % Coding for multiple
transmissions and multiple receptions
T = 4;
curr = 1;
sendin = reshape(temp, 4, n/4); % n is the length of the code after
channel coding and T is the number of transmitting antennas
send = zeros(4, n/4);
for t = 1: 4: n/4
   send(:, t) = sendin(:, t);
end
j = -1;
for t = 2: n/4
   j = j + 2;
   if t == 4 * curr + 1
      curr = curr + 1;
      continue;
   else
      i = 1;
      for i = 1: T
```

```
            m = mod(t + i + j, 4);
            if m == 0
              m = 4;
            end
           send(i, t) = sendin(m, t);
         end
      end
   end
   for i = 1: 2: n/4
      change24(:, (i + 1) / 2) = 2 * send(:, i) + send(:, i + 1) + 1;
   end
```

5.3.4 Performance Comparison and Analysis of H-BLAST, D-BLAST, and T-BLAST

The performance of the horizontal-, turbo-, and diagonal-layered space-time codes in weak and strong turbulence is compared and analyzed, when the maximum-likelihood detection algorithm is selected.

Figure 5.32 shows the simulation results of the maximum-likelihood detection algorithm for H-BLAST, T-BLAST, and D-BLAST with different antenna numbers in weak turbulence. It can be seen that the performance of the diagonal-layered space-time code is the best of the three codes, followed by the turbo-layered space-time code, and the horizontal-layered space-time code is the worst. This is consistent with the previous theoretical analysis.

Figure 5.33 shows the comparison of the bit-error rate of H-BLAST, T-BLAST, and D-BLAST when the scintillation index SI = 1 (medium turbulence). It can be seen that among the three codes, the diagonal-layered space-time coding is the best,

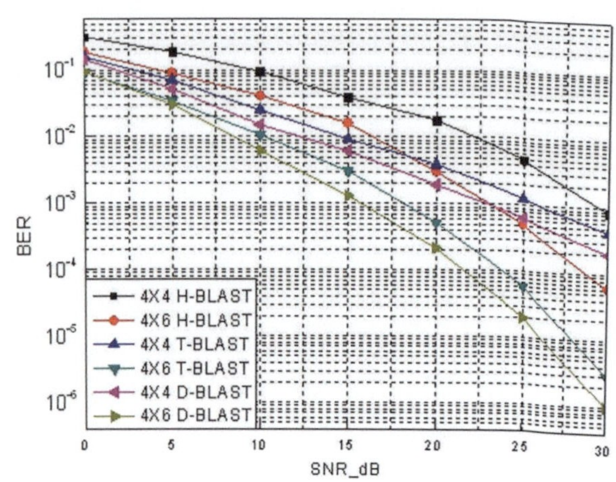

Fig. 5.32 Three layered space-time codes. 4 × 4 and 4 × 6 simulation diagram (SI = 0.6)

the turbo-layered space-time coding is next, and the horizontal-layered space-time coding is the worst.

The bit-error rate of the diagonal-layered space-time coding does not change greatly when the turbulence is enhanced, and the greatest change is in the horizontal layered space-time coding, which also shows that the diagonal-layered space-time coding system can suppress the impact of medium atmospheric turbulence.

Figure 5.34 shows the simulation results of the maximum-likelihood detection algorithm for H-BLAST, T-BLAST, and D-BLAST under strong turbulence with different antenna numbers. In strong turbulence, the performance of H-BLAST, T-BLAST, and D-BLAST is affected, and the bit-error rate increases. However, in the case of four transmitters and multiple receivers, increasing the signal-to-noise ratio can also reduce the bit-error rate of the system and improve its anti-fading performance. As in the case of weak turbulence, diagonal-layered space-time coding is the best method.

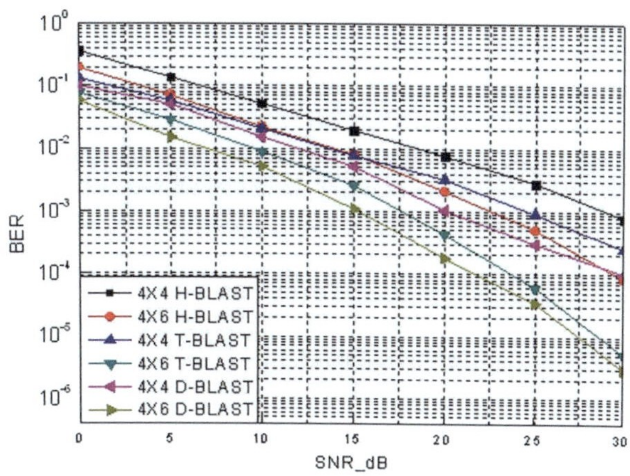

Fig. 5.33 Three layered space-time codes. 4 × 4 and 4 × 6 simulation diagram (SI = 1)

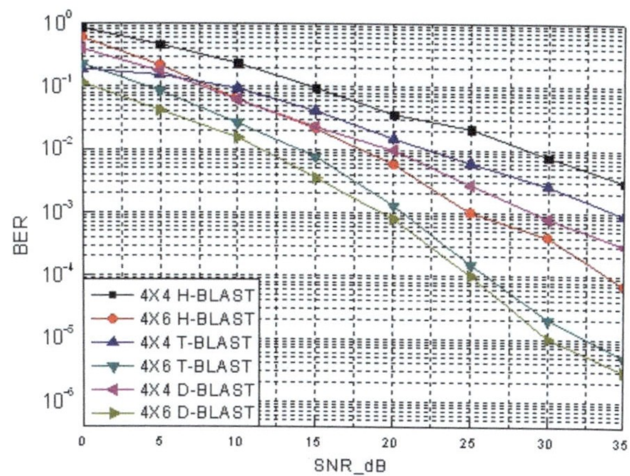

Fig. 5.34 Three layered space-time codes with 4 × 4 and 4 × 6 simulation diagram (SI = 3)

The results show that diagonal-layered space-time coding, turbo-layered space-time coding, and horizontal-layered space-time coding have the best performance when the number of transmit and receive antennas increases gradually, whether in weak or strong turbulence. Horizontal-layered space-time coding is not coded in the spatial domain; hence, its performance is poor. Compared with diagonal-layered space-time coding, turbo-layered space-time coding has the advantages of lower coding complexity, no redundancy, and better use of space and time. Although diagonal layered space-time coding has the best performance and makes full use of the resources in the space and time domains, it has a redundancy of $n(n-1)/2$; thus, the coding increases its redundancy and obtains higher coding performance.

5.3.5 Simulation Analysis of Cascaded-Layered Space-Time Codes

Figure 5.35 shows the basic structure of the low-density parity-check (LDPC)-BLAST system. The basic concept is that after the source sends out a message C, we will apply LDPC to the message and change C into a codeword x with a better code distance. The coded codeword x is subjected to a serial–parallel conversion and layered space-time coding; then, the codeword is PPM modulated and transmitted through an array composed of N lasers.

At the receiving end, the signal is received by M detectors and the received signal is PPM demodulated. The demodulated signal is detected and decoded to obtain the codeword. After the parallel–serial conversion, a codeword y is obtained, and then the final information C' is obtained using the belief propagation (BP) decoding algorithm. Finally, the information C sent by the source is compared with the information C' obtained at the receiving end, the corresponding bit-error rate is calculated, and the performance of the system is judged according to the bit-error rate.

The maximum-likelihood, zero-forcing, and minimum mean-squared error detection algorithms are used to simulate the D-BLAST- and T-BLAST-coded LDCP + BLAST cascaded system. The simulation conditions are as follows:

(1) Assuming that the total power E_S remains unchanged, the value is 1 (normalized);

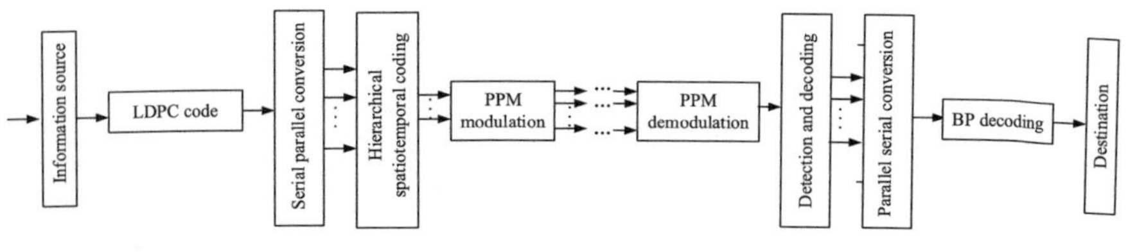

Fig. 5.35 LDPC-BLAST system model

(2) The channel characteristics and noise can be accurately estimated by the receiver;
(3) The photoelectric-conversion efficiency is 0.6;
(4) The atmospheric channel follows a lognormal distribution under weak turbulence, and the scintillation index SI is 0.6. Under medium and strong turbulence, the atmospheric channel follows a gamma–gamma distribution, and the scintillation index SI is 1 and 3, respectively;
(5) The system adopts 4-PPM pulse-position modulation;
(6) LDPC code with binary domain rules is selected for coding, with a code rate of 1/2 and a code length of 2014 bits. The probabilistic BP decoding algorithm is adopted, and the number of iterations is 100.
(7) The number of antenna pairs is 2×2, 2×4, 4×4, and 4×6.

5.3.5.1 Simulation of the LDPC + D-BLAST System

Figures 5.36 and 5.37 show the system-simulation diagrams when the scintillation index SI = 0.6; that is, the curves of the bit-error rate and signal-to-noise ratio of the LDPC + D-BLAST (cascaded LDPC code and D-BLAST) system under weak turbulence. It can be seen that when the signal-to-noise ratio of the system with four transmitters and multiple receivers is 25 dB, the bit error rate is 10^{-6}, while when the signal-to-noise ratio of the system with two transmitters and multiple receivers is increased to 30 dB, the bit-error rate can reach 10^{-6}. Compared with the D-BLAST system without LDPC coding, the LDPC + D-BLAST system has better performance and lower requirements for the signal-to-noise ratio in the case of weak turbulence.

Figures 5.38 and 5.39 show the system-simulation results when the scintillation index SI = 1; that is, the curves of the bit-error rate and signal-to-noise ratio of the LDPC + D-BLAST system under medium turbulence. It can be seen that the enhancement of the scintillation index has a great impact on the bit-error rate. The LDPC + D-BLAST system needs to increase the signal-to-noise ratio of the system

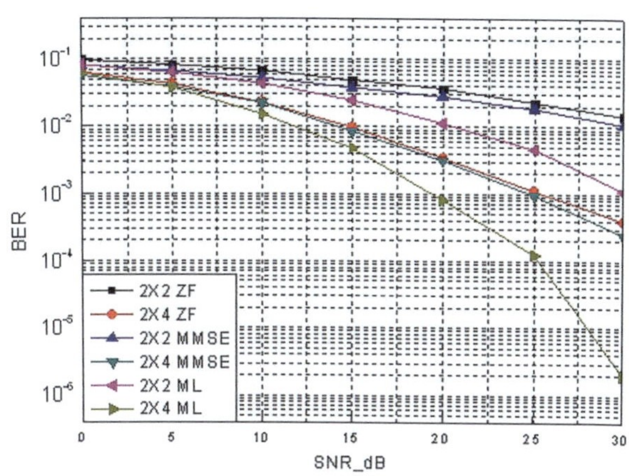

Fig. 5.36 LDPC + D-BLAST system. 2×2 and 2×4 simulation diagram (SI = 0.6)

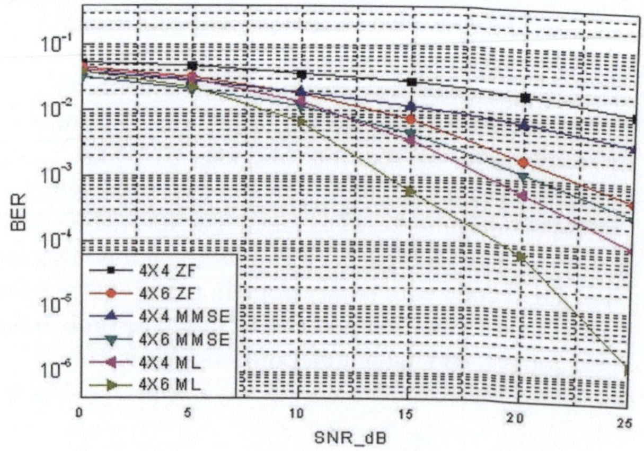

Fig. 5.37 LDPC + D-BLAST system. 4 × 4 and 4 × 6 simulation diagram (SI = 0.6)

to obtain the bit-error rate it achieves under weak turbulence; that is, the bit-error rate of the two-transmitter multiple-receiver system can be close to 10^{-6} when the signal-to-noise ratio is 35 dB. This increases the power requirements of the transmitter.

Figures 5.40 and 5.41 show the system-simulation results when the scintillation index SI = 3; that is, the curves of the bit-error rate and signal-to-noise ratio of the LDPC + D-BLAST system under strong turbulence. It can be seen that under the same SNR, the BER of the system under strong turbulence is much higher than that under weak turbulence (when the SNR = 25 dB, the BER under weak turbulence is 10^{-6} and that under strong turbulence is 10^{-4}). In the case of strong turbulence, it is necessary to increase the transmitting power at the transmitting end and increase the signal-to-noise ratio to 40 dB to reduce the BER of the system to close to 10^{-6}.

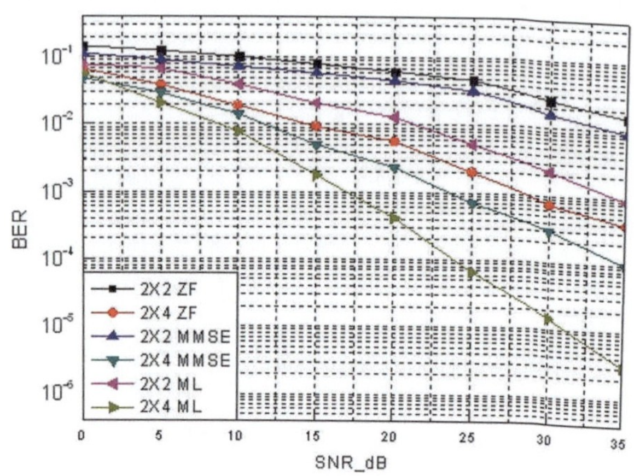

Fig. 5.38 LDPC + D-BLAST system. 2 × 2 and 2 × 4 simulation diagram (SI = 1)

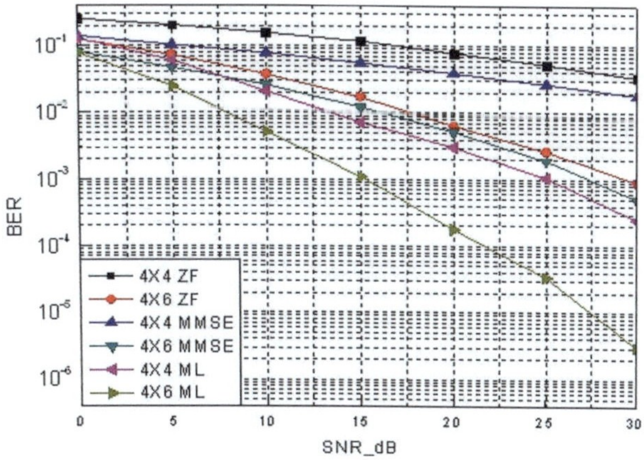

Fig. 5.39 LDPC + D-BLAST system. 4 × 4 and 4 × 6 simulation diagram (SI = 1)

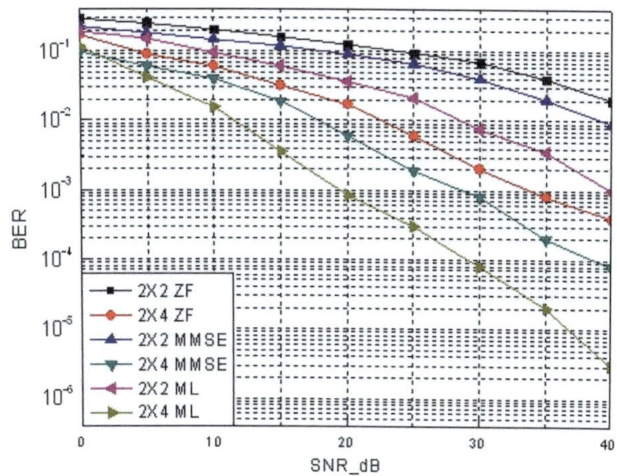

Fig. 5.40 LDPC + D-BLAST system. 2 × 2 and 2 × 4 simulation diagram (SI = 3)

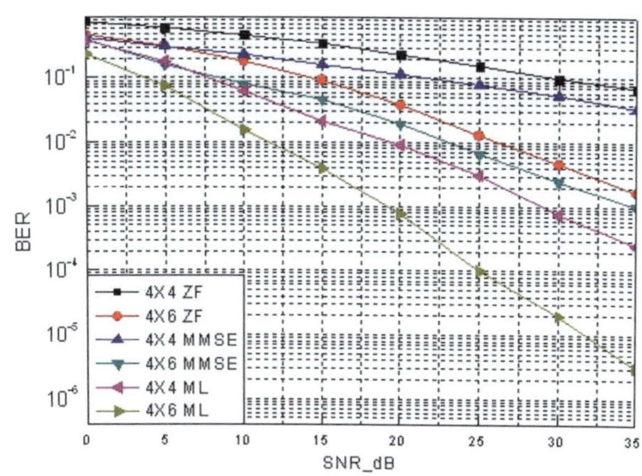

Fig. 5.41 LDPC + D-BLAST system. 4 × 4 and 4 × 6 simulation diagram (SI = 3)

5.3.5.2 LDPC + T-BLAST System

Figures 5.42 and 5.43 show the system-simulation results when the scintillation index SI = 0.6; that is, the curves of the bit-error rate and signal-to-noise ratio of the LDPC + T-BLAST system under weak turbulence. It can be seen that, in the case of weak turbulence, the bit-error rate of the system with two transmitters and multiple receivers is close to 10^{-6} when the signal-to-noise ratio is 30 dB, which is slightly less than that of the LDPC + D-BLAST system. This is because the performance of diagonal-layered space-time coding is better than that of turbo-layered space-time coding. However, in a system with four transmitters and multiple receivers, a bit-error rate close to 10^{-6} can be obtained as long as the signal-to-noise ratio is 25 dB. This shows that increasing the number of transmit and receive antennas can significantly improve the system performance.

Figures 5.44 and 5.45 show the system-simulation results when the scintillation index SI = 1; that is, the curves of the bit-error rate and signal-to-noise ratio of the

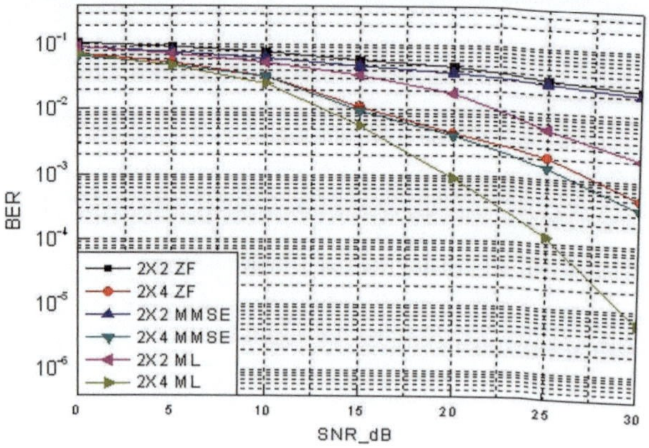

Fig. 5.42 LDPC + T-BLAST system. 2 × 2 and 2 × 4 simulation diagram (SI = 0.6)

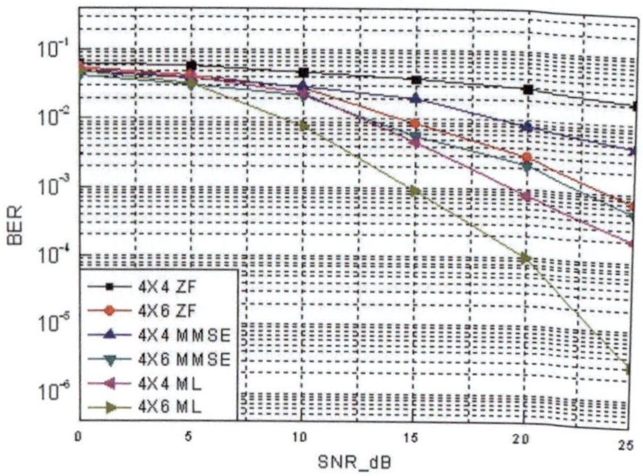

Fig. 5.43 LDPC + D-BLAST system. 4 × 4 and 4 × 6 simulation diagram (SI = 0.6)

LDPC + T-BLAST system under medium turbulence. It can be seen that the intensity of the atmospheric turbulence will affect the performance of the system to a certain extent.

In the system with two transmitters and multiple receivers, it is necessary to increase the signal-to-noise ratio to obtain the minimum bit-error rate under weak turbulence (Fig. 5.44). The bit-error rate of the system is close to 10^{-6} when the signal-to-noise ratio is 35 dB, under medium turbulence. In the system with four transmitters and multiple receivers, the bit-error rate also increases. When the signal-to-noise ratio is increased to 30 dB, the bit-error rate can be reduced to approximately 10^{-6}.

Figures 5.46 and 5.47 show the system-simulation results when the scintillation index SI = 3; that is, the curves of the bit-error rate and signal-to-noise ratio of the LDPC + T-BLAST system under strong turbulence. It can be seen that the performance of the system becomes much worse in the case of strong turbulence,

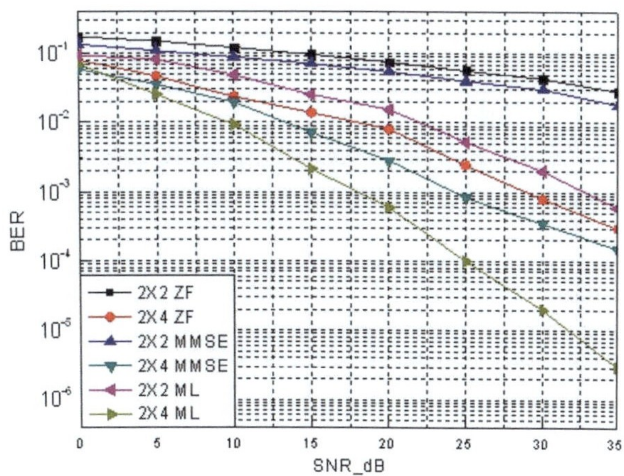

Fig. 5.44 LDPC + T-BLAST system. 2 × 2 and 2 × 4 simulation diagram (SI = 1)

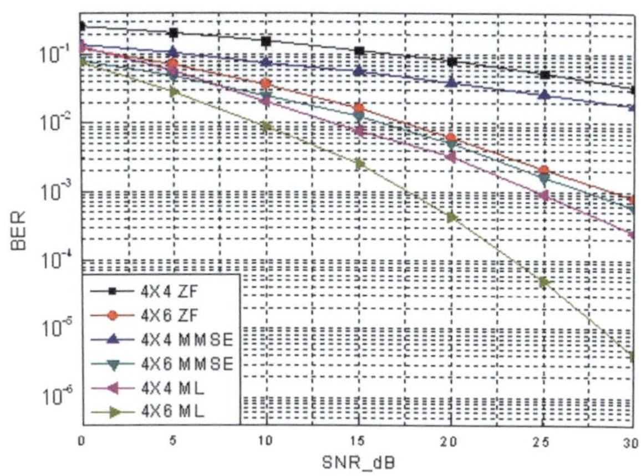

Fig. 5.45 LDPC + T-BLAST system. 4 × 4 and 4 × 6 simulation diagram (SI = 1)

and the bit-error rate increases under the same signal-to-noise ratio. When the signal-to-noise ratio is 25 dB, the bit-error rate of the two-transmitter multi-receiver system increases from 10^{-4} to 10^{-3}, while the bit-error rate of the four-transmitter multi-receiver system increases from 10^{-5} to 10^{-4}.

To improve the influence of strong turbulence on the system, it is necessary to increase the transmission power of the system; that is, to increase the signal-to-noise ratio. When the signal-to-noise ratio of the two-transmitter multi-receiver system needs to be increased to 40 dB, the bit-error rate of the system is close to 10^{-6}, while the signal-to-noise ratio of the four-transmitter multi-receiver system needs to be increased to 35 dB.

Adding the LDPC code increases the complexity of the system; however, the bit-error rate of the LDPC + BLAST system decreases significantly. It shows that the LDPC code can obviously improve the bit-error rate of the system, making it better than that of a single BLAST system. In weak turbulence, when the signal-to-noise ratio is 30 dB, the bit-error rate of the two-transmitter multiple-receiver system

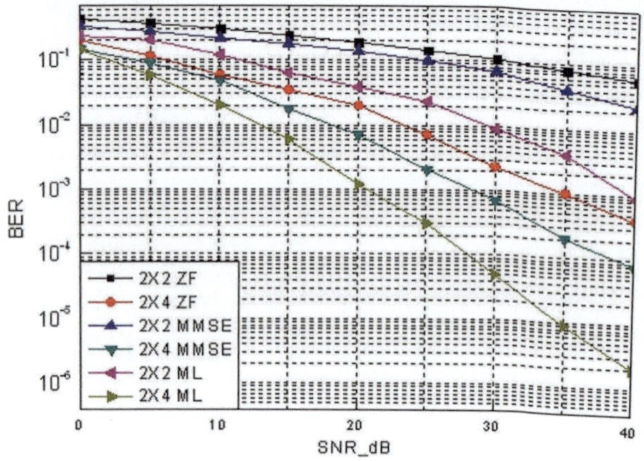

Fig. 5.46 LDPC + T-BLAST system. 2 × 2 and 2 × 4 simulation diagram (SI = 3)

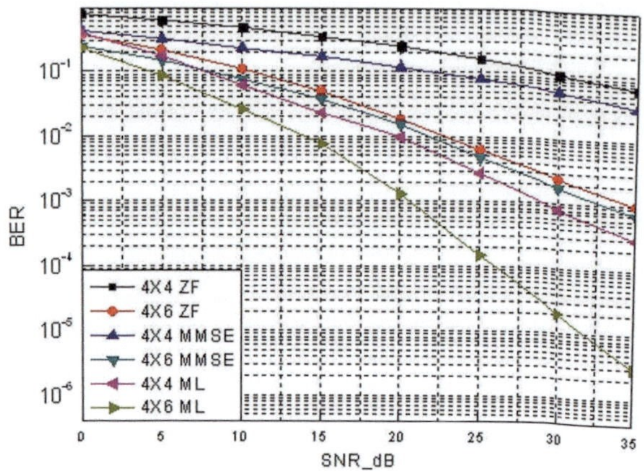

Fig. 5.47 LDPC + T-BLAST system. 4 × 4 and 4 × 6 simulation diagram (SI = 3)

decreases from 10^{-4} to 10^{-6}, indicating that the system with LDPC coding can better overcome the noise interference on the system and improve the reliability of the system information transmission. Moreover, the BER increases with the signal-to-noise ratio, and the slope of the BER curve of the LDPC-BLAST system is very steep, indicating that the BER decreases rapidly.

5.4 Performance Analysis of V-BLAST

To concretely analyze the BER performance of different decoding algorithms in atmospheric laser communication after using V-BLAST code, the Monte Carlo method is used to simulate the system. The effects of the number of transmit and receive antennas, noise, and atmospheric flicker on the BERs are analyzed respectively, and compared with an uncoded (i.e., 1 × 1) system. The performance of the atmospheric laser-communication system is compared.

The simulation conditions are as follows:

(1) The total transmission power E_s is unchanged and $E_s = 1$ (normalized value);
(2) The channel characteristics and noise can be accurately estimated;
(3) The photoelectric-conversion efficiency $\eta = 0.5$;
(4) The light-intensity scintillation follows a lognormal distribution, and the scintillation index is $S \cdot I = 0.6$ and 1.0;
(5) The signal is fully synchronized;
(6) The system adopts Q-PPM modulation with $Q = 4$;
(7) The number of antennas is 1 × 1, 1 × 2, 1 × 4, 2 × 2, 2 × 3, 4 × 4, and 4 × 5.

5.4.1 *Performance Analysis of the Maximum-Likelihood Decoding Algorithm*

To better describe the BER performance of layered space-time coding when the maximum-likelihood (ML) decoding algorithm is adopted, the BER performance of the algorithm is simulated and analyzed; the results are shown in Figs. 5.48 and 5.49.

Figure 5.48 shows the BER and SNR curves of the maximum-likelihood decoding algorithm at $S \cdot I = 0.6$. The following can be seen from the figure:

(1) Relative to 1 × 1, the BER performance of other systems has significantly improved. This shows that layered space-time coding can effectively overcome the scintillation effect caused by atmospheric turbulence.
(2) The diversity gain is related to the number of detectors; the more detectors, the greater the diversity gain, and the better the BER performance. When the bit-error rate is 10^{-3}, relative to the 1 × 1 system, the signal-to-noise ratios of the 1 × 2 and 1 × 4 systems are improved by approximately 6 dB and 14 dB, respectively. When the bit-error rate is 10^{-3}, compared with the 1 × 1 system,

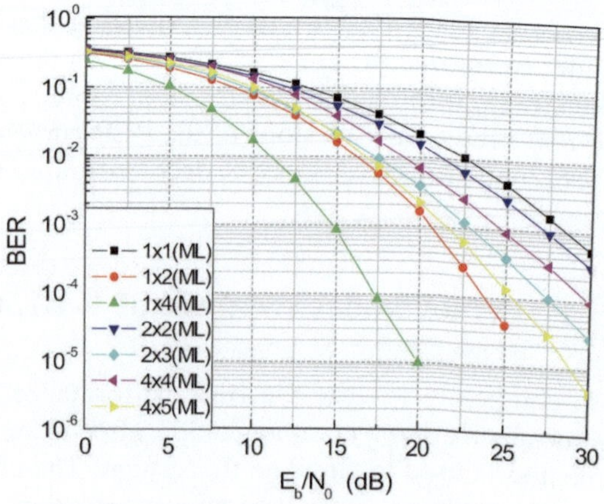

Fig. 5.48 Maximum likelihood at $S \cdot I = 0.6$. BER and SNR curves

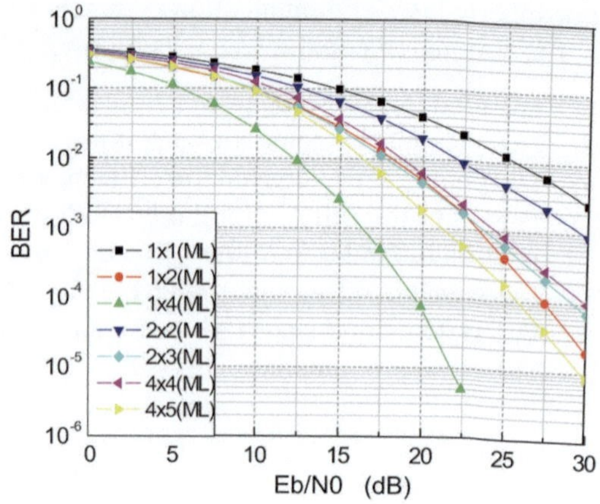

Fig. 5.49 Maximum likelihood at $S.I = 1.0$. BER and SNR curves

the signal-to-noise ratios of the 2 × 3 and 2 × 2 systems were improved by approximately 4.5 dB.

(3) When the number of detectors is the same, the system with more lasers has worse performance than the system with fewer lasers. This is because the vertical-layered space-time coding is not based on transmit diversity and does not obtain transmit-diversity gain; and, more importantly, because the spatial interference increases.

Figure 5.49 shows the BER and SNR curves of the maximum-likelihood decoding algorithm at SI = 1. It can be seen from the figure that the rule reflected at this time is similar to that in Fig. 5.48. Comparing Figs. 5.48 and 5.49, it can be seen that with the increase of the scintillation index relative to the 1 × 1 system, the improvement of the signal-to-noise ratio of the same system increases significantly. When the bit-error rate is 3×10^{-2}, relative to the 1 × 1 system, the signal-to-noise ratios of the 1 × 4 and 1 × 2 systems are improved by approximately 8 dB and 15 dB, respectively.

The 2 × 2, 2 × 3, 4 × 4, 4 × 5, and other systems especially obtain a greater diversity gain. This shows that the stronger the scintillation, the more obvious the diversity gain obtained by the MIMO system, resulting in the obvious suppression of light-intensity fluctuation.

5.4.2 *Performance Analysis of Linear Decoding Algorithms*

The linear decoding algorithms include the zero-forcing and least mean-squared error algorithms. We simulate the error performance of these two decoding algorithms.

5.4.2.1 Zero-Forcing Decoding Algorithm

To better describe the bit-error performance of time-layered space-time coding using the zero-forcing decoding algorithm (ZF), we have simulated and analyzed its bit-error performance, and the results are shown in Figs. 5.50 and 5.51.

Figure 5.50 shows the BER and SNR curves of the ZF and ML decoding algorithms at $S.I. = 0.6$. It can be seen from the figure that after the ZF decoding algorithm is adopted, with the increase in the number of lasers and detectors, the bit-error rate of the system gradually decreases with the increase of the signal-to-noise ratio. This shows that it is feasible to apply the ZF decoding algorithm to V-BLAST code in atmospheric laser communication. However, there are some differences between the rule presented at this time and that of the maximum-likelihood decoding algorithm (ML):

(1) After the ZF decoding algorithm is adopted, the BER performance caused by the increase in the number of detectors is significantly improved. When the bit-error rate is 6×10^{-2}, relative to the 4 × 4 system, the signal-to-noise ratio of

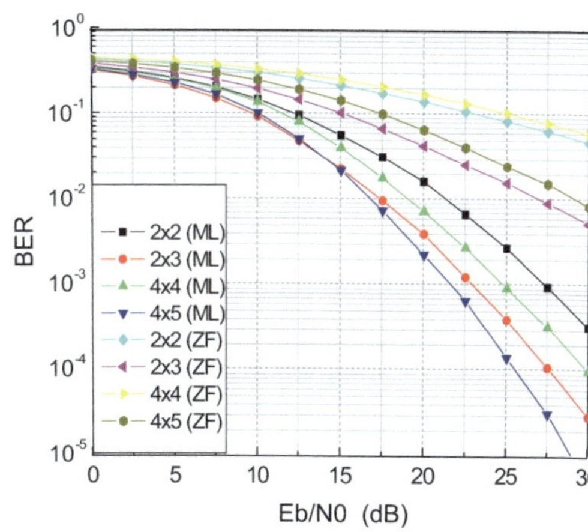

Fig. 5.50 Bit error rate and signal-to-noise ratio curves of the ZF and ML decoding algorithms at time 1, $S \cdot I = 0.6$

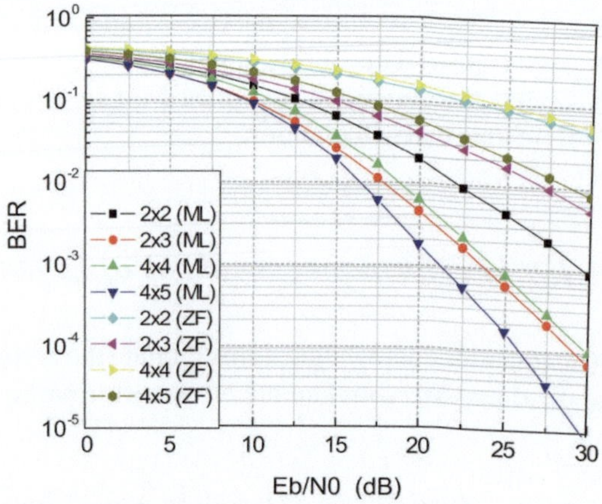

Fig. 5.51 ZF and ML decoding at $S.I = 1.0$. BER and SNR curves of the algorithm

the 4×5 system is improved by about 7.5 dB. Relative to the signal-to-noise ratio of the 2×2 system, that of the 2×3 system is improved by about 11.5 dB. This is because increasing the number of detectors is equivalent to increasing the receiving aperture, which indicates that the system can effectively suppress the scintillation effect caused by atmospheric turbulence.

(2) After using the ZF decoding algorithm, the bit-error performance ratio of the 4×5 system was worse than that of the 2×3 system. Moreover, the bit-error performance ratio of the 2×2 system was similar to that of the 4×4. The poor bit-error performance of the system is due to the increase of spatial interference among the signals after the multiplexing technology was adopted by the originator.

Figure 5.51 shows the BER and SNR curves of the ZF and ML decoding algorithms at $S.I = 1.0$. It can be seen from the figure that the rule reflected at this time is the same as that in Fig. 5.50; however, the difference is only for 2×2 and 4×4. The performance gap between the 2×3 and 4×5 systems is significantly reduced. This again shows that when the scintillation is stronger, the diversity gain obtained by the MIMO system is more obvious.

5.4.2.2 Least Mean-Squared Error Decoding Algorithm

To analyze the influence of the number of transmit and receive antennas, noise, and atmospheric scintillation on the system bit-error rate (BER), when the scintillation index is 0.6 and 1.0, and the minimum mean-squared error decoding algorithm is adopted, the curves of the system BER with the signal-to-noise ratio are simulated, as shown in Figs. 5.52 and 5.53.

Figure 5.52 shows the BER curve of the MMSE and ML decoding algorithms when the scintillation index is 0.6. It can be seen from the figure that after the MMSE decoding algorithm is adopted, with the increase of the number of lasers

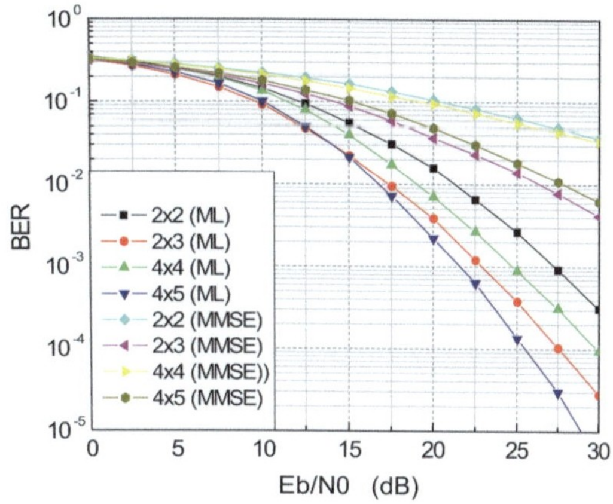

Fig. 5.52 MMSE and ML decoding at $S.I. = 0.6$. Bit-error rate curve of the algorithm

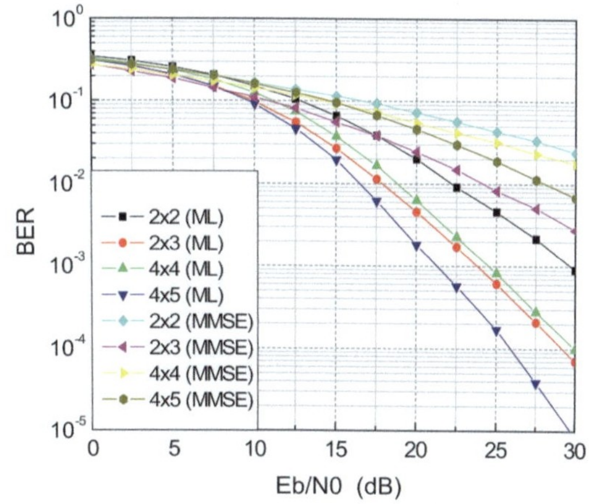

Fig. 5.53 MMSE and ML decoding at $S.I = 1.0$. BER and SNR curves of the algorithms

and detectors, the bit-error rate of the system gradually decreases with the increase of the signal-to-noise ratio, which is closer to the performance of the ML decoding algorithm compared with Fig. 5.50. However, there are some differences between the rule presented at this time and that presented in Fig. 5.50.

Specifically, after the MMSE decoding algorithm is adopted, the 4×5 bit-error performance ratio was worse than that of the 2×3 system; however, the 4×4 bit-error performance ratio was similar to that of the 2×2 system. The error performance of the MMSE system is better because the MMSE decoding algorithm considers the influence of noise on the error performance.

Figure 5.53 shows the BER and SNR curve of the MMSE and ML decoding algorithms when $S.I = 1.0$. The rule reflected in this figure is very similar to that in Fig. 5.52. The difference between them is that the improvement of the system bit-error rate is different; especially in the case of strong flicker, compared with the single-input single-output (SISO) system, when the minimum mean-squared error

Fig. 5.54 ZF and MMSE at $S.I. = 0.6$. BER curves of the algorithms

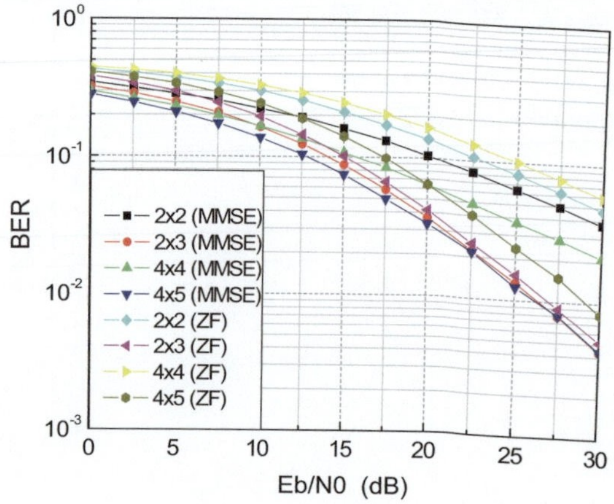

decoding algorithm is used. The improvement of the system bit-error rate for 2×2 and 4×4 is greater.

5.4.2.3 Comparison of the ZF and MMSE Decoding Algorithms

When the total transmission power is constant and the channel-fading coefficient is determined, the Monte Carlo method is used to simulate the bit error rate of 4×4 and 4×5 V-BLAST systems. The simulation parameters are $Q = 4$, $S.I. = 0.6$, and $\eta = 0.5$. The simulation results are shown in Fig. 5.54.

Figure 5.54 shows the performance comparison of the ZF and MMSE decoding algorithms. As can be seen from the figure, the MMSE algorithm has significantly better BER performance than the ZF algorithm, considering the influence of noise. However, when the signal-to-noise ratio is small, the performance difference between the 2×2 and 4×4 systems is large. It shows that the MMSE algorithm performs better at low SNRs. However, at high SNRs, the performance of the 2×2 and 4×4 systems is almost the same; hence, the MMSE decoding algorithm is more suitable for low SNRs.

5.4.3 *Performance Analysis of the Successive Interference-Cancellation Decoding Algorithm*

The BER performance of 4×4 and 4×5 vertical-layered space-time coding (V-BLAST) systems is simulated using the Monte Carlo method when the total transmission power is constant and the channel-gain coefficient is determined. The simulation parameters are $Q = 4$ and $S.I. = 0.6$. The simulation results are shown in Figs. 5.55 and 5.56.

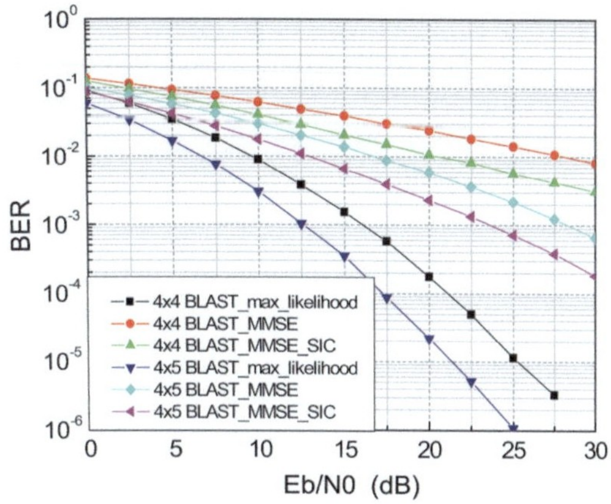

Fig. 5.55 Bit-error performance of the SIC-MMSE decoding algorithm

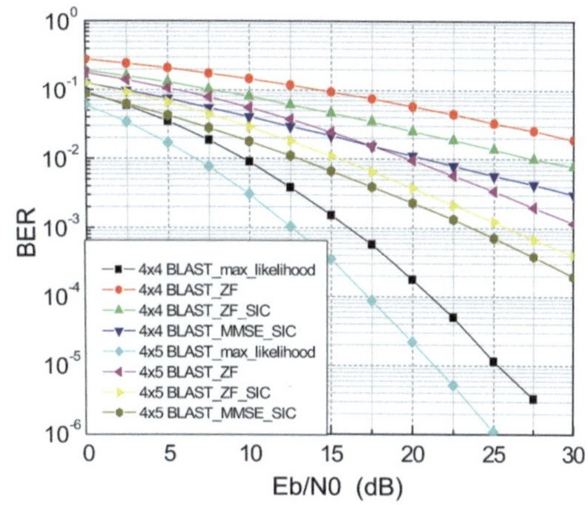

Fig. 5.56 Bit-error performance of the SIC-ZF and SIC-MMSE decoding algorithms

Figure 5.55 shows the bit-error performance of the BLAST code when the SIC-MMSE decoding algorithm is adopted. The following can be seen from Fig. 5.55:

(1) After the successive interference-cancellation algorithm is adopted, the bit-error performance of the system is closer to the theoretical value (that is, the bit-error performance when the maximum-likelihood decoding algorithm is adopted), indicating that the bit-error performance has been significantly improved. When the bit-error rate is $BER = 10^{-2}$ in the 4×4 system, compared with the MMSE detection algorithm, the bit-error performance of the SIC-MMSE detection algorithm is improved by approximately 7.5 dB. In the 4×5 system, the bit-error performance of the SIC-MMSE detection algorithm is improved by approximately 4 dB.

When the bit error rate is $BER = 10^{-3}$, in the 4×5 system, the bit-error performance of the SIC-MMSE detection algorithm is improved by approximately 4.5 dB, compared with that of the MMSE detection algorithm.

(2) The slope of the BER curve is steeper when the successive interference-cancellation algorithm is used. It shows that subtracting the decision component of the detected signal from the received vector can avoid intersymbol interference and improve the diversity of the detection algorithm without losing the multiplexing gain. The simulation results are consistent with the theoretical derivation (Eqs. 5.31 or 5.32).

Figure 5.56 shows the bit-error performance of the BLAST code when the SIC-MMSE and SIC-ZF decoding algorithms are used. The following can be seen from the figure:

(1) After the successive interference-cancellation algorithm is adopted, the bit-error performance of the system is significantly improved. When the bit-error rate is $BER = 2 \times 10^{-2}$ in the 4×4 system, compared with the ZF detection algorithm, the bit-error performances of the SIC-ZF and SIC-MMSE detection algorithms are improved by approximately 7.5 dB and 14.5 dB, respectively. When the bit-error rate is $BER = 10^{-3}$ in a 4×5 system, compared with the ZF detection algorithm, the bit-error performances of the SIC-ZF and SIC-MMSE detection algorithms are improved by approximately 3.5 dB and 6.5 dB, respectively. The simulation results are consistent with the theoretical derivation.
(2) In the same system, although the performance of the SIC-MMSE detection algorithm is better than that of the SIC-ZF detection algorithm, the slopes of their BER curves are almost the same, indicating that they have the same diversity gain. The SIC-MMSE detection algorithm outperforms the SIC-ZF detection algorithm because the former considers the influence of noise on the signal.

5.5 Summary

Space-time layered code is one of the earliest space-time coding methods. Its basic principle is to decompose the input information bit stream into multiple bit streams, encode and modulate them independently, and map them to multiple transmitting antennas. At the receiving end, special processing techniques are used to separate the signals that arrive together at the receiving antenna, and then send them to the corresponding decoder.

Comparing the four BLAST coding schemes, the horizontal-layered space-time coding is easy to implement; however, it is rarely used in practice because it has the worst space-time characteristics. Although diagonal-layered space-time coding has a good hierarchical structure and space-time characteristics, it can reach the theoretical capacity of MIMO systems. However, owing to its $\frac{M(M-1)}{2} bit$ transmission redundancy, its encoding and decoding algorithms are more complex, which also limit its application. Compared with diagonal-layered space-time coding, vertical-layered space-time coding has poorer space-time characteristics and hierarchical structure; however, it has no transmission redundancy and a simple coding structure, so it is widely used in practice.

Layered space-time codes cannot achieve maximum diversity, and their performance is relatively poor. They can be considered as a spatial-reuse technology. Their advantage is that the rate change is flexible, and the rate increases linearly with the number of transmitting antennas. These methods are often used in combination with binary-coding methods close to the channel capacity, such as concatenated code, to improve the coding performance.

References

1. Tao L, Guangxin L (2005) Principle and application of multi antenna wireless communication. Beijing University of Posts and Telecommunications Press, Beijing
2. Wolniansky PW, Foshini GJ, Golden GD (1998) V-BLAST: an architecture for realizing very high data rates over the rich-scattering wireless channel. ISSSE, Pisa, Italy, pp 295–300
3. Foschini GJ, Gans M (1998) On limits of wireless communication in a fading environment when using multiple-element antennas. Wireless Pers Commun 3:311–335
4. Telatar E (1999) Capacity of multi-antennas Gaussian channels. Eur Trans Telecomm 10(11):585–595
5. Vucetic B, Yuan J (2004) Space time coding technology. Translated by Wangxiaohai et al. China Machine Press, Beijing 2004
6. Wangxueping, Wen L, Yongzhao Z (2010) A compromise layered space-time coding scheme. Manuf Autom 32(09):35–36
7. Sellathurai M, Haykin S (2003) T-BLAST for wireless communications: first experimental results. IEEE Trans Veh Technol 3(52):530–531
8. Ankiraman M (2004) Space-time codes and MIMO systems. Artech House, Boston, London
9. Jafarkhani H (2005) Space-time coding theory and practice. Cambridge University Press, New york
10. Ying L, Jia L, Wangxinmei (2003) An improved layered space-time code detection algorithm. J Commun 24(3):113–118

Chapter 6
Mixed Space-Time Encoding

Layered space-time coding (LSTC) [1–3] and space-time block coding (STBC) [4–6] are two important space-time coding methods. They have been widely studied in radio-frequency (RF) communication and have been adopted by 3G. LSTC achieves high spectral efficiency despite its poor error-code performance. STBC codes use orthogonality to achieve transmission diversity in the system and their performance is significantly better than that of LSTC codes. However, owing to the redundant information in the encoding matrix, their transmission efficiency is low [7, 8]. Therefore, improving the transmission rate of STBC codes is the first problem that needs to be solved.

A hybrid space-time encoding scheme, suitable for intensity modulation/direct detection (IM/DD) optical communication, was constructed by making a reasonable compromise between multiplex gain and diversity gain. Orthogonal space-time coding is used to provide diversity gain for the first information symbol to be translated, and the reliability of the system is improved by ensuring the accuracy of the first-layer detection.

6.1 Encoding Mixed Space-Time Codes in Optical Communications

Owing to their good performance, space-time block codes and layered space-time codes have been widely recognized in the field of wireless optical communication [1, 4, 5, 9, 10]. To effectively suppress the influence of atmospheric turbulence, space-time block code schemes for intensity modulation were proposed in [4, 5, 11]. Although these schemes differ, they have characteristics similar to STBC codes in the RF domain; that is, they have maximum diversity gain.

X. Ke, *Space-Time Coding Theory in Optical Wireless MIMO Systems*, Optical Wireless Communication Theory and Technology,
https://doi.org/10.1007/978-981-97-6449-5_6

To fully utilize the channel capacity of a multiple-input multiple-output (MIMO) system, a layered space-time encoding scheme suitable for intensity-modulated atmospheric laser communication is presented in references [1, 10, 12–14]; it is decoded by interference cancellation (IC). In the decoding process, the information symbols decoded in the first layer do not obtain diversity gain; that is, the first decoded codeword may be wrong, resulting in error propagation in the process of successive interference cancellation (SIC), which seriously affects the error-code performance of the system.

The idea of hybrid space-time coding is to introduce a space-time block code into the lower layer of the layered space-time coding to improve the diversity of the first translated code. Figure 6.1 shows the principle of hybrid space-time coding in atmospheric laser communication.

For a hybrid space-time $M \times N$ coding system (M represents the number of lasers and N represents the number of detectors), the basic principle is to decompose the serial-parallel transformed data into a layer-G packet data stream and a layer-M_{L} direct-transmission data stream. Each direct-transmission data stream is directly transmitted by the corresponding laser. After $M \times N$ orthogonal space-time coding, the packet data streams of each layer are sent out by the corresponding m lasers. The number of lasers meets $G_{\mathrm{m}} + M_{\mathrm{L}} = M$, if the original STBC code is directly introduced into the first layer of layered space-time coding.

In hybrid space-time coding, there will be complex conjugates and negative numbers of characters, owing to the STBC coding. However, in IM/DD atmospheric laser communication, the optical signal has only two states: "*on*" and "*off*," and it

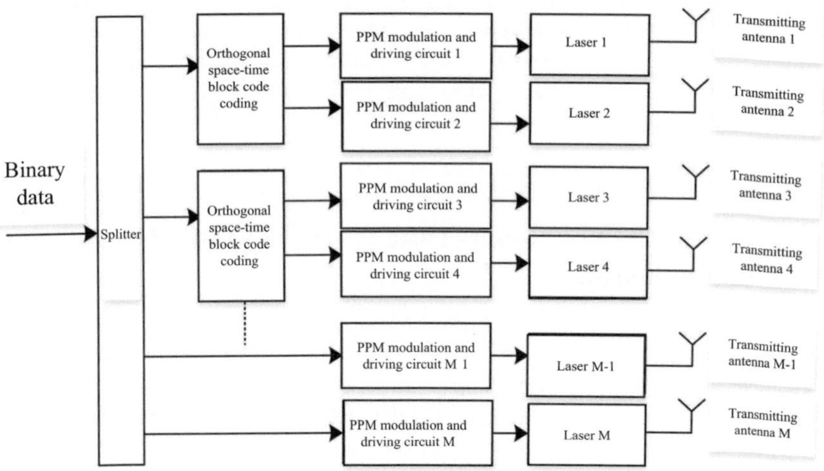

Fig. 6.1 Principle of hybrid space-time coding in optical communication

is impossible to express complex conjugates and negative numbers of symbols with these two states.

Therefore, we introduce bit-complement cross-space-time coding into the first layer of layered space-time coding to improve the diversity gain of the first layer. In this manner, the coding matrix is transformed into a real matrix, which can be transmitted through IM/DD optical communication [15].

6.2 Decoding Mixed Space-Time Codes in Optical Communication

In hybrid space-time encoding, because of the special encoding structure, the received signal contains not only the grouped data with the STBC layer but also the direct data with the Bell Laboratories layered space-time (BLAST) layer. Therefore, the received signal cannot be directly decoded by STBC. Interference from other layers must be eliminated first. We used group-interference suppression (GIS) to eliminate interference from other layers.

At the receiving end, decoding one set while suppressing the effects of other sets of signals (considered interference) reduces the decoding complexity at the cost of losing the diversity gain. The GIS method not only suppresses intergroup interference, but also simplifies the decoding process. STBC codes are detected for each set of signals after GIS using merge criteria, and the original signal can be restored after hierarchical space-time decoding. The decoding principle of mixed space-time encoding is shown in Fig. 6.2.

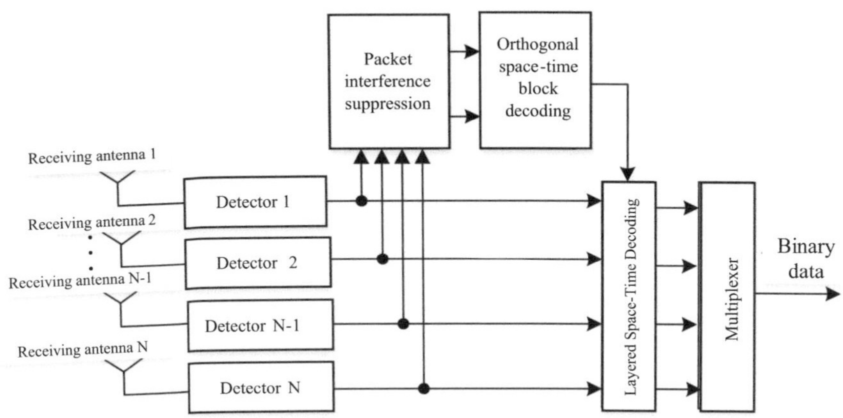

Fig. 6.2 Decoding principle of hybrid space-time coding in optical communication

6.2.1 Grouping Interference-Suppression Principle

GIS is a signal-processing algorithm based on antenna grouping. This method divides M transmitting antennas at the transmitter into $q = G + M_L$ groups, and $G_1, G_2, \ldots, G_q$ groups are used to represent q groups of antennas. We assume that they contain $m_1, m_2, \ldots, m_q$ antennas, where $m_1+m_2+\ldots+m_q = M$. We assume that the corresponding transmission-symbol vector for group q is C_i, where $1 \leq i \leq q$.

Generally, the transmission symbol C_1, corresponding to the first group of transmission antennas, $C_1 = (c_1^1, c_1^2, \ldots, c_1^m)^T$, is selected as the expected signal $C_1 = (c_1^1, c_1^2, \ldots, c_1^m)^T$ and uses the GIS method for decoding. There is a total of $M - m$ interference signals, assuming that the number of receiving antennas N satisfies $N \geq M - m + 1$ and the receiver knows the channel-state information (CSI), that is, H. Channel-state information H can then be expressed as $H = H_1 + H_2$, where H_1 represents the channel-gain matrix from the first set of transmit antennas, that is,

$$H_1 = \begin{bmatrix} h_{1,1} & \cdots & h_{1,m} \\ h_{2,1} & \cdots & h_{2,m} \\ \vdots & \vdots & \vdots \\ h_{N,1} & \cdots & h_{N,m} \end{bmatrix}. \tag{6.1}$$

H_2 represents the channel-gain matrix of the interference signal and is defined as follows:

$$H_2 = \begin{bmatrix} h_{1,m+1} & \cdots & h_{1,M} \\ h_{2,m+1} & \cdots & h_{2,M} \\ \vdots & \vdots & \vdots \\ h_{N,m+1} & \cdots & h_{N,M} \end{bmatrix}. \tag{6.2}$$

Because H_2 is an $N \times (M - m)$ matrix, the rank of H_2 is less than or equal to the number of columns in the matrix, *rank* $[A_2] \leq M - m$. Defining the zero space of H_2 as Ω, Ω is made up of all the line vectors that satisfy the following conditions:

$$xH_1 = (0,0,\cdots,0)\text{, 且 } \dim(\Omega) + rank(H_2) = N \tag{6.3}$$

where the dimension $\dim(\Omega) \geq N - M + m$ because *rank* $(H_2) \leq M - m$. Therefore, a set of non-unique orthogonal normalization vectors $\{v_1, v_2, \ldots, v_{N-M+m}\}$ can be found in the zero-space Ω of H_2, where i is the row vector of $1 \times N$, and $j = 1, 2, \ldots, N - M + m$. V is defined as an $(N - M + m) \times N$ matrix, where j behaves as v_j; then, V is structured as follows:

$$V = [v_1, v_2, \ldots, v_{N-M+m}]^T. \tag{6.4}$$

Because v_j is an orthogonal normalized vector, $v_j v_l^{\mathrm{T}} = \begin{cases} 1 & j = l \\ 0 & j \neq l \end{cases}$; then,

$$\mathrm{VV}^{\mathrm{T}} = I_{N-M+m}, \tag{6.5}$$

where I_{N-M+m} represents a unit matrix of order $N - M + m$ and $(\bullet)^T$ represents the transposition of the matrix.

Assume that the Gaussian channel model is still used, that is

$$y(t) = \eta Hx(t) + n_0(t). \tag{6.6}$$

Equation (6.6) is multiplied by V to obtain

$$Vy(t) = \eta VHx(t) + Vn_0(t). \tag{6.7}$$

If we set $\tilde{r}(t) = Vy(t), \tilde{H} = VH$, and $\tilde{n}(t) = Vn_0(t)$, Eq. (6.7) can be simplified to

$$\tilde{r}(t) = \eta\tilde{H}x(t) + \tilde{n}(t). \tag{6.8}$$

From Eq. (6.8), the signals from the group laser are retained, whereas the signals from other lasers are suppressed.

6.2.2 Decoding STBC in an Optical Channel

We can see that Eqs. (6.8) and (6.6) have the same structure, so we can use the bit-complement orthogonal space-time decoding method. Introducing STBC into layered space-time coding can increase the diversity gain of the first layer, which can obtain the diversity gain of the STBC codes, and thereby improve the reliability of the lower layer of the BLAST codes.

In traditional layered space-time coding, each laser is emitted at equal power; that is, the power of each laser is equal to the total power of the transmitter. In the decoding process, because the diversity gain obtained by the former decoding is lower than that of the latter, we can further guarantee the correctness of the first decoding by distributing the transmission power, according to the size of the diversity gain in the hybrid space-time coding system.

6.2.3 Decoding Algorithm in Hierarchical Space-Time Coding

After decoding the STBC layer, the influence of the STBC code on other symbols is eliminated from the original system, that is,

$$r_0 = r - \eta \begin{bmatrix} h_{11} & h_{12} \\ h_{21} & h_{22} \end{bmatrix} \bullet \begin{bmatrix} \hat{x}_1 \\ \hat{x}_2 \end{bmatrix}. \tag{6.9}$$

Hierarchical space-time decoding is performed on the symbols after the STBC decoding interference is eliminated. In this case, any of the decoding algorithms mentioned in Chap. 5 can be used, which will not be repeated here.

6.3 Performance Analysis of Hybrid Space-Time Coding

To evaluate the performance of hybrid space-time coding, we analyzed the error-code performance, diversity, and spectral efficiency.

6.3.1 Error-Code Performance Analysis

Theorem 6.1 [16]: For a MIMO ($M \times$ N) system, a space-time coded data block with a length of B bits is divided into q strings, each of which is counted as $B_1, B_2, \ldots, B_q$ and ($B_1 + B_2 + \ldots + B_q = B$). The M transmission antennas are divided into q groups, which are recorded as $G_1, G_2, \ldots, G_q$, respectively. Each group contains m_1,$m_2, \ldots, m_q$ antennas and $m_1 + m_2 + \ldots + m_q = M$. We assume that the number of receiving antennas N satisfies the following criterion: $N \geq \mathrm{M} - \mathrm{m}_1 + 1$. Each B_i substring C_i is encoded by an encoder and then transmitted by a group of G_i m_i antennas. The average power of this group of transmitting antennas is recorded as follows.

The receiver detects the bit-error rate of symbol C_1 after using the group-interference suppression method. Simultaneously, in another $(m_1, N - M + m_1)$ MIMO system, data of length B_1 enters encoder C_1, and the output signal of the encoder is simultaneously transmitted through M_1 antennas. Assuming that the average transmission power of antenna M_1 is E_1 and P_2 is defined as the error rate of the system, then $P_1 = P_2$. Refer to [16] for detailed evidence.

We consider the 4 × 4 system as an example and analyze the error rate of STBC-BLAST codes according to Theorem 6.1. In the system, it is assumed that the grouping method is ($\mathrm{G}_1 + \mathrm{G}_2 + \mathrm{G}_2$), and the transmitting antenna is divided into three groups, corresponding to one STBC encoder and two sublayered space-time encoders. After the group-interference suppression, a 2×2 STBC encoding subsystem is obtained.

According to Theorem 6.1, the error rate of any subsystem is the same as that of a traditional 2×2 STBC system when the average transmission power is the same. In other words, to achieve the same error-rate level, the total transmission power of STBC-BLAST codes grouped by $G_1 + G_2 + G_2$ should be three times that of traditional 2×2 STBC systems. Therefore, the error rate of the STBC-BLAST system with grouping ($G_1 + G_2 + G_2$) is higher than that of the 2×2 STBC system under the same average transmission power. It is worth noting that with ($G_1 + G_2 + G_2$), the transmission rate of the STBC-BLAST system in grouping mode is three times that of the 2×2 STBC system.

6.3.2 Comparison of Diversity and Spectrum Efficiency in Several Encoding Methods

For a BLAST system, the spectral efficiency is

$$\eta = Rkn_T \, bit/s/Hz, \tag{6.10}$$

where k represents the number of bits in each modulation symbol, M is the number of transmitting antennas, R is the code rate, and $R = 1$ for an uncoded vertical (V)-BLAST system. For an (m, n, G) STBC-VBLAST system, the spectral efficiency is

$$\eta = k(M - mG + GR_{STBC}) \, bit/s/Hz, \tag{6.11}$$

where R_{STBC} represents the code rate of STBC encoding, m represents the number of transmit antennas encoded by the STBC, n represents the number of receiving antennas encoded by the STBC, and G denotes the number of STBC encoding blocks used. For $M \times N$ STBC systems, $R_{STBC} = \frac{M}{n} \leq 1$. Compared with uncoded BLAST systems, the minimum diversity gain of STBC-BLAST codes is achieved at the expense of a $\frac{G(m-R_{STBC})}{M}$ reduction in spectral utilization. When M is larger, the $\frac{G(m-R_{STBC})}{M}$ value is smaller.

To clearly represent the tradeoff between diversity and spectral efficiency, Table 6.1 shows the diversity and spectral efficiency using the STBC-BLAST and V-BLAST codes, respectively. M is the number of transmitting antennas and N is the number of receiving antennas.

Table 6.1 Comparison of minimum diversity and spectral efficiency between STBC-BLAST and VBLAST codes

	Using (*m*, *n*, *G*) STBC-BLAST	Unencoded BLAST
Diversity (d)	$\min(m(N-M)+m^2, nN-nM+Gm+1)$	$N-M+1$
Spectral efficiency (η)	$K(M-Gm+GR_{STBC})$	KM

6.4 Simulation Analysis

The Monte Carlo method was used to simulate the error-code performance of the system, where the total transmission power is constant and the channel-attenuation coefficient is determined. The simulation parameters are $Q=4$, scintillation index $S.I.=0.6$, and $\eta=0.5$, using one 2×2 STBC block in the mixed encoding scheme. The simulation results are shown in Figs. 6.3 and 6.4.

Figure 6.3 shows the error-code performance when different systems use different encoding methods. From Fig. 6.3, we can see the following:

(1) Contrasting an STBC system and a (4) × zero-forcing (ZF)_4 BLAST system, it can be seen that there is a wide error-rate range between STBC and BLAST, and the error-rate curve of the STBC system is steeper than that of the BLAST system. This shows that the STBC system can obtain a larger diversity gain at the cost of some multiplexing gain; that is, the performance of the STBC system is significantly better than that of the BLAST system.

(2) The error-code performance of the hybrid space-time coding system is between a 2 × 2 STBC system and a 4 × 4 ZF-BLAST system. It can be seen that the error performance of mixed space-time codes is significantly better than that of traditional BLAST codes.

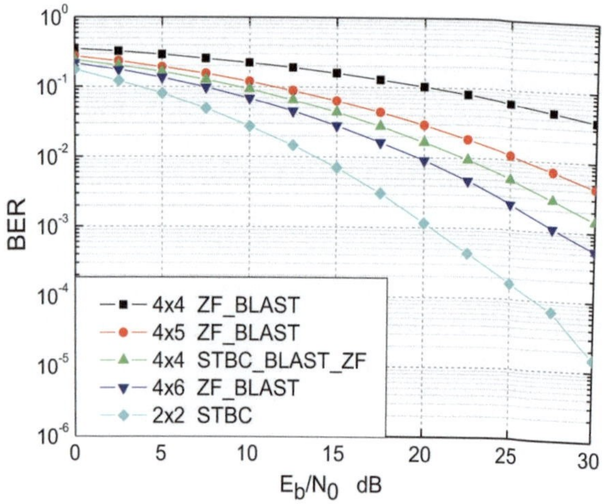

Fig. 6.3 Bit error performance in different encoding systems

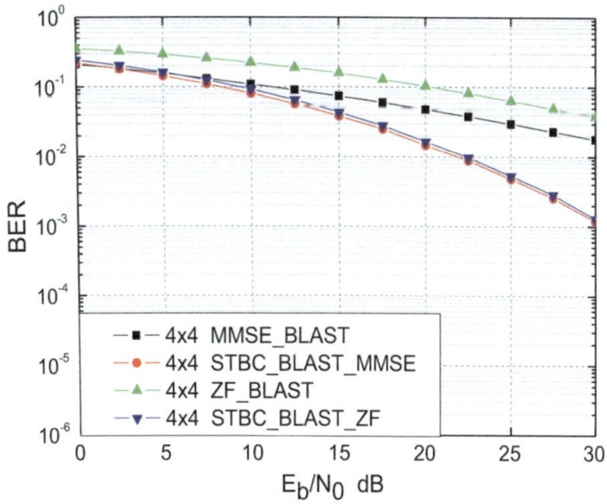

Fig. 6.4 Error performance of two systems with different decoding algorithms

Relative to ZF, when the error rate is shown for BLAST systems, the performance of hybrid space-time coding systems is improved approximately. This shows that introducing the STBC code into layered space-time coding increases the diversity of the first layer, ensures the correctness of the first layer, and reduces error propagation.

(3) 2×2 and 4×4 ZF-BLAST systems were compared. The error code performance of the hybrid space-time coding system was between them. Note that a 4×4 hybrid space-time coding system achieves both receive and send diversity gains.

To further illustrate the good error-code performance of the hybrid space-time encoding system, minimum mean-squared error (MMSE) and ZF detection algorithms were used to analyze the traditional BLAST system and the hybrid space-time encoding system, and their error-code performance is shown in Fig. 6.4. The following can be observed:

(1) When the MMSE decoding algorithm was used in a hybrid space-time coding system, the error performance was significantly better than that of the traditional BLAST system.
(2) In the traditional BLAST system, the performance of the MMSE algorithm was significantly better than that of the ZF algorithm because the influence of noise on the signal was considered.

However, in hybrid space-time coding systems, the error performance is very close when using the two algorithms. When the error rate is 2×10^{-3}, the error performance of the two algorithms is less than 0.5 dB. This shows that after ensuring the correctness of the first-layer BLAST decoding, the noise has little effect on the signal, which further shows that the error in the BLAST system is mainly caused by the first-layer error. It also reflects that in a hybrid space-time coding system, the ZF detection algorithm is more suitable for engineering applications to reduce receiver complexity.

Because both space-time block encoding and layered space-time encoding pursue a single performance index, they have some drawbacks. The anti-fading performance of layered space-time encoding is poor, whereas the STBC code matrix has redundant information, which reduces the transmission efficiency of the system. For this reason, this scheme first combines the two codes and presents a hybrid space-time coding scheme suitable for IM/DD optical communication.

6.5 Summary

This scheme used bit-complement orthogonal space-time encoding to provide diversity gain for the first information symbol to be translated and improved the reliability of the system by ensuring the accuracy of the first-layer detection. The essence of this encoding method was to make a reasonable compromise between multiplexing and diversity gains. Second, the error-code performance of this method was analyzed, and the diversity and spectral efficiency of several space-time coding methods were compared.

References

1. Huiqin W, Zheng K (2008) Free-space optical communication based on vertical layered space-time codes. China Laser 35(6):874–878
2. Wolniansky PW, Foshini GJ, Golden GD (1998) V-BLAST: an architecture for realizing very high data rates over the rich-scattering wireless channel. 1998 ISSSE, Pisa, Italy, pp 295–300
3. Ying L, Jia L, Xinmei W (2003) An improved hierarchical space-time code detection algorithm. J Commun 24(3):113–118
4. Brandt-Pearce M, Wilson S, Cao Q et al (2004) Code design for optical MIMO systems over fading channels. In: Proceedings of the 38th Asilomar conference on signals, systems & computers, Monterey, CA, pp 871–875
5. Simon MK, Vilnrotter V (2005) Alamouti-type space-time coding for free-space optical communication with direct detection. IEEE Trans Wirel Commun 4(1):35–39
6. Zaidi SAR, Afees M (2008) Cross layer design for orthogonal space time block coded optical MIMO systems. In: WOCN '08. Wireless and optical communications networks, pp 1–5
7. Wilson SG, Brandt-Pearce M, Cao Q, Baedke M (2005) Optical repetition MIMO transmission with multipulse PPM. IEEE J Sel Areas Commun 23(9):1901–1910
8. Tao L, Guangxin L (2005) Principle and application of multi-antenna wireless communication. Beijing University of Posts and Telecommunications Press, Beijing
9. Haas SM et al (2001) Space-time codes for wireless optical channels. IEEE Transactions on Information Theory, Washington, DC, p 244
10. Wongpooml J, Sripimanwat K, Sangaroon O (2006) multilevel photon communication with binary pulse position modulation. In: Communications, circuits and systems proceedings, 2006 international conference on, Guilin, pp 1904–1907
11. Huiqin W, Zheng K (2009) Orthogonal space-time block coding in atmospheric laser communication. J Opt 36(S2):63–68

12. Anguita JA, Neifeld MA, BV Vasic (2006) Multi-beam space-time coded systems for optical atmospheric channels. SPIE, pp 63041B-63041–63041B-63049
13. Huiqin W, Zheng K (2010) Serial interference cancellation algorithm for hierarchical space-time encoding in FSO. Photoelectr Eng 37(2):1–6
14. Djordjevic IB, Vasic B, Neifeld MA (2006) Multilevel coding in free space optical MIMO transmission with Q-ary PPM over the atmospheric turbulence channel. IEEE Photonics Technol Lett 18(14):1491–1493
15. Huiqin W, Zheng K (2009) Hybrid space-time encoding in free space optical communication. J Opt 29(1):132–137
16. Tarokh V, Ngauib A, Seshadr N (1999) Combined array processing and space-time coding. IEEE Trans Inf Theory 45(4):1121–1128

Chapter 7
Space-Time Trellis Coding

Space-time trellis coding (STTC) is a combination of coding and modulation, which does not reduce the frequency-band or power utilization, but improves coding at the expense of increasing the system complexity using gain technology. STTC can be regarded as a finite-state transition device. The value of the latest set of data streams can determine the transition relationship between the current and next states. The result of this transition is the space-time symbol transmission process.

Space-time trellis coding combines transmit diversity with channel coding, thereby improving the anti-fading performance of the system. This chapter studies space-time trellis code technology in a turbulent atmospheric channel, and analyzes the transmit and receive diversity schemes of space-time trellis codes in atmospheric turbulence.

7.1 Space-Time Trellis Coding

In 1998, Tarokh et al. [1] proposed space-time trellis codes by jointly designing error-control coding, modulation, and transmit and receive diversity techniques and produced a channel based on flat slow and fast fading. The space-time trellis code comprehensively considers coding and diversity gains to achieve a reasonable compromise between coding complexity, system performance, and spectrum utilization.

The Viterbi decoding algorithm is generally used to decode space-time trellis codes, and its decoding complexity increases exponentially with an increase in the transmission rate. Therefore, finding a more-efficient decoding algorithm is a key issue. In [2], the bit-error rate and frame-error rate performance of space-time trellis codes were compared. In [3], a space-time trellis code–based impulse-radio ultra-wideband (STTC-IR-UWB) system was proposed that adopted a bi-orthogonal

X. Ke, *Space-Time Coding Theory in Optical Wireless MIMO Systems*,
Optical Wireless Communication Theory and Technology,
https://doi.org/10.1007/978-981-97-6449-5_7

modulation scheme based on pulse-position modulation (PPM). Its performance was better than that of impulse-radio ultra-wideband (IR-UWB).

In [4], an IR-UWB multiple-input multiple-output (MIMO) system combining PPM and STTC was studied and its performance in slow-fading and fast-fading multichannel communication systems was analyzed. In [5], space-time block coding (STBC) was applied to an improved multicarrier code-division multiple-access (MC-CDMA) system, and extended to space-time trellis-code technology. The performance of the two systems was compared.

In [6], an STTC-orthogonal frequency-division multiplexing (OFDM) system with a maximum-combining ratio was proposed for quasi-static flat fading and frequency-selective fading environments. In [7], Altera's Stratix EP1S25 field-programmable gate array (FPGA) chip was used to build a system that could be configured to use STBC and STTC, and the performance of different configurations was measured. In [8], an optimal *MN*-ary pulse-position amplitude modulation (PPAM) space-time trellis code was introduced based on a UWB MIMO system.

In [9], space-time trellis codes for optimal *M*-ary pulse-amplitude modulation (M-PAM)-based multi-antenna ultra-wideband communication systems in fast Rayleigh fading were discussed, and the upper bound of the pairwise error probability (PEP) was given. In [10], a multi-carrier code-division multiple-access (MC-CDMA) system based on space-time trellis coding was proposed and the bit-error rates in the maximum-ratio combining (MRC) and equal-gain combining modes were analyzed.

In [11], for backward compatibility with space-time transmit diversity (STTD), a solution for space-time trellis codes in wideband CDMA was proposed, and a set of STTC suitable for this scheme was given. It not only had the same backward-compatibility characteristics as STTD, but also obtained better anti-fading performance. In [12], a new super-quasi-orthogonal space-time trellis code (SQO-STTC) was proposed, which is a twiddle factor–based quasi-orthogonal space-time grouping. It can achieve full rate and full diversity without increasing the encoding and decoding complexity.

In [13], combining the advantages of space-time trellis codes and hierarchical space-time codes and using multiple high-rate convolutional codes, a class of generalized recursive space-time trellis codes (G-RSTTC) suitable for serial concatenated structures was designed. In [14], a joint iterative maximum-likelihood estimation and decoding method suitable for space-time trellis codes was proposed, which could obtain better system performance than traditional non-iterative maximum-likelihood decoding. In [15], the advantages of MIMO systems and ultra-wideband were combined to achieve high-speed data transmission. Three different methods, based on pulse-position modulation, pulse-amplitude modulation, and bi-level quadrature modulation, were discussed.

7.2 MIMO-PPM Analysis

Atmospheric optical communication systems generally adopt an intensity modulation/direct detection (IM/DD) system [16]; the commonly used modulation methods include on–off keying (OOK), pulse-position modulation (PPM), etc.

Figure 7.1 shows a MIMO-PPM system in which we use M lasers and N photodetectors (PD). After the signal is transmitted through an atmospheric channel, the scintillation caused by atmospheric turbulence causes the signal to fade. It is assumed that each PD can ideally receive optical signals incoherently. After the source is modulated by PPM, it is emitted by M lasers, assuming that the laser beam is sufficiently narrow to cover all the photodetectors.

It is assumed that the equal-gain combining (EGC) method is adopted at the receiving end, and that thermal noise and shot noise are not considered. When a single photodetector is used at the receiving end, we assume that the total incident power is Pr and the average number of photoelectrons received by the photodetector is

$$n_s = \frac{\eta P_r T}{hf} = \frac{\eta E_s}{hf} \text{ photoelectrons/time slot.} \tag{7.1}$$

In Eq. (7.1), h represents Planck's constant and η is the quantum efficiency (photoelectric-conversion efficiency) of the detector. E_s is the symbol energy transmitted by all antennas simultaneously, f is the frequency of the light wave, and T is the time-slot cycle time where the information is located.

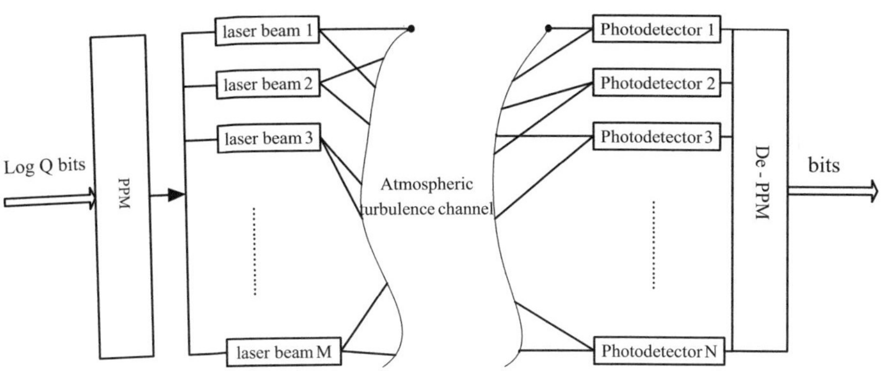

Fig. 7.1 Block diagram of a MIMO pulse-position modulation system

7.2.1 No Fading Caused by Atmospheric Scintillation

When no fading is caused by flicker in the system, and assuming that there are no other losses in the system, all the signals emitted by the lasers, that is, "1," appear in the first time slot. Hence, the only case considered to be an error is if "0" appears in the first time-slot position; thus, considering that the PD receives photoelectrons with Poisson distribution properties and independent channels, we can obtain the symbol-error rate [17]:

$$P_S = \frac{Q-1}{Q}\left[e^{-\frac{M\eta(\frac{P_T}{M})T_S}{hfQ}}\right]^N = \frac{Q-1}{Q}e^{-\frac{\eta E_S N}{hf}}. \tag{7.2}$$

Equation (7.2) can be converted into the bit-error rate:

$$P_b = \frac{Q}{2(Q-1)}P_S = \frac{1}{2}e^{-\frac{\eta E_S N}{hf}}. \tag{7.3}$$

In Eq. (7.3), Q represents the modulation order, h represents Planck's constant, and η is the quantum efficiency (i.e., photoelectric conversion efficiency) of the detector. E_s is the symbol energy emitted by all antennas simultaneously, EEE is the frequency of the light wave, and f is the time-slot cycle time where the information is located.

It can be seen from Eq. (7.3) that when the energy per symbol is constant, the bit-error rate (BER) is unrelated to Q. We used $\eta = 0.7$ as the calibration point for $E_s = 10^{-16}$, which corresponds to 80 photoelectrons per symbol detected by the photodetector.

From Fig. 7.2, it can be seen that increasing the number of detectors at the receiving end significantly improves the bit-error rate. When the detector was 2 at Es $= -175$dBj, it was two orders of magnitude higher than when the detector was 1. As can be seen from Fig. 7.3, when using PPM modulation with different Q values, the bit-error rate is not significantly improved; however, with the increase in Q, the bit-error rate worsens because the total energy is fixed.

7.2.2 Fading Caused by Atmospheric Scintillation

When fading caused by atmospheric scintillation is introduced into the system, the effect of fading on the received signal must be considered. In a MIMO system, the fading at each photodetector is different and time varying, and can be expressed as $\{H_{mn}\}$ With N detectors, the number of effective photoelectrons per time slot is

$$n_s \cdot \sum_{m=1}^{M} H_{mn}^2/M. \tag{7.4}$$

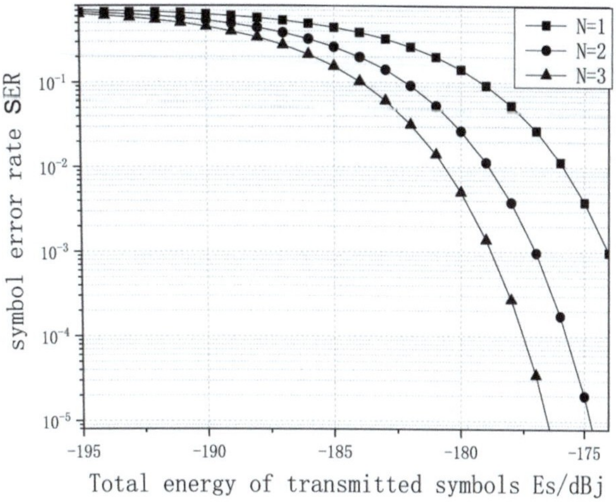

Fig. 7.2 Relationship between different numbers of PDs and the bit-error rate when there is no channel fading

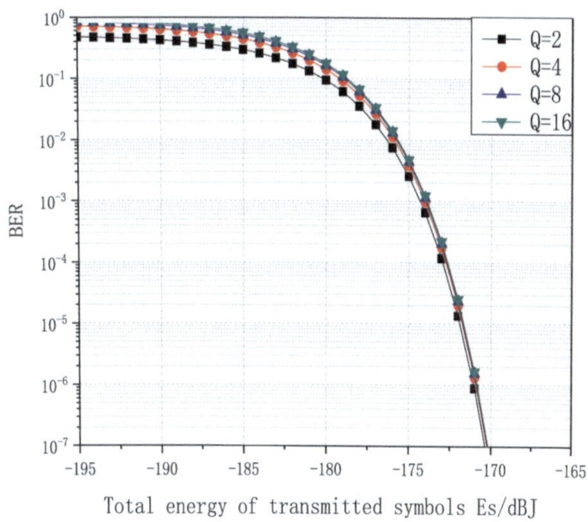

Fig. 7.3 Relationship between different Q and the bit-error rate

Then, at the n th detector, the probability of "0" appearing in the first time slot can be expressed as

$$P[L_{n1} = 0/slot1, H] = e^{-\sum_{m=1}^{M} h_{mn}^2 \frac{\eta P_r T_s}{M h f Q}}. \tag{7.5}$$

Because the received signal at each detector is relatively independent, the probability that the first time slot of all detectors at a certain moment is "0" can be expressed as

$$P\left[\sum_{n=1}^{N} l_{n1} = 0/slot1, H\right] = e^{-\sum_{n=1}^{N}\sum_{m=1}^{M} h_{mn}^2 \frac{\eta P_r T_s}{MhfQ}}. \tag{7.6}$$

Assuming that h_{mn} is independent of the others and has the same distribution in practice, the average symbol-error rate can be expressed as follows [18]:

$$\begin{aligned} P_s &= \int P_{s/H} f_H(h)dh \\ &= \frac{Q-1}{Q}\left\{\left[\int e^{-h_{mn}^2 \frac{\eta P_r T_s}{MhfQ}} f_H(h)dh\right]^N\right\}^M \\ &= \frac{Q-1}{Q}\left[\int e^{-h_{mn}^2 \frac{\eta E_s}{Mhf}} f_H(h)dh\right]^{MN}. \end{aligned} \tag{7.7}$$

Equation (7.7) can be converted into the bit-error rate:

$$P_b = \frac{1}{2}\left\{\left[\int e^{-h_{mn}^2 \frac{\eta E_s}{Mhf}} f_H(h)dh\right]^N\right\}^M. \tag{7.8}$$

As shown in Fig. 7.4a, the symbol-error rate of the system increases with a continuous increase in the scintillation index. Figure 7.4b demonstrates the influence of diversity on the system when the same Q value and a scintillation index of 0.5 are used.

It can be seen from the figure that when the receiving end remains unchanged and the transmitting end increases the power of the laser beam, the bit-error rate of the system is significantly improved. When the number of laser beams and photodetectors is increased, the bit-error rate of the system can also be improved. At Es = −175dBj,

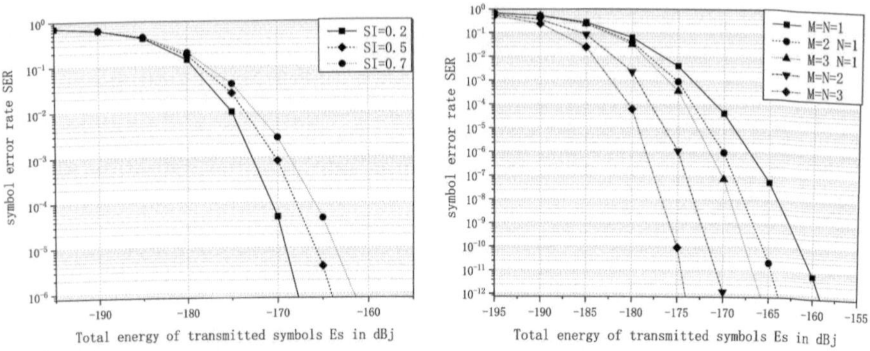

(a) M = N =2, Q = 4, with different scintillation indices (S.I.)

(b) Q = 4, S.I. = 0.5 for different *M* and *N* values

Fig. 7.4 Symbol-error rate under different conditions

the bit-error rate of a 3 × 3 system is increased by three orders of magnitude compared with a 2 × 2 system; it can also be seen that diversity is indeed present. The fading caused by atmospheric scintillation can be suppressed and the bit-error rate performance can be improved.

The results show that the fading caused by atmospheric turbulence will have a great impact on a wireless optical communication system; however, after using MIMO technology, the bit-error rate of the system can be effectively reduced by increasing the number of transmitter lasers and photodetectors.

7.3 Wireless Optical MIMO Space-Time Trellis Codes

In Fig. 7.5, the binary data of the source are transformed into multiple data streams through a serial-to-parallel transformation into the encoder of the space-time trellis code, and the symbols output by the encoder are sent out by laser after PPM modulation. The optical signal passing through the atmospheric channel is affected by atmospheric turbulence, which causes the signal to fade.

In this chapter, a lognormal fading model is used to observe the effect of the proposed method at different scintillation-index values [17]. At the receiving end, a group of PDs detects the optical signal and sends it to a maximum-likelihood Viterbi decoder for joint demodulation and decoding. The codeword is decoded and then output through a parallel–serial converter.

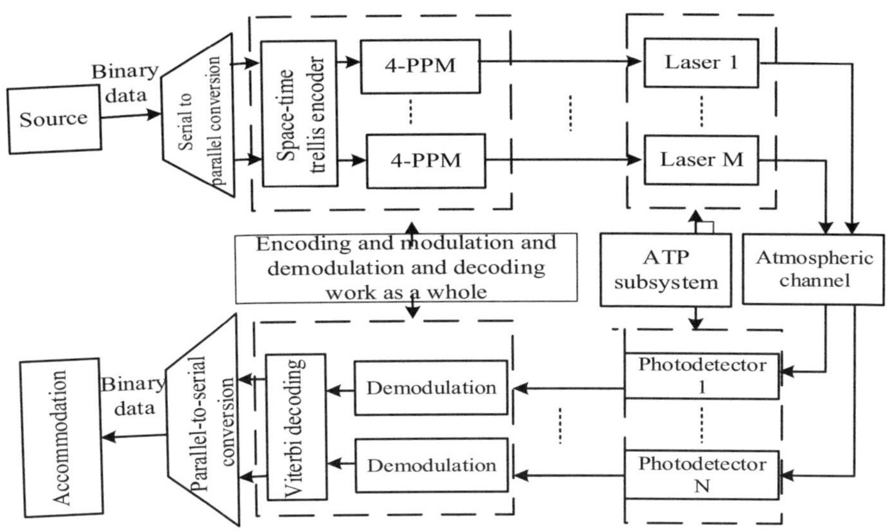

Fig. 7.5 Block diagram of a wireless optical space-time trellis code system

7.3.1 Transmit-Signal Model

In a wireless optical channel, the channel-fading period is much slower than the signal period; therefore, we can assume that the optical channel is a non-ergodic slow-fading channel. The scintillation index in a given block signal B of length L can be considered to remain unchanged; therefore, the received signal at the N receiving antennas (photodetectors) is expressed as

$$y_b^j[l] = \sum_{i=1}^{M} h_b^{i,j} \sqrt{p_b} x_b[l] + n_b^j[l]. \tag{7.9}$$

In Eq. (7.9), $\sqrt{p_b}$ represents the electrical-signal power at each photodetector of the block signal B (without the flickering effect); $y_b^j[l]$ and $n_b^j[l]$ respectively represent the received and noise elements of the signal at time l in the block signal B at detector n. The noise $n_b^j[l]$ is independently distributed and follows $n \sim N(0, 1)$; $x_b[l]$ represents the transmission signal; $h_b^{i,j}$ represents the scintillation fading coefficient from laser m to detector n. Each transmission symbol $x_b[l]$ is obtained from the PPM mapping table, $x_b[l] = \{e_1 \ldots e_Q\}$, where e_q is a normalized unit vector. The value is nonzero at position q; the other slot positions are "0".

7.3.2 4-PPM-STTC Modulation and Decoding

In space-time trellis coding (STTC), the encoder adopts a convolutional code with a generated sequence; thus, coding redundancy is introduced in the channel coding, and a good error-correction capability is obtained. By merging the mapping part of the L-PPM with the mapping part of the space-time trellis code, coding and modulation can be combined; that is, coding and diversity gains can be provided simultaneously.

As shown in Fig. 7.6, the space-time trellis-code 4-PPM encoder has two transmit antennas. The input message flow is represented as

$$c = (c_0, c_1, \ldots c_t). \tag{7.10}$$

m = 2 when L = 4. c_t is a set of $m = \log_2 L$ information bits at time t, as shown in Eq. (7.11):

$$c_t = (c_t^1, c_t^2). \tag{7.11}$$

The encoder maps sequence c_t to a 4-PPM modulated signal sequence:

$$x = (x_0, x_1 \cdots x_t). \tag{7.12}$$

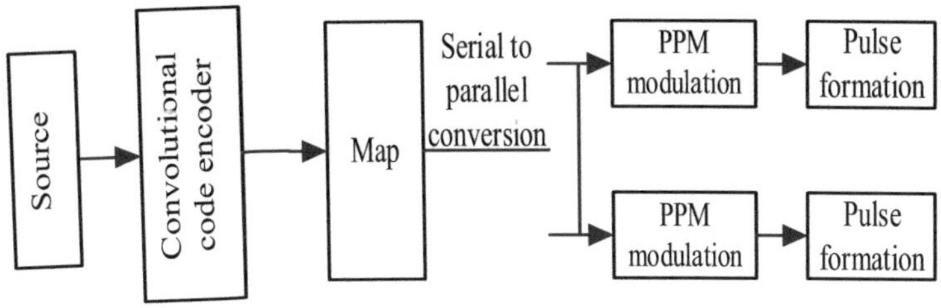

Fig. 7.6 System block diagram of the encoding part

Fig. 7.7 4-PPM encoder with two transmit antennas and two memory slots

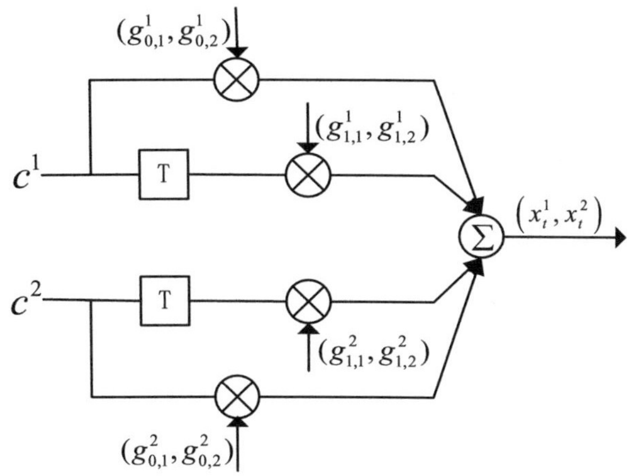

In Eq. (7.12), $x_t = (x_t^1, x_t^2)$ is transmitted through two transmitting antennas.

Figure 7.7 shows the structure of an STTC encoder with two-antenna transmission, where the encoder length is v = 2, and the total number of trellis-encoder states is $2^v = 4$.

The two input information bit streams of input sequences c^1 and c^2 are respectively multiplied by the following coefficients:

$$g^1 = \left[\left(g_{0,1}^1, g_{0,2}^1\right), \left(g_{1,1}^1, g_{1,2}^1\right)\right] \tag{7.13}$$

$$g^2 = \left[\left(g_{0,1}^2, g_{0,2}^2\right), \left(g_{1,1}^2, g_{1,2}^2\right)\right]. \tag{7.14}$$

In this manner, the mathematical expression of the encoder is

$$x_t^i = \sum_{k=1}^{2}\sum_{j=0}^{1} g_{j,i}^k c_{2-j}^k \text{ mod } 4\,(i = 1, 2). \tag{7.15}$$

Different coding coefficients are used in systems with three transmit antennas:

$$g^1 = \left[\left(g_{0,1}^1, g_{0,2}^1, g_{0,3}^1\right), \left(g_{1,1}^1, g_{1,2}^1, g_{1,3}^1\right)\right] \tag{7.16}$$

$$g^2 = \left[\left(g_{0,1}^2, g_{0,2}^2, g_{0,3}^2\right), \left(g_{1,1}^2, g_{1,2}^2, g_{1,2}^2\right)\right]. \tag{7.17}$$

At this time, the mathematical expression of the encoder is

$$x_t^i = \sum_{k=1}^{2}\sum_{j=0}^{2} g_{j,i}^k c_{2-j}^k \bmod 4\,(i = 1, 2, 3). \tag{7.18}$$

In the design process of Eq. (7.18), 0 is output when (00) is input, 1 is output when (01) is input, 2 is output when (10) is input, and 3 is output when (11) is input.

Combined with the x_t^i obtained in the above equation in the previous 4 - PPM encoding process, the calculation result of the mapping can be replaced with the time-slot position of the 4 - PPM modulated pulse. Because the encoder input may only receive [(00) (01) (10) (11)], it can be seen that the correspondence is one-to-one, which satisfies the uniqueness requirement of modulation. The following mapping assumptions are used during simulation:

When $l = 0$, a transmitted 1000 indicates that the signal is in the first time slot.

When $l = 1$, a transmitted 0100 indicates that the signal is in the second time slot.

When $l = 2$, a transmitted 0010 indicates that the signal is in the third time slot.

When $l = 3$, a transmitted 0001 indicates that the signal is in the fourth time slot.

For the 4-PPM STTC decoder, the Viterbi algorithm, based on maximum likelihood (ML), can be used and averaged, and then the quadratic Euclidean distance with the corresponding time slot of the theoretically received frame can be calculated, and expressed as

$$\sum_{j=1}^{M}\sum_{l=1}^{4}\left|r_{t,x}^j - \sum_{i=1}^{N} h_{j,i}^t x_{t,x}^i\right|^2. \tag{7.19}$$

Equation (7.19) represents the square of the Euclidean distance between a set of modulation symbols, sent by the transmitting antenna at time t after passing through an atmospheric fading channel and arriving at the receiving end, and the theoretical received symbols, which is called the branch metric (BM). M and N represent an $M \times N$ system; $x_{t,x}^i$ represents the average value of N samples from the xth time slot of the information frame received at time t; $h_{j,i}^t$ represents the channel gain at time t; and $r_{t,x}^j$ represents each time in the theoretically received information frame.

The calculated branch-metric result is used by the Viterbi algorithm to select a set of sequences with the smallest path metric as the decoding sequence. The Viterbi algorithm for the space-time trellis code is different from that of other decoders. It uses the Euclidean distance between the signal points as the decoding sequence. A

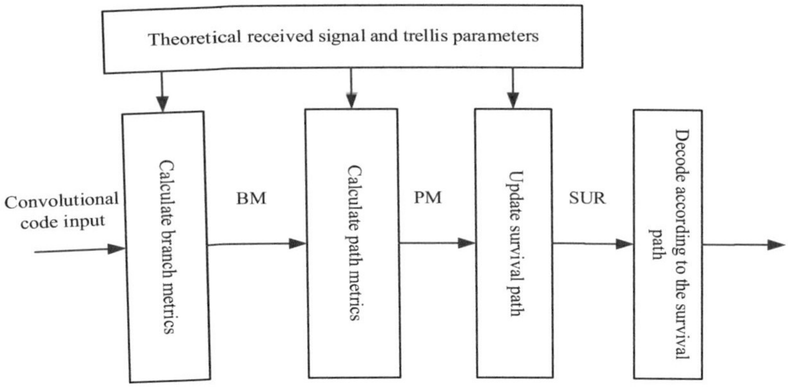

Fig. 7.8 Viterbi decoding-process block diagram

parallel-transfer path in trellis-coded modulation must be considered when designing a Viterbi decoder. The decoder can be designed based on the following four aspects: branch metrics, path metrics, survival path, and decoding. A block diagram of Viterbi decoding is shown in Fig. 7.8, and a flowchart is shown in Fig. 7.9.

The first step is to calculate the branch metric (BM); that is, the Euclidean distance between the received signal R at time n and the signal subset corresponding to the jth possible state transition to the ith state. The parallel-transition path metric can be expressed as

$$BM_n = \min_{i \in \text{subset}} \left\{ \mathrm{d}_{\mathrm{f}}^2 [R_n(t), S_i(t)] \right\}. \tag{7.20}$$

BM_n represents the Euclidean distance between the received signal and two signal points at time n, and S_i represents different signal points.

The second step is to update the state path metrics (PM). Over time, each small branch is connected to a path, and the path metric is equal to the sum of the metrics of each branch. The equivalent of updating the path metric of the ith state is to calculate the sum of the first two states, the PM at moment $\mathrm{n} - 1$, and the branch metric corresponding to this transition, and select the smallest one as the path metric at this moment.

The third step is to update the surviving path (SUR) of each state. The path traversed on the trellis map required for Viterbi decoding is recorded. The survivor path is actually a sequence consisting of the output convolutional codes of L-trellis convolutional encoders, which can be expressed as

$$SUR_n(i) = \{C_n(i), C_{n-1}(i), \cdots C_{n-L+1}(i)\}. \tag{7.21}$$

$C_n(i)$ represents the convolutional code corresponding to the transition from a certain state to state i at time n.

Fig. 7.9 Viterbi coding flow chart

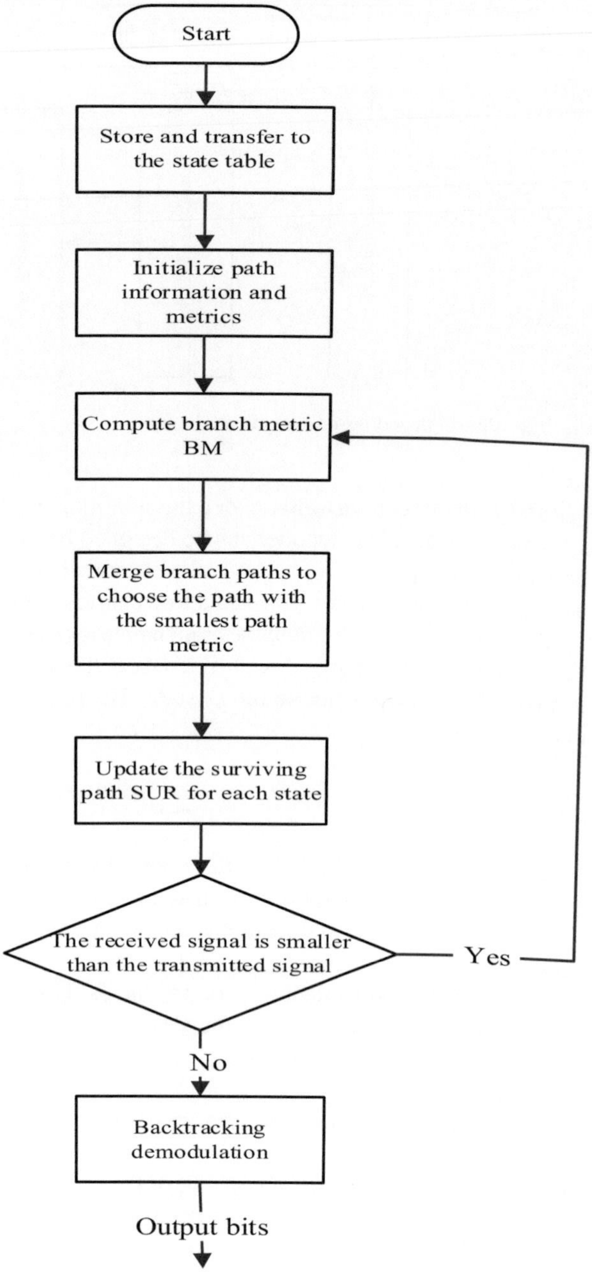

The fourth step is to output a decision. The output of the decoding sequence is generated, based on the signals stored by the survivor paths.

7.3.3 *Simulation and Result Analysis*

The system was simulated and verified using the Monte Carlo method, and the variation in bit-error rate with the signal-to-noise ratio is analyzed when two transmit antennas are used with different numbers of receive antennas. The simulation conditions were as follows:

(1) The channel characteristics and noise at the receiving end can be accurately estimated;
(2) The photoelectric-conversion efficiency is set to 0.8, and the scintillation index S.I. is between 0.7 and 0.9 (S.I. is usually between 0.4 and 1);
(3) The transmitted signals of one antenna all arrive at the receiving antenna synchronously and simultaneously;
(4) 4-PPM is used for system modulation;
(5) The number of antennas is 2×2, 2×3, and 2×4 when analyzing the bit-error rate;
(6) $g_1 = [0\ 2;\ 2\ 0]$; $g_2 = [0\ 1;\ 1\ 0]$.

In Fig. 7.10, the BER curves of two transmitting antennas, different receiving antennas, and different scintillation indices are simulated. As shown in Fig. 7.10, with an increase in the number of receiving antennas, the bit-error rate is significantly improved. In addition, the 4-PPM-STTC system can achieve a better bit-error rate than 4-PPM modulation with the same noise added. Combined with Fig. 7.10, it can be seen that the bit-error rate of the wireless optical communication system using the space-time trellis code is significantly improved under different scintillation indices. For example, when S.I. = 0.9 and SNR = 20 dB, the bit-error rate of the 2×4 system is an order of magnitude higher than that of the 2×3 system.

In a wireless optical STTC system with three transmitting antennas, the influence of the signal-to-noise ratio on the bit-error rate for different numbers of receiving antennas was analyzed. The simulation conditions were as follows:

(1) The photoelectric-conversion efficiency is set to 0.8, and the scintillation index S.I. is set to 0.7;
(2) The transmitted signals of each antenna arrive at the receiving antenna synchronously and simultaneously;
(3) The system modulation adopts 4 - PPM pulse-position modulation;
(4) The number of antennas when analyzing the bit-error rate is 3×2, 3×3, and 3×4;
(5) $g_1 = [0\ 2\ 2;\ 1\ 2\ 3]$; $g_2 = [2\ 3\ 3;\ 2\ 0\ 2]$.

It can be seen from Fig. 7.11a that the three-antenna system has the same advantages as the two-antenna system. A direct comparison is made between the two and

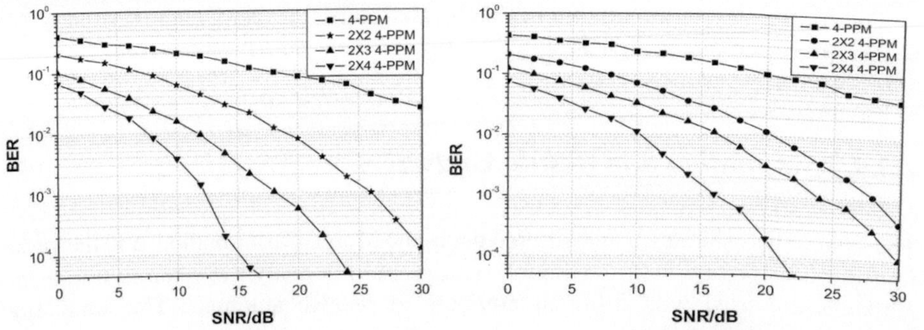

(a) S.I. = 0.7 (b) S.I. = 0.9

Fig. 7.10 Relationship between the bit-error rate and signal-to-noise ratio for different scintillation indices

three transmit-antenna systems in Fig. 7.11b. It can be seen that when the number of receiving antennas is equal to three, the bit-error rate of the 3 × 3 system is improved by two orders of magnitude, compared with the 2 × 3 system at SNR = 20 dB, indicating that the 3 × 3 system has achieved a better bit-error rate. This indicates that when the number of receive antennas is fixed, the bit-error rate decreases with an increase in the number of transmit antennas.

The results show that after using MIMO technology, the bit-error rate of the system can be effectively reduced by increasing the number of lasers and photodetectors. After using STTC in a wireless optical communication system, the bit-error rate of the system is significantly improved, compared with the system without STTC. Finally, when the number of transceiver antennas is different, the bit-error rate also decreases with the increase in the number of transceiver antennas.

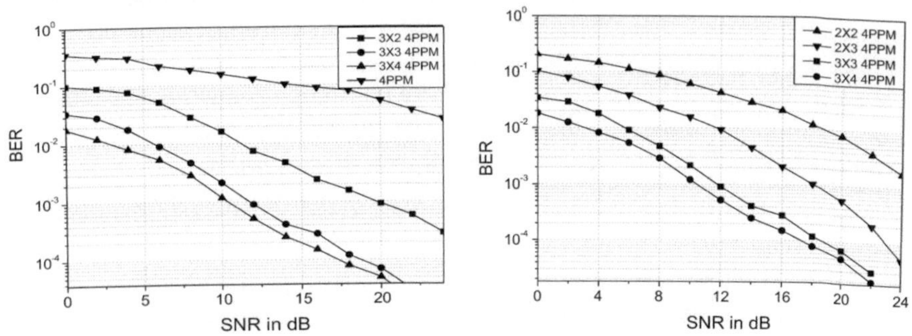

(a) Curve of the bit-error rate and signal-to-noise ratio of the transmitting antenna (b) Comparison of the bit-error rate between three- and two-antenna systems

Fig. 7.11 Bit-error rate for different combinations of transmit and receive antennas (S.I. = 0.7)

7.4 Summary

Space-time trellis coding is combined with the trellis-coding idea so that the modulated signal can obtain the maximum Euclidean free distance. Space-time trellis codes introduce a coding gain under the condition of a given diversity gain and transmission rate, and their performance is better than that of space-time block codes.

The codewords designed by space-time trellis codes can provide the maximum coding and diversity gains without losing bandwidth efficiency. The maximum diversity gain is equal to the number of transmit antennas. However, when the number of transmit antennas is fixed, the decoding complexity of space-time trellis codes increases exponentially with an increase in the diversity gain and transmission rate, which limits its application.

References

1. Tarokh V, Seshadri N, Calderbank AR (1998) Space-time codes for high data rate wireless communication performance criterion and code construction. IEEE Trans Inform Theory 44(2):744–765
2. Rassool BA, Allen B, Nakhai R, Roberts R, Sweeney P (2003) Error statistics of optimal and sub-optimal space-time trellis codes: concatenation requirements. IEEE Conf 2:1007–1011
3. Ohtsuki T (2004) Space-time trellis coding for UWB-IR. IEEE Conf 2:1054–1058
4. Tyagi A, Bose R (2008) Pulse position modulated space-time trellis coding for ultra-wideband impulse radio multiple-input multiple-output communication systems. IET J 2:366–371
5. Kumaratharan N, Jayapriya S, Dananjayan P (2009) STTC based site diversity technique for MC-CDMA system. IEEE Conf 35:1–5
6. Kim YJ, Li X, Park NY (2007) Generalized suboptimum decoding for STTC-OFDM systems in MIMO Rayleigh fading channels. IEEE Conf ISCIT'07 42:194–197
7. Fadera GND, Ignacio LRT, Nastor MBR, Urriza PIM, Marciano JJS (2007) FPGA implementation of space-time encoders. In: IEEE conferences on intelligent and advanced systems. ICIAS 2007, pp 368–371
8. Tyagi A, Bose R (2007) A new distance notion for PPAM space–time trellis codes for UWB MIMO communications. IEEE J 55(7):1279–1282
9. Tyagi A, Bose R (2005) M-PAM space time trellis codes for multi-antenna ultra-wide band communications. In: IEEE conferences ICPWC 2005, pp 308–311
10. Bansal LK, Trivedi A (2008) Performance study of space-time trellis coded MC-CDMA system employing different detection techniques. In: IEEE fourth international conference on wireless communication and sensor networks, 2008, vol 23. WCSN, pp 137–140
11. Ying L, Xinmei W (2003) The solution of trellis space-time code in WCDMA. Chin J Electron 31(5):772–775
12. Junhui D, Jun W, Shouyin L (2016) A new superquasi-orthogonal space-time trellis code and its performance analysis. J Cent China Norm Univ (Nat Sci Ed) 41(3):358–363
13. Ying L, Xudong G, Xinmei W (2007) Design of generalized recursive trellis space-time code suitable for serial cascade structure. Chin J Electron Inf 29(8):1872–1876
14. Zhuo L, Jian-Dong L, Wei-Ying L (2006) Jointly iterative maximum-likelihood estimation and decoding for space-time trellis coding 28(8):1350–1353
15. Yang L, Giannakis GB (2004) Analog space-time coding for multi-antenna ultra-wideband transmissions. IEEE Trans Commun 52(3):507–517

16. Ghassemlooy Z, Rajbhandari S (2007) Performance of diffused indoor optical wireless links employing neural and adaptive linear equalizers. In: IEEE information, communications & signal processing, 2007 6th international conference on (ICICS), pp 1–6
17. Wilson SG, Brandt-Pearce M, Cao Q (2005) Free-space optical MIMO transmission with Q-ary PPM. IEEE Trans Wirel Commun 6(53):1402–1412
18. Winters JH (1987) On the capacity of radio communication systems with diversity in a Rayleigh fading environment. IEEE J Sel Areas Commun 5(5):868–875

Chapter 8
Differential Space-Time Coding

In multiple-input multiple-output (MIMO) systems, because signals from different transmission antennas overlap each other at each receiving antenna, channel estimation is no longer feasible, owing to computational complexity when the fading is fast or the number of transmission antennas is large. Differential space-time modulation avoids complex MIMO channel estimation and achieves a higher diversity gain.

Inspired by differential modulation technology in conventional single-send and single-receive wireless communication systems, researchers have attempted to extend differential modulation methods to MIMO systems. Differential space-time codes have attracted considerable attention because they do not require channel estimation. Based on group codes in the radio-frequency (RF) domain, this chapter analyzes differential space-time codes suitable for intensity modulation/direct detection (IM/DD) in wireless optical communications.

8.1 Research Status of Differential Space-Time Codes

The concept of differential space-time codes comes from differential phase-shift keying (DPSK) in radio-frequency communication. Assuming that the channel response is approximately constant from one symbol period to the next, differential encoding [1] encodes the transmission information as the phase difference between two consecutive symbols. Reference symbols are sent first, followed by differential phase-shift symbols. The receiver decodes by comparing the phase difference between the current and previous symbols.

In 2000, Tarokh and Jafarkhani proposed a two-antenna differential space-time block code [2] based on an orthogonal design, which is an improvement of the Alamouti orthogonal space-time block code [3]. However, when the number of transmitting antennas is more than two, the space-time block code cannot achieve a full rate. Compared with the traditional two-antenna space-time block code (Alamouti

X. Ke, *Space-Time Coding Theory in Optical Wireless MIMO Systems*,
Optical Wireless Communication Theory and Technology,
https://doi.org/10.1007/978-981-97-6449-5_8

encoding), this scheme has a 3-dB performance loss [3], and quadrature phase-shift keying (QPSK) modulation still has signal-expansion problems. Tarokh and Jafarkhani generalized the results in the literature [2] such that they could be applied to multiple transmission antennas (differential space-time coding for multiple transmission antennas) [4].

In 2000, Hochwald and Hughes proposed differential unitary space-time modulation [5] and differential space-time modulation [6]. Differential unitary space-time modulation introduces a type of diagonal matrix in which the symbols on the diagonal line are PSK-modulated signals (that is, only one antenna works at any time). A differential space-time modulation scheme for multiple transmit antennas, based on group codes, was proposed in [6].

In 2002, Hassibi and Hochwald proposed the Cayley differential unitary space-time code [7]. This scheme uses the Cayley transformation to map a higher-order nonlinear Stiefel-manifold unitary matrix to a skew-Hermitian matrix in linear space, which can effectively encode and decode any transmit antenna and decoding rate.

In 2003, Shao proposed a new differential space-time block code [8] that uses a recursive structure and a Viterbi decoding algorithm at the receiver. Because the redundant information introduced by the recursive structure provides coding gain, it improves the error performance by 1–1.5 dB [2] compared with the schemes proposed by Tarokh and Jafarkhani. In the same year, Hwang proposed a differential space-time block code for irregular constellations [9] that used quadrature amplitude modulation (QAM), which is different from the traditional differential space-time block codes using quadrature phase-shift keying (QPSK).

Chen et al. proposed an amicable-orthogonal-design-based differential space-time block code (ADSTBC) [10, 11] based on a mutual orthogonal design. Prior to this, most differential space-time codes used PSK modulation, which limited the improvement in spectral efficiency. ADSTBC has no restrictions on signal constellation maps; thus, efficient modulation techniques can be used, such as multi-amplitude and multi-phase signal constellation maps (such as QAM and amplitude and phase-shift keying (APSK)). Other space-time codes can further improve the performance of communication systems, such as cascaded differential space-time codes with space-time trellis codes or space-time turbo codes, etc. [12–15].

8.2 Differential Space-Time Codes Based on Cluster Codes

Suppose that a wireless communication system has t transmitting antennas and r receiving antennas. At the transmitting end, the source is sent to t parallel encoders after a serial–parallel conversion and the output data of each encoder are transmitted by the corresponding antenna. The encoded signals are mapped to a normalized constellation, modulated at time T and sent to a wireless channel. $X_{jk} \in X$ represents the coded constellation points, $j = 1, \ldots, t$ represents the corresponding transmitting antenna, and $k = 1, \ldots, n$ is the corresponding launch time.

The signal received by each of the r receiving antennas is the superposition of t fading-transmission signals and noise. Assuming that the delay spread of multiple paths is small, the receiver can receive signals from t transmit antennas simultaneously. Similarly, assuming that each symbol period is T, the continuous n-symbol time nT is much less than the channel-coherence time, and the fading coefficient of the channel is approximately constant. The received signal of each receiving antenna is synchronously demodulated and sampled to obtain r decision statistical values, as follows:

$$y_{ik} = \sqrt{p_t}\sum_{j=1}^{t} h_{ij}x_{jk} + n_{ik}, i = 1, \ldots r; k = 1, \ldots n, \tag{8.1}$$

where h_{ij} is the complex fading-path gain from transmitting antenna j to receiving antenna i, n_{ik} is a complex Gaussian random variable, and ρ_t is the signal-to-noise ratio of each receiving antenna. It is assumed that the path from each transmitting antenna to the receiving antenna is independent of the others, so the path gain h_{ij} and the noise variable n_{ik} are independently and identically distributed. The coding matrix X is defined as

$$\mathrm{X}=\begin{bmatrix} x_{11} & x_{12} & \cdots & x_{1n} \\ x_{21} & x_{22} & \cdots & x_{2n} \\ \vdots & \cdots & \ddots & \vdots \\ x_{t1} & x_{t2} & \cdots & x_{tn} \end{bmatrix}. \tag{8.2}$$

The received signal is written as a matrix in the form of

$$Y = \sqrt{p_t}HX + N, \tag{8.3}$$

where $Y = \{y_{ik}\}$ is an $r \times n$ receive matrix, H is an $r \times t$ path-decay matrix, and N is an $r \times n$ noise matrix.

8.2.1 Coding of Differential Space-Time Codes Based on Group Codes

In 2000, Hughes [6] proposed a differential modulation method for multitransmitter antennas based on group codes. For a system with t transmitting antennas and a modulated constellation A, for any $n \geq t$, let g be the set of unitary matrices of an $n \times n$ group; for any $G \in g$, we have

$$G^H G = GG^H = I. \tag{8.4}$$

Suppose there exists a $t \times n$ matrix D, such that any of the unitary matrices G in group DG produces a $t \times n$ matrix, and the non-zero entries of this matrix are the elements of the signal constellation set Λ. $(\forall G \in g)$, we have

$$DG \in \wedge^{t \times n}. \tag{8.5}$$

Then, the set of matrices

$$Dg \triangleq \{DG : G \in g\}. \tag{8.6}$$

Then, it constitutes a space-time group code. Each matrix in the set specifies a null-time transmission codeword of length n on signal constellation Λ for t transmitting antennas. Signal constellation Λ is modulated using $M - PSK$, where $M = 2^m$. In the kth coded grouping, $\log_2 M$ bits of information are mapped into group code g and matrix G_1 is selected, where $G_k \in g$. To initialize the differential transmission, we initially send $X_0 = D$. The differential coding guidelines are as follows [6]:

$$X_k = X_{k-1} G_k, k = 1, \ldots, K. \tag{8.7}$$

Because $G_k \in g$, and g is a unitary matrix of an $n \times n$ group, group-code g ensures that when $X_{k-1} \in \Lambda^{t \times n}$, $X_k \in \Lambda^{t \times n}$, $X_k \in \Lambda^{t \times n}$; for example, if transmission antenna $t = 2$ is used to send two modulated symbols $x_1, x_2 \in \Lambda$, the encoding matrix is

$$X = \begin{bmatrix} x_1 & -x_2^* \\ x_2 & x_1^* \end{bmatrix}. \tag{8.8}$$

The process of sending signals for this code matrix is as follows: At some point t_0, antenna 1 sends symbol x_1 and antenna 2 sends symbol x_2; if the character cycle is T, at the next moment, $t_0 + T$, antenna 1 sends symbol $-x_2^*$ and antenna 2 sends symbol x_1^*. If the BPSK constellation mapping $\Lambda = \{1, -1\}$ is selected, the group code g is

$$\text{g}=\left\{\begin{bmatrix} 1 & 0 \\ 0 & 1 \end{bmatrix}, \begin{bmatrix} 0 & 1 \\ -1 & 0 \end{bmatrix}\right\}.$$

The initially sent reference matrix is $D : D = \begin{bmatrix} 1 & -1 \\ 1 & 1 \end{bmatrix}$.

8.2.2 Decoding Differential Space-Time Codes Based on Group Codes

When the number of transmitting antennas $t = n$, in the absence of channel-state information (CSI), the maximum-likelihood detection method is based on the full sequence of received signals $Y = [Y_0 : \ldots : Y_k]$; according to Eq. (8.3), it is known that

$$Y_k = \sqrt{p_t} H X_k + N_k, k = 0, 1, \ldots, K, \tag{8.9}$$

where ρ_t represents the signal-to-noise ratio of each receiving antenna, H is the channel-fading coefficient matrix, X_k is the kth encoding matrix, and N_k is the noise matrix. Only the last two receive blocks are required to estimate G_k:

$$\overline{Y}_k \triangleq \left[Y_{k-1} : Y_k\right] \tag{8.10}$$

When $G_k = G$, the following encoding matrix affects $\overline{Y}_k$:

$$\overline{C}_G \triangleq \left[X_{k-1} : X_{k-1} G\right] \tag{8.11}$$

The initial launch matrix D satisfies $DD^H = nI$; when $n = t$, $DD^H = tI$. Thus, we can conclude that for all $X \in Dg$, there is $XX^H = X^H X = tI$. For $X \in Dg$, there is a $t \times 2t$ matrix $\left\{\overline{C}_G : G \in g\right\}$ satisfying $\overline{C}_G \overline{C}_G^H = 2tI$; therefore, it can be considered a null-time transmission codeword of length $\overline{n} = 2t$. If the transmitted x_{k-1} is known in advance at the receiver, the best decoding is based on the product of the matrices of block codes of length $\overline{n} = 2t$.

$$\overline{C}_G^H \overline{C}_G = \begin{bmatrix} tI & tG \\ tG^H & tI \end{bmatrix}. \tag{8.12}$$

In Fig. 8.1, the Z^{-1} in the figure indicates a block delay. To obtain the current information, it is necessary to know the currently received information block and the previous information block. The differential decoding equation is [6]

$$\begin{aligned}
\hat{G} &= \arg\max_{G \in g} Tr\left\{\overline{Y}_k \overline{G}_G^H \overline{G}_G \overline{Y}_k^H\right\} \\
&= \arg\max_{G \in g} Tr\left\{\left[Y_{k-1} : Y_k\right] \begin{bmatrix} tI & tG \\ tG^H & tI \end{bmatrix} \left[Y_{k-1} : Y_k\right]^H\right\} \\
&= \arg\max_{G \in g} \mathrm{Re} Tr\left\{Y_{k-1} G Y_k^H\right\} \\
&= \arg\max_{G \in g} \mathrm{Re} Tr\left\{G Y_k^H Y_{k-1}\right\}
\end{aligned} \tag{8.13}$$

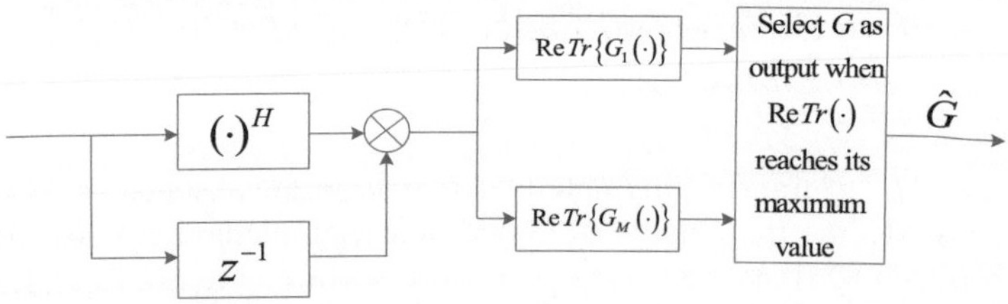

Fig. 8.1 Diagram of a differential receiver

where arg max$f$ (•) denotes taking a certain subset in the defined domain that maximizes$f$ (•), $(\bullet)^H$ denotes the conjugate transpose of a matrix, G is the unitary matrix (see Eq. 8.4), g is the set of a group of unitary $n \times n$ matrices, Y_k represents the received signal matrix in the kth block time, and Y_{k-1} represents the matrix of received signals in the $K - 1^{st}$ block time. ReTv(•) represents the real part of a matrix trace, and the last step is when matrices A, B are square matrices of order n and $Tr(AB) = Tr(BA)$.

8.3 Differential Space-Time Codes in Optical Communications

Because differential space-time coding in RF communications generally uses PSK or QAM modulation, the unitary matrix in the transmit-code matrix of Eq. (8.8) and the unitary matrix used to map information bits to group codes exist in negative and complex forms. In IM/DD free-space optical (FSO) communication, only positive real signals are used; i.e., the channel transmits optical signals in only two states, "*on*" and "*off*." This method needs to be improved to find a differential space-time code (DSTC) suitable for IM/DD optical communications.

8.3.1 Coding of Differential Space-time Codes Based on 2-PPM Modulation

In RF communication, space-time block codes can be divided into space-time block codes with real signals and space-time block codes with complex signals, according to the signal constellation [16]. When the numbers of transmitting antennas are two and four, the transmitting-code matrix forms of their real signal space-time block codes are

$$C^{(2)} = \begin{bmatrix} x_1 & -x_2 \\ x_2 & x_1 \end{bmatrix} \tag{8.14}$$

$$C^{(4)} = \begin{bmatrix} x_1 & -x_2 & -x_3 & -x_4 \\ x_2 & x_1 & x_4 & -x_3 \\ x_3 & -x_4 & x_1 & x_2 \\ x_4 & x_3 & -x_2 & x_1 \end{bmatrix}. \tag{8.15}$$

The upper corners of the matrices in Eqs. (8.14) and (8.15) are used to distinguish the systems with different numbers of antennas at the transmitting end.

From Eqs. (8.14) and (8.15), it can be seen that RF communications have a negative number of elements in the transmitting-code matrix because PSK or QAM modulation is used, while in IM/DD optical communication, the transmitted optical signal has only two states, "*on*" and "*off*." There is no corresponding form for representing negative numbers in the symbols.

Therefore, we adopt binary pulse-position modulation (BPPM), based on the orthogonal space-time coding scheme proposed by Chadi Abou-Rjeily [17]. When the numbers of transmitting antennas are two and four, the transmitting-code matrices in the null-time coding scheme of the orthogonal design proposed by Chadi Abou-Rjeily [17] respectively take the form of

$$C_k^{(2)} = \begin{bmatrix} x_{k,1} & x_{k,2}\Omega \\ x_{k,2} & x_{k,1} \end{bmatrix} \tag{8.16}$$

$$C_k^{(4)} = \begin{bmatrix} x_{k,1} & x_{k,2}\Omega & x_{k,3}\Omega & x_{k,4}\Omega \\ x_{k,2} & x_{k,1} & x_{k,4} & x_{k,3}\Omega \\ x_{k,3} & x_{k,4}\Omega & x_{k,1} & x_{k,2} \\ x_{k,4} & x_{k,3} & x_{k,2}\Omega & x_{k,1} \end{bmatrix}, \tag{8.17}$$

where $C_k^{(m)}$ denotes the code matrix sent at the kth time when the number of transmitting antennas is m. $x_{k,1}, x_{k,2}, \ldots x_{k,m}$ is the modulated signal, and the period occupied by this modulated signal is T. Assuming that BPPM is used, the set of constellation mappings $A = \{[\,1\ 0\,], [\,0\ 1\,]\}$, where a scalar 0 indicates no light pulse at the corresponding position and a scalar 1 indicates a light pulse at the corresponding position. The corresponding bit-to-constellation mapping relationships are listed in Table 8.1.

Ω is a circular 2×2 substitution matrix:

$$\Omega = \begin{bmatrix} 0 & 1 \\ 1 & 0 \end{bmatrix}. \tag{8.18}$$

Table 8.1 Information bits to 2-PPM mapping relationships

Binary information flow ($Dg \triangleq \{DG : G \in \mathbf{g}\}$)	2-PPM constellation mapping (Λ)
0	1 0
1	0 1

It is easy to verify that, when $x_{k,1}, x_{k,2}, \ldots x_{k,m} \in A$, $x_{k,i}\Omega \in A$ $i = 1, \ldots \in, m$, and A is the set of 2-PPM constellation mappings. For example, if $X_{k,i} = \begin{bmatrix} 1 & 0 \end{bmatrix}$, $x_{k,i}\Omega = \begin{bmatrix} 0 & 1 \end{bmatrix}$; $x_2\Omega$ means that the "*on*" and "*off*" states of the signal waveform are reversed, allowing transmission in FSO communications with IM/DD.

In Eqs. (8.16) and (8.17), the rows of the matrix represent the corresponding transmit antennas, and the columns represent the signals at the corresponding transmitting moments. Taking a transmitting-code matrix with four transmitting antennas as an example, the operation process is shown in Fig. 8.2.

The send-code matrix in Fig. 8.2 (Eq. 8.17) operates by sending signals $x_{k,1}, x_{k,2}, x_{k,3}, x_{k,4}$ from antennas 1, 2, 3, and 4 at time 1; signals $x_{k,2}\Omega, x_{k,1}, x_{k,4}\Omega x_{k,3}$ are sent from antennas 1, 2, 3, and 4 at time 2; signals $x_{k,3}\Omega x_{k,4}, x_{k,1,x_{k,2}}\Omega$ are sent from antennas 1, 2, 3, and 4 at time 3; and signals $x_{k,4}\Omega, x_{k,3}\Omega, x_{k,2}, x_{k,1}$ are sent from antennas 1, 2, 3, and 4 at time 4. Combining the orthogonal space-time coding scheme [17] proposed by Chadi Abou-Rjeily and the differential concept from the literature [18], a differential space-time coding scheme for laser communication based on 2-PPM modulation is obtained (as shown in Fig. 8.3).

The differential encoding process is as follows. The transmitter initially transmits a reference modulation matrix, $C_0^{(m)}$, which is similar to the process shown in Fig. 8.2 (the rows of the matrix represent the signals sent by different antennas, and the columns of the matrices represent the signals at the corresponding sending time).

Fig. 8.2 Send-code matrix operation (four transmit lasers as an example)

Transmitting antenna	*Time* 1	2	3	4
1	$x_{k,1}$	$x_{k,2}\Omega$	$x_{k,3}\Omega$	$x_{k,4}\Omega$
2	$x_{k,2}$	$x_{k,1}$	$x_{k,4}$	$x_{k,3}\Omega$
3	$x_{k,3}$	$x_{k,4}\Omega$	$x_{k,1}$	$x_{k,2}$
4	$x_{k,4}$	$x_{k,3}$	$x_{k,2}\Omega$	$x_{k,1}$

Transmitting antenna	Time 1	2	3	4
1	$x_{k,1}$	$x_{k,2}\Omega$	$x_{k,3}\Omega$	$x_{k,4}\Omega$
2	$x_{k,2}$	$x_{k,1}$	$x_{k,4}$	$x_{k,3}\Omega$
3	$x_{k,3}$	$x_{k,4}\Omega$	$x_{k,1}$	$x_{k,2}$
4	$x_{k,4}$	$x_{k,3}$	$x_{k,2}\Omega$	$x_{k,1}$

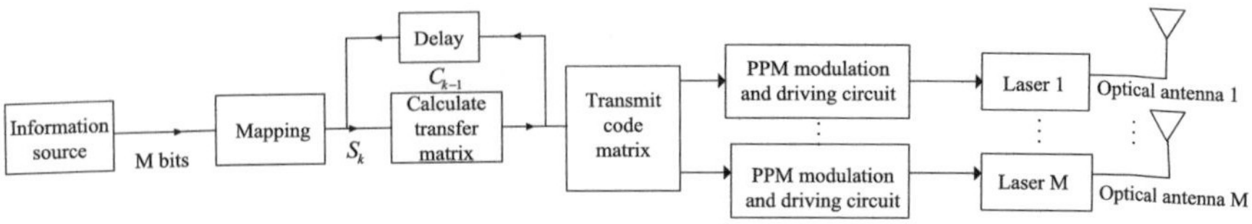

Fig. 8.3 Block diagram for differential space-time coding in optical communications

Element $x_{o,i}$ (where $i = 1$) in the matrix is the reference-modulation symbol, and this m transfer does not carry any data information. The transmitter encodes the data sequence in differential mode and sends it as follows.

Suppose that the $K - 1^{st}$ send-code matrix $C_{k-1}^{(m)}$ has sent the modulation signal through m transmission times. At the next moment, m information bits arrive at the encoder, which is used to construct a mapping matrix $S_k^{(m)}$. Matrix $S_k^{(m)}$ is combined with reference matrix C_{k-1}^{m} sent in the previous moment and calculates the code matrix $C_k^{(m)}$ sent at the next moment, based on the differential encoding criteria in Eqs. (8.19) and (8.20). Each sent matrix (i.e., the delay unit in Fig. 8.3) is stored as the previous reference matrix for the next sent-code matrix. This process is repeated until the end of the data frame:

$$\overline{C}_k = C_{k-1}^{(m)} S_k^{(m)} \tag{8.19}$$

$$C_k^{(m)} = \mathrm{sgn}\left(\overline{C}_k - \left[\min\left(\overline{C}_k\right) \otimes u_1\right]\right). \tag{8.20}$$

When m antennas are used at the transmitter end of the system, $C_k^{(m)}$ represents the code matrix sent at the kth moment, $S_k^{(m)}$ represents the mapping matrix at the kth moment, $u_1 = \begin{pmatrix} 1 & 1 \end{pmatrix}$, and $\otimes$ represents the Kronecker product. sgn(X) represents a matrix giving the symbolic values of the elements of X; if an element is zero or positive, the symbolic value is equal to 0 or 1, respectively; if an element is negative, the symbolic value is -1. I_2 indicates that the matrix gives the smaller of two consecutive elements in a row of matrix X.

Because $\min\left(\overline{C}_k\right) \otimes u$ operates on a matrix of the same size as matrix $\overline{C}_k$, each element in the matrix is always less than or equal to the element at its corresponding position in matrix$\overline{C}_k$; that is, the element in $\overline{C}_k - \left[\min\left(\overline{C}_k\right) \otimes u_1\right]$ is always greater than or equal to zero. It can be seen that only 0 and 1 are in the sent-code $C_k^{(m)}$, indicating the presence or absence of an optical pulse. The specific method for constructing $X_{BCOSTBC}$ bits of information for map matrix $S_k^{(m)}$ is as follows.

When $m = 2$, that is, when two binary information bits, b_1 and b_2, arrive at the encoder each time, the mapping matrices $S_k^{(2)}$, based on 2-PPM constellation mappings (see Table 8.1), are $b_1 \to s_1$, $b_2 \to s_2$, s_1, and s_2:

$$S_k^{(2)} = \begin{bmatrix} s_1 & s_2 \\ s_1\Omega & s_2\Omega \\ s_2\Omega & s_1 \\ s_2 & s_1\Omega \end{bmatrix}. \tag{8.21}$$

When $m = 4$, that is, when four binary bits $b_1, b_2, b_3, b4$ arrive at the encoder each time, the 2-PPM constellation is mapped to $b_1 \to s_1$, $b_2 \to s_2$, $b_3 \to s_3$, and $b_4 \to s_4$.s_1, s_2, s_3, s_4 are used to construct the mapping matrix $S_k^{(4)}$:

$$S_k^{(4)} = \begin{bmatrix} s_1 & s_2 & s_3 & s_4 \\ s_1\Omega & s_2\Omega & s_3\Omega & s_4\Omega \\ s_2\Omega & s_1 & s_4\Omega & s_3 \\ s_2 & s_1\Omega & s_4 & s_3\Omega \\ s_3\Omega & s_4 & s_1 & s_2\Omega \\ s_3 & s_4\Omega & s_1\Omega & s_2 \\ s_4\Omega & s_3\Omega & s_2 & s_1 \\ s_4 & s_3 & s_2\Omega & s_1\Omega \end{bmatrix}. \tag{8.22}$$

When a nonzero element in C_{k-1}^m belongs to 2-PPM constellation mapping set $A = \{[1\ 0], [0\ 1]\}$, the differential coding criteria in Eqs. (8.19) and (8.20) can guarantee that the nonzero element in the next transmit-code matrix $C_k^{(m)}$ also belongs to modulation constellation mapping set C. A zero-vector $[0\ 0]$ is included within a character cycle T in the sent-coding matrix $C_k^{(m)}$, which represents not sending symbols at a certain time. At least one antenna of code matri x $C_k^{(m)}$, which is sent each time, sends an optical pulse signal (i.e., at least one antenna sends information every time).

8.3.2 *Differential Space-Time Decoding Based on 2-PPM Modulation*

For a MIMO optical communication system with M transmitting antennas and N receiving antennas, it is assumed that the background light at the receiving end can be completely removed by a filter, considering only the thermal noise from the device. The signal sent by the ith transmitting antenna is represented by x_i and the photocurrent received by the ith detector is represented by r_j. h_{ji} represents the intensity decay factor from the ith emitted laser to the jth detector. Thus, they satisfy each other.

$$r_j = \frac{\eta I_s}{M}\left(\sum_{i=1}^{M} h_{ji}x_i + v_j\right), i = 1, \ldots, M, j = 1, \ldots N, \tag{8.23}$$

where η is the photoelectric-conversion efficiency, I_s is the maximum received light intensity without fading, v_j is Gaussian white noise with mean 0 and variance $N_0/2$ (N_0 is the average power-spectral density of the noise, which is a positive constant, and the noise is generated by the detector and its attached electronic circuit), and h_{ji} denotes the channel-attenuation coefficient from the ith transmitting antenna to the jth receiving antenna.

In a wireless optical atmospheric channel, h_{ji} approximately obeys the lognormal distribution model [19]. The optical-intensity undulation characteristics under weak

Table 8.2 Values of each parameter corresponding to different degrees of turbulence

Parameter	Degree of turbulence		
	Weak	Medium	Strong
σ_R^2	0.2	1.6	3.5
α	11.6	4.0	4.2
β	10.1	1.9	1.4
SI	0.19	0.9	1.12

turbulence satisfy the lognormal distribution, whereas the optical-intensity undulation characteristics under strong turbulence satisfy the gamma–gamma distribution model [20]. The probability-density function of the gamma–gamma distribution is

$$f(I) = \frac{2(\alpha\beta)^{\alpha+\beta/2}}{\Gamma(\alpha)\Gamma(\beta)} I^{(\alpha+\beta)/2-1} K_{\alpha-\beta}\left(2\sqrt{\alpha\beta I}\right), I > 0, \tag{8.24}$$

where I is the light intensity at the receiving end, $\Gamma()$ is the gamma function, $K_{\alpha-\beta}$ is a type-II modified Bessel function, and α and β are the two parameters characterizing the magnitude of the scintillation-index intensity, which are defined as follows:

$$\alpha = \left[\exp\left(\frac{0.49\sigma_R^2}{\left(1 + 1.11\sigma_R^{12/5}\right)^{7/6}}\right) - 1\right]^{-1} \tag{8.25}$$

$$\beta = \left[\exp\left(\frac{0.51\sigma_R^2}{\left(1 + 0.69\sigma_R^{12/5}\right)^{5/6}}\right) - 1\right]^{-1}, \tag{8.26}$$

where $\sigma_R^2 = 1.23C_n^2 k^{7/6} L^{11/6}$, k is the number of optical waves, λ is the wavelength, and L is the communication-link distance (in meters). C_n^2 represents the atmospheric refractive-index structure constant. The scintillation index is defined as m, and the relationship between SI and α and β is defined as $SI = 1/\alpha + 1/\beta + 1/(\alpha\beta)$ [20]. The gamma–gamma channel model can describe atmospheric undulations from weak to strong turbulence [21]: $\sigma_R^2 > 1$ indicates weak turbulence; $\sigma_R^2 \cong 1$ indicates moderate turbulence; and $\sigma_R^2 > 1$ indicates strong turbulence. The different turbulence levels are presented in Table 8.2.

8.3.2.1 Decoding of a Single Receive Aperture

From the encoding method, we can see that matrix $S_k^{(m)}$ represents the emitted data information, $S_k^{(m)}$ is decoded, and then the matrix is inverse mapped to obtain the information bits.

(1) Two transmit antennas and one receive antenna

When the number of transmitting antennas is two and the number of receiving antennas is one, the received signals of moments 1 and 2 of the receiving antennas in the first block of time can be expressed by Eq. (8.23) as

$$r_{k,1} = \frac{\eta I_s}{2}\left(h_{11}x_{k,1} + h_{12}x_{k,2}\right) + v_{k,1} \tag{8.27}$$

$$r_{k,2} = \frac{\eta I_s}{2}\left(h_{11}x_{k,2}\Omega + h_{12}x_{k,1}\right) + v_{k,2}. \tag{8.28}$$

We rewrite Eqs. (8.27) and (8.28) into matrix form as

$$R_k^{(2)} = \frac{\eta I_s}{2} H^{(2)} C_k^{(2)} + V_k^{(2)}, \tag{8.29}$$

where $R_k^{(2)} = \left[r_{k+1}\ r_{k+2}\right]$, $C_k^{(2)} H^{(2)} = [h_{11}\ h_{12}]$, $V_k^{(2)} = [v_{k,1}\ v_{k.2}]$, and see Eq. (8.16) for $SNR1$.

(2) Four transmit antennas and one receive antenna

For the 4 × 1 system, the received signal at moments 1, 2, 3, and 4 in the kth block time of the receiving antenna can be expressed by Eq. (8.23) as

$$r_{k,1} = \frac{\eta I_s}{4}\left(h_{11}x_{k,1} + h_{12}x_{k,2} + h_{13}x_{k,3} + h_{14}x_{k,4}\right) + v_{k,1} \tag{8.30}$$

$$r_{k,2} = \frac{\eta I_s}{4}\left(h_{11}x_{k,2}\Omega + h_{12}x_{k,1} + h_{13}x_{k,4}\Omega + h_{14}x_{k,3}\right) + v_{k,2} \tag{8.31}$$

$$r_{k,3} = \frac{\eta I_s}{4}\left(h_{11}x_{k,3}\Omega + h_{12}x_{k,4} + h_{13}x_{k,1} + h_{14}x_{k,2}\Omega\right) \tag{8.32}$$

$$r_{k,4} = \frac{\eta I_s}{4}\left(h_{11}x_{k,4}\Omega + h_{12}x_{k,3}\Omega + h_{13}x_{k,1} + h_{14}x_{k,1}\right) + v_{k,4}. \tag{8.33}$$

Rewriting Eqs. (8.30)–(8.33) into matrix form yields

$$R_k^{(4)} = \frac{\eta I_s}{4} H^{(4)} C_k^{(4)} + V_k^{(4)}, \tag{8.34}$$

where $R_k^{(4)} = \left[r_{k,1}\ r_{k,2}\ r_{k,3}\ r_{k,4}\right]$, $H_k^{(4)} = \left[h_{11}\ h_{12}\ h_{13}\ h_{14}\right]$, $V_k^{(4)} = \left[v_{k,1}\ v_{k,2}\ v_{k,3}\ v_{k,4}\right]$, and see Eq. (8.17) for $C_k^{(4)}$.

When the number of transmitting antennas is m, $R_k^{(m)}$ denotes the kth receive matrix, $H^{(m)}$ denotes the channel-fading coefficient matrix, and $V_k^{(m)}$ denotes the yth noise matrix.

For $S_k^{(m)}$ decoding, the current received information block $R_k^{(m)}$ and the previously received information block $R_{k-1}^{(m)}$ are decoded jointly, and the optical-intensity fading coefficient matrix $H^{(m)}$ is invariant within two adjacent information blocks (i.e., the channel is a slow flat fading channel). The matrix of messages received within the previous time block, $R_{k-1}^{(m)}$, is

$$R_{k-1}^{(m)} = \frac{\eta I_s}{m} H^{(m)} C_{k-1}^{(m)} + v_{k-1}^{(m)}. \tag{8.35}$$

Because $C_k^{(m)} = C_{k-1}^{(m)} S_k^{(m)}$ at encoding and assuming that the channel $H^{(m)}$ remains constant within two adjacent blocks, the received signal matrix $R_k^{(m)}$ in the kth time block is

$$R_k^{(m)} = \frac{\eta I_s}{m} H^{(m)} C_{k-1}^{(m)} S_k^{(m)} + V_k^{(m)}. \tag{8.36}$$

Substituting Eq. (8.35) into Eq. (8.36) yields

$$\begin{aligned} R_k^{(m)} &= \left(R_{k-1}^{(m)} - V_{k-1}^{(m)}\right) S_k^{(m)} + V_k^{(m)} \\ &= R_{k-1}^{(m)} S_k^{(m)} + \overline{V}_k^{(m)}, \end{aligned} \tag{8.37}$$

where $\overline{V}_k^{(m)} = -V_{k-1}^{(m)} S_k^{(m)} + V_k^{(m)}$, $\overline{V}_k^{(m)}$ is the noise term, and $S_k^{(m)}$ is given by Eqs. (8.21) and (8.22). The decoding equation based on the current received signal matrix $R_k^{(m)}$ and previously received signal matrix $R_{k-1}^{(m)}$ is

$$\{\hat{s}_1, \ldots \hat{s}_m\} = \arg \min_{s_1, \ldots s_m \in A} \left\| R_k^{(m)} - R_{k-1}^{(m)} S_k^{(m)} \right\|_F^2, A = \{[\,1\ 0\,], [\,0\ 1\,]\} \tag{8.38}$$

where $\|\bullet\|_F^2$ denotes the Frobenius parameterization (also called Euclidean parameterization) of finding a matrix with the following properties:

$$\|X\|_F^2 = trace\left(X^H X\right) = trace\left(XX^H\right), \tag{8.39}$$

where *trace* indicates the operation for taking the trace of a matrix of order N. The trace of the matrix is equal to the sum of all eigenvalues of a matrix of order N, which is the sum of the main diagonal elements of the matrix. Equation (8.38) then becomes

$$\begin{aligned} \{\hat{s}_1, \ldots \hat{s}_m\} &= \arg \min_{s_1, \ldots s_m \in A} \left\| R_k^{(m)} - R_{k-1}^{(m)} S_k^{(m)} \right\|_F^2 \\ &= \arg \min_{s_1, \ldots s_m \in A} trace\left\{ \left(R_k^{(m)} - R_{k-1}^{(m)} S_k^{(m)}\right) \left(R_k^{(m)} - R_{k-1}^{(m)} S_k^{(m)}\right)^H \right\} \end{aligned} \tag{8.40}$$

Here, s_i and $\overset{\frown}{\underset{i}{S}}$ belong to the set of BPPM constellation mappings; $R_{YY} = E[Y_k Y_k^H]$, $R_{HY} = E[H_k Y_k^H]$ is the element of mapping matrix $S_k^{(m)}$ and $\overset{\frown}{\underset{i}{s}}$ is the value of the final estimated binary information bits from the transmitter side to the BPPM constellation mapping. By expanding the product term in Eq. (8.40), because $S_k^{(m)}$ represents the emitted data, for different $s_1, \ldots s_m \in A$, the $S_k^{(m)}\left(S_k^{(m)}\right)^H$ term is a definite matrix; thus, eliminating $S_k^{(m)}\left(S_k^{(m)}\right)^H$ and the terms that are not related to $S_k^{(m)}$ gives

$$\{\hat{s}_1, \ldots \hat{s}_m\} = \arg \max_{s_1,\ldots s_m \in A} trace\left[R_{k-1}^{(m)} S_k^{(m)} \left(R_k^{(m)}\right)^H\right], A = \{[\,1\ 0\,], [\,0\ 1\,]\} \quad (8.41)$$

The estimated $\overset{\frown}{\underset{1}{s}}, \ldots \overset{\frown}{\underset{m}{s}}$ inverse mapping yields the sent message bits.

8.3.2.2 Decoding Multiple Receive Apertures

(1) Two transmit antennas and N receive antennas

For a $2 \times N$ system, signals $\tilde{r}_{k,1}$ and $\tilde{r}_{k,2}$ from the N receiving apertures at moments 1 and 2 of the kth block time at the receiving end are combined using equal gain (Here, equal-gain combination means that the signals received by each receiving antenna are summed by equal gain; here, the gain value is taken as a constant, 1):

$$\tilde{r}_{k,1} = \sum_{n=1}^{N} r_{k,1}^n = \frac{\eta I_s}{2}\left(\sum_{n=1}^{N} h_{n1} x_{k,1} + \sum_{n=1}^{N} h_{n2} x_{k,2}\right) + \sum_{n=1}^{N} v_{k,1}^n \quad (8.42)$$

$$\tilde{r}_{k,2} = \sum_{n=1}^{N} r_{k,2}^n = \frac{\eta I_s}{2}\left(\sum_{n=1}^{N} h_{n1} x_{k,2}\Omega + \sum_{n=1}^{N} h_{n2} x_{k,1}\right) + \sum_{n=1}^{N} v_{k,2}^n, \quad (8.43)$$

where $r_{k,1}^n$ denotes the received signal of the nth receive antenna at moment 1, $r_{k,2}^n$ denotes the received signal of the nth receive antenna at moment 2, and h_{n1}, h_{n2} denote the channel gain from transmit antennas 1 and 2 to the nth receive antenna, respectively. Rewriting Eqs. (8.42) and (8.43) into matrix form yields

$$\tilde{R}_k^{(2)} = \frac{\eta I_s}{2} \tilde{H}^{(2)} C_k^{(2)} + \tilde{V}_k^{(2)}, \quad (8.44)$$

where $\tilde{R}_k^{(2)} = \left[\,\tilde{r}_{k,1}\ \tilde{r}_{k,2}\,\right]$, $\tilde{H}^{(2)}$ is the channel-gain matrix, $\tilde{H}^{(2)} = \sum_{n=1}^{N} H_n^{(2)}, H_n^{(2)} = [\,h_{n1}\ h_{n2}\,]$, $C_k^{(2)} \tilde{V}_k^{(2)} = [\,\sum_{n=1}^{N} v_{k,1}\ \sum_{n=1}^{N} v_{k,2}\,]$, and see Eq. (8.16) for $C_k^{(2)}$.

(2) Four transmit antennas and N receive antennas

For a $4 \times N$ system, signals $\tilde{r}_{k,1}, \tilde{r}_{k,2}, \tilde{r}_{k,3}, \tilde{r}_{k,4}$ from the N receiver apertures are combined by equal gain at moments 1, 2, 3, and 4 at the kth block time at the receiver end as

$$\begin{aligned}\tilde{r}_{k,1} &= \sum_{n=1}^{N} r_{k,1}^{n} \\ &= \frac{\eta I_s}{4}\left(\sum_{n=1}^{N} h_{n1}x_{k,1} + \sum_{n=1}^{N} h_{n2}x_{k,2} + \sum_{n=1}^{N} h_{n3}x_{k,3} + \sum_{n=1}^{N} h_{n4}x_{k,4}\right) + \sum_{n=1}^{N} v_{k,1}^{n}\end{aligned} \tag{8.45}$$

$$\begin{aligned}\tilde{r}_{k,2} &= \sum_{n=1}^{N} r_{k,2}^{n} \\ &= \frac{\eta I_s}{4}\left(\sum_{n=1}^{N} h_{n1}x_{k,2}\Omega + \sum_{n=1}^{N} h_{n2}x_{k,1} + \sum_{n=1}^{N} h_{n3}x_{k,4}\Omega + \sum_{n=1}^{N} h_{n4}x_{k,3}\right) + \sum_{n=1}^{N} v_{k,2}^{n}\end{aligned} \tag{8.46}$$

$$\begin{aligned}\tilde{r}_{k,3} &= \sum_{n=1}^{N} r_{k,3}^{n} \\ &= \frac{\eta I_s}{4}\left(\sum_{n=1}^{N} h_{n1}x_{k,3}\Omega + \sum_{n=1}^{N} h_{n2}x_{k,4} + \sum_{n=1}^{N} h_{n3}x_{k,1} + \sum_{n=1}^{N} h_{n4}x_{k,2}\Omega\right) + \sum_{n=1}^{N} v_{k,3}^{n}\end{aligned} \tag{8.47}$$

$$\begin{aligned}\tilde{r}_{k,4} &= \sum_{n=1}^{N} r_{k,4}^{n} \\ &= \frac{\eta I_s}{4}\left(\sum_{n=1}^{N} h_{n1}x_{k,4}\Omega + \sum_{n=1}^{N} h_{n2}x_{k,3}\Omega + \sum_{n=1}^{N} h_{n3}x_{k,2} + \sum_{n=1}^{N} h_{n4}x_{k,2}\right) + \sum_{n=1}^{N} v_{k,4}^{n},\end{aligned} \tag{8.48}$$

where $r_{k,j}^{n}$ denotes the received signal of the nth receive antenna at moment $i(i = 1, 2, 3, 4))$. $h_{n1}, h_{n2}, h_{n3}, h_{n4}$ denote the channel gain from transmit antennas 1, 2, 3, and 4 to the nth receive antenna, respectively. Equations (8.45)–(8.48) are rewritten in matrix form as

$$\tilde{R}_k^{(4)} = \frac{\eta I_s}{4}\tilde{H}^{(4)}C_k^{(4)} + \tilde{V}_k^{(4)}, \tag{8.49}$$

where $\tilde{R}_k^{(4)} = \left[\tilde{r}_{k+1}\ \tilde{r}_{k+2}\ \tilde{r}_{k+3}\ \tilde{r}_{k+4}\right]$, $\tilde{H}^{(4)}$ is the channel-gain matrix, $\tilde{H}^{(4)} = \sum_{n=1}^{N} H_n^{(4)}$, $H_n^{(\pi 4)} = \left[h_{n1}\ h_{n2}\ h_{n3}\ h_{n4}\right]$, $V_k^{(4)} = \left[\sum_{N=1}^{n} v_{k,1}\ \sum_{N=1}^{n} v_{k,2}\ \sum_{N=1}^{n} v_{k,3}\ \sum_{N=1}^{n} v_{k,4}\right]$, and see Eq. (8.17) for k.

Note that when the number of transmit antennas is m and the number of receive antennas is N, $\tilde{R}_{k-1}^{(m)}$ denotes the kth receive matrix, $\tilde{H}^{(m)}$ denotes the channel-fading coefficient matrix, and $\tilde{V}_k^{(m)}$ denotes the kth noise matrix. It is also assumed that the channel-gain coefficient $H^{(m)}$ is constant within two adjacent information blocks, that is, the channel is a slow flat fading channel. Then, the decoding equation is

$$\{\hat{s}_1, \ldots \hat{s}_m\} = \arg \max_{s_1, \ldots s_m \in A} trace\left[\tilde{R}_{k-1}^{(m)} S_k^{(m)} \left(\tilde{R}_k^{(m)}\right)^H\right], A = \{[\,1\ 0\,], [\,0\ 1\,]\} \tag{8.50}$$

where $S_k^{(m)}$ is the previous Eq. (8.21); Eq. (8.22), which maps the estimated $\widehat{s}_1, \ldots \widehat{s}_m$, is mapped inversely to obtain the sent message bits.

8.3.3 Simulation Analysis

We define the received energy per symbol per detector in the absence of turbulence as $E_s = (\eta I_s)^2 T_s$, where $T_s = T/Q$ is the time-slot length, Q is the modulation order, and T is the character period; the electrical signal-to-noise ratio per symbol per detector in the absence of turbulence is μ: $\mu = E_s/N_0$; and the electrical signal-to-noise ratio per bit in the absence of turbulence is $SNR = \mu/(R \log_2 Q)$, with R being the transmission rate. The total power E_s of the system is assumed to remain constant and is normalized to 1, and the transmission rate $R = 1$. The channel coefficients are assumed to remain constant within the two adjacent transmit-code matrices, according to Taylor's frozen turbulent-atmosphere theory.

The bit-error rates of the 2 × 1, 2 × 2, 2 × 3, 4 × 1, 4 × 2, 4 × 3, and 4 × 4 systems were simulated using the Monte Carlo method. The simulation parameter $\eta = 0$ and binary pulse-position modulation (BPPM) was used; that is, $Q = 2$. The sample value of the binary data stream is 1 million. The turbulence strength is characterized by the magnitude of the scintillation index SI, as shown in Table 8.2.

Figure 8.4 shows the bit-error rate of differential space-time coding with a lognormal-distribution turbulence model, scintillation index SI = 0.4, with two transmit antennas. It can be seen from the figure that the system performance gradually improves as the number of receive apertures increases; at a 15-dB signal-to-noise ratio, the bit-error rate performance of the 2 × 3 system improves by one order of magnitude, relative to the 2 × 2 system, and the bit-error rate performance of the 2 × 3 system improves by two orders of magnitude, relative to the 2 × 1 system.

Figure 8.5 shows the bit-error rate curves of differential space-time coding with two transmit antennas using the lognormal turbulence model and scintillation index SI = 0.9. At an 18-dB signal-to-noise ratio, the 2 × 3 system exhibits a one order

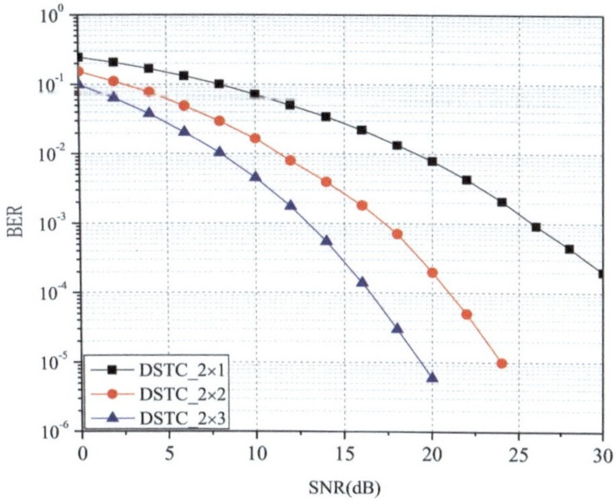

Fig. 8.4 Coding BER for differential space-time coding with BPPM modulation for SI = 0.4

of magnitude improvement in the bit-error rate performance relative to the 2 × 2 system, and a two order of magnitude improvement relative to the 2 × 1 system. Figure 8.5 shows the same pattern as that presented by the curves in Fig. 8.4, where the system performance improves as the number of receive apertures increases.

Figure 8.6 shows the bit-error rate of differential space-time coding for two transmit antennas with a gamma–gamma-distribution turbulence model and scintillation index SI = 1.12. At a 27-dB signal-to-noise ratio, the 2 × 3 system exhibited a one order of magnitude improvement in the bit-error rate performance relative to the 2 × 2 system, and a two order of magnitude improvement relative to the 2 × 1 system.

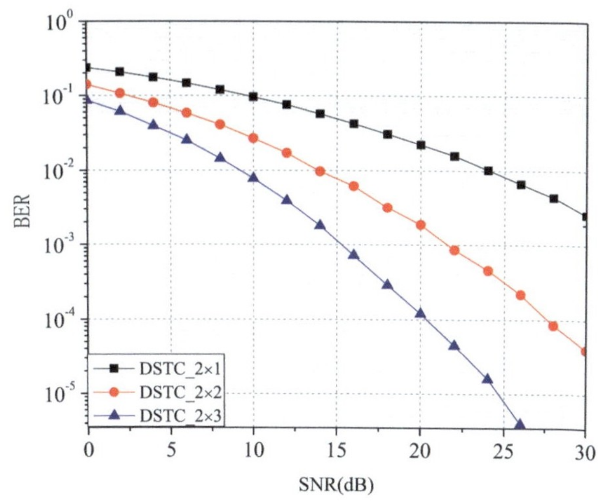

Fig. 8.5 Coding BER for differential space-time coding with BPPM modulation for SI = 0.9

Fig. 8.6 Bit-error rate of differential space-time coding with BPPM modulation for SI = 1.12

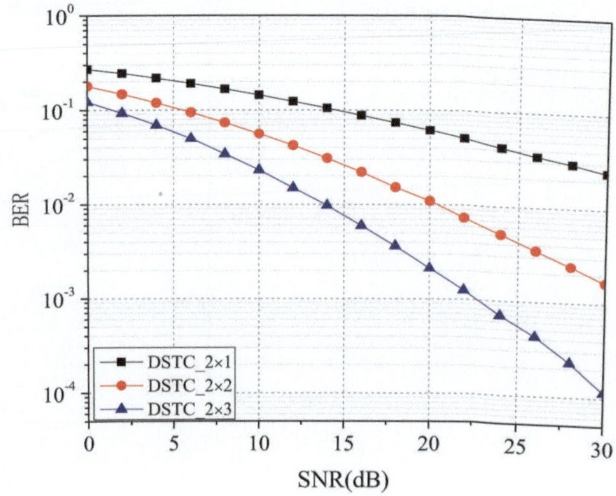

Figure 8.7 shows the bit-error rate comparison between differential space-time coding and the improved Alamouti code [22] by Simon and Vilnrotter when the lognormal-distribution turbulence model is used and the scintillation index $SI = 0.9$. The performance curves of the differential space-time coding method and the modified Alamouti coding method are approximately parallel, as shown in the figure, which indicates that the differential scheme can achieve complete transmit diversity with the modified Alamouti code [22].

Because the differential scheme does not require channel estimation, the performance is inferior to that of the improved Alamouti coding; at a 26-dB signal-to-noise ratio, there is an approximately one order of magnitude difference in the bit-error

Fig. 8.7 Bit-error rate comparison of differential space-time coding and Alamouti code with BPPM modulation for SI = 0.9

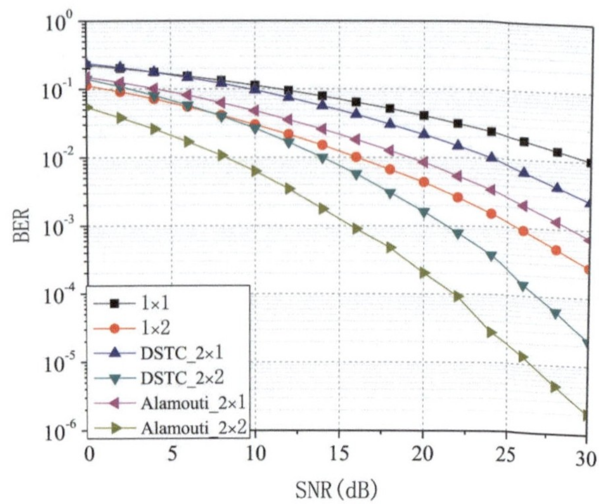

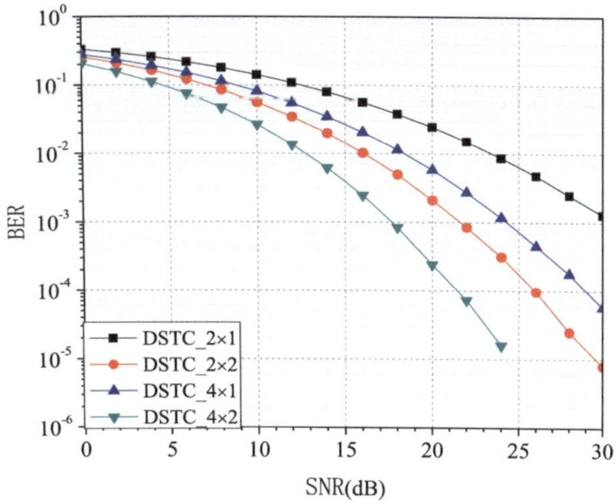

Fig. 8.8 Comparison of the bit-error rate of differential space-time codes with two and four transmitting antennas for SI = 0.4

rate performance between the 2 × 2 system with differential coding and the 2 × 2 system with improved Alamouti coding; however, the 2 × 1 and 2 × 2 systems with the differential coding scheme are significantly better than the uncoded 1 × 1 and 1 × 2 systems.

Figure 8.8 shows a comparison of the bit-error rate of the differential space-time code with two and four transmit antennas using the lognormal-distribution turbulence model with scintillation index $S = 0.4$. From the figures, we can see the following:

(1) The differential space-time code scheme with four transmit antennas is significantly better than that with two transmit antennas because the diversity gain provided by four transmit antennas at the transmitting end is significantly higher than that provided by two transmit antennas. At a 25-dB signal-to-noise ratio, the bit-error rate performance of the 4 × 1 system was improved by one order of magnitude relative to the 2 × 1 system, and at a 20-dB signal-to-noise ratio, the bit-error rate performance of the 4 × 2 system was improved by one order of magnitude relative to the 2 × 2 system.
(2) The system performance tends to improve as the number of receive antennas increases, owing to the effect of the receive diversity. At a 20-dB signal-to-noise ratio, the bit-error rate performance of the 2 × 2 system improves by one order of magnitude compared to that of the 2 × 1 system; at an 18-dB signal-to-noise ratio, the bit-error rate performance of the 4 × 2 system was improved by one order of magnitude compared to that of the 4 × 1 system.
(3) For two different MIMO systems, the bit-error rate performance of the 2 × 2 system is better than that of the 4 × 1 system when the product of the number of transmit antennas and the number of receive antennas is equal. At a 30-dB signal-to-noise ratio, the bit-error rate performance of the 2 × 2 system was improved by approximately one order of magnitude with respect to the 4 × 1 system.

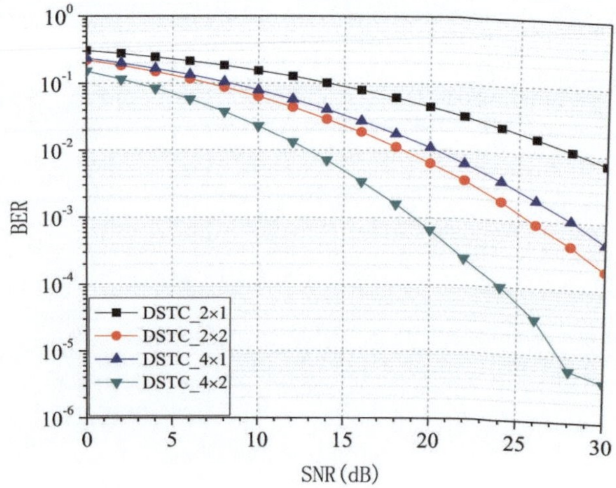

Fig. 8.9 Comparison of the bit-error rate of differential space-time codes with two and four transmitting antennas for SI = 0.9

Figure 8.9 shows a comparison of the differential space-time code bit-error rate for two and four transmit antennas using a lognormal-distribution turbulence model with a scintillation index $SI = 0.9$. The performance shown in Fig. 8.9 is significantly degraded relative to that in Fig. 8.8, owing to the intensity of the atmospheric turbulence. Three conclusions similar to those in Fig. 8.8 were drawn from Fig. 8.9:

(1) The differential space-time code scheme with four transmit antennas is significantly better than that with two transmit antennas because the diversity gain provided by four transmit antennas at the transmitter end is significantly higher than that provided by two transmit antennas. At a 28-dB signal-to-noise ratio, the bit-error rate performance of the 4×1 system was improved by one order of magnitude relative to the 2×1 system, and at a 20-dB signal-to-noise ratio, the bit-error rate performance of the 4×2 system was improved by one order of magnitude relative to the 2×2 system.

(2) The performance of the system improves as the number of receiving antennas increases, owing to the effect of receive diversity from the receiver side. At a 22-dB signal-to-noise ratio, the bit-error rate performance of the 2×2 system is improved by one order of magnitude compared to that of the 2×1 system; at an 18-dB signal-to-noise ratio, the bit-error rate performance of the 4×2 system is improved by one order of magnitude compared with that of the 4×1 system.

(3) For two different MIMO systems, the bit-error rate performance of the 2×2 system is better than that of the 4×1 system when the product of the number of transmit antennas and the number of receive antennas is equal. The combined effect of transmit and receive diversity suppresses turbulence better than the system with transmit diversity alone because the 2×2 system has the combined effect of transmit diversity at the transmitter and receive diversity at the receiver,

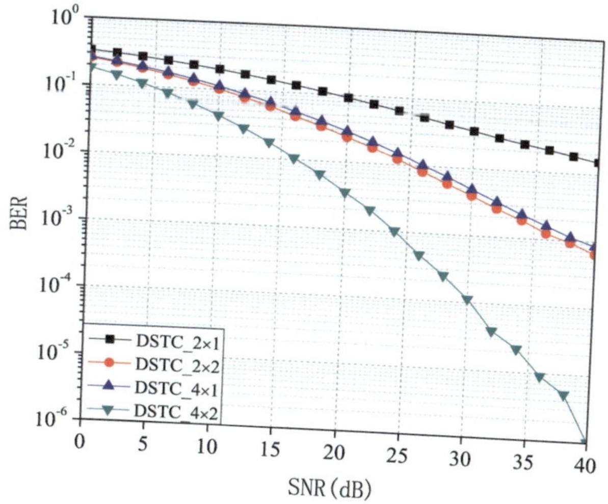

Fig. 8.10 Comparison of the bit-error rate of differential space-time codes with two and four transmitting antennas for SI = 1.12

whereas the 4 × 1 system has only the effect of transmit diversity at the transmitter. At a 30-dB signal-to-noise ratio, the 2 × 2 system has a half-order of magnitude improvement in the bit-error rate relative to the 4 × 1 system.

Figure 8.10 shows a comparison of the differential space-time code bit-error rate for two and four transmit antennas using the gamma–gamma-distribution turbulence model with scintillation index $SI = 1.12$. The performance shown in Fig. 8.10 is significantly worse than that in Figs. 8.8 and 8.9, because of the atmospheric-turbulence intensity.

From Fig. 8.10, it is obvious that the difference in differential space-time codes using the 4 × 2 system under strong turbulence conditions is significantly larger than that of the 2 × 2, 4 × 1, and 2 × 1 systems, and the effect of strong turbulence can be effectively suppressed using a 4 × 2 system. At a 20-dB signal-to-noise ratio, the bit-error rate performance of the 4 × 2 system was improved by one order of magnitude compared with that of the 2 × 2 system, while at a 40-dB signal-to-noise ratio, the bit-error rate performance of the 4 × 2 system was improved by three orders of magnitude compared with that of the 2 × 2 system.

Figure 8.11 shows the plot of the bit-error rate versus the signal-to-noise ratio for differential space-time codes with a scintillation index SI = 0.6 for four transmit antennas and one, two, three, and four receive antennas, under the lognormal-distribution turbulence model. It can be seen from the graph that the performance of the system tends to improve as the number of receive antennas increases. At a 16-dB signal-to-noise ratio, the bit-error rate performance of the 4 × 4 system has a three order of magnitude improvement relative to the 4 × 1 system, the bit-error rate performance of the 4 × 4 system has a two order of magnitude improvement relative to the 4 × 2 system, and the bit-error rate performance of the 4 × 4 system has a one order of magnitude improvement relative to the 4 × 3 system.

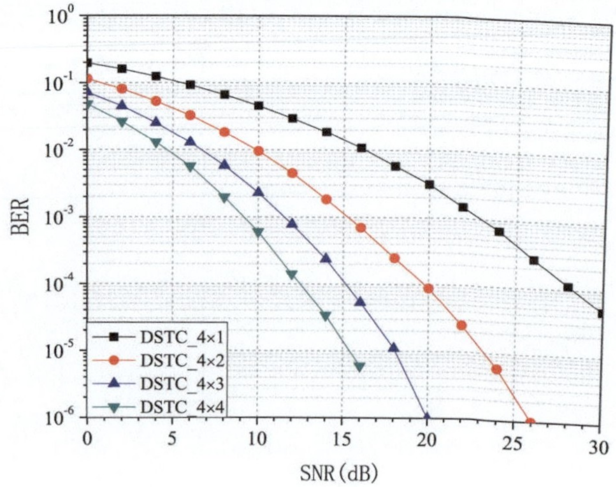

Fig. 8.11 Bit-error rate of differential space-time codes with four transmit antennas and different numbers of receive antennas for SI = 0.6

8.4 Summary

Because free-space optical communications are different from RF communications, some improvements in differential space-time codes are needed to make them suitable for IM/DD-type free-space optical communication. This chapter focused on the compiled code method for differential space-time codes in free-space optical communications.

The analysis results showed the following:

(1) The differential scheme had the same diversity gain as the improved Alamouti code proposed by Simon and Vilnrotter [4]. Although the performance was worse, it was also effective in reducing the BER performance of the system compared with an uncoded system, and it reduced the system complexity because it did not require channel estimation.
(2) As the number of transmit and receive antennas increased (e.g., with differential space-time codes for a 4 × 4 system), the system became more capable of suppressing turbulence.

References

1. Vucetic B, Yuan J, Wang XH, Huang KZ et al (2004) Space-time coding technology. Machinery Industry Press, Beijing, p 198
2. Tarokh V, Jafarkhani H (2000) A differential detection scheme for transmit diversity. IEEE J Sel Areas Com 18(07):1169–1174
3. Alamouti SM (1998) A simple transmit diversity scheme for wireless communication. IEEE J Sel Areas Com 16(08):1451–1458

4. Tarokh V, Jafarkhani H (2001) Multiple transmit antenna differential detection from generalized orthogonal designs. IEEE Trans Inf Theory 47(06):2626–2631
5. Hochwald MB (2000) Differential unitary space-time modulation. IEEE Trans Com 48(12):2041–2052
6. Hughes BL (2000) Differential space-time modulation. IEEE Trans Inf Theory 46(O7):2567–2578
7. Hassibi B, Hochwald BM (2002) Cayley differential unitary space-time codes. IEEE Trans Inf Theory 48(6):1485–1503
8. Shao X (2003) A new differential space-time block coding scheme. IEEE Com Letters 7(9):437–439
9. Hwang C-S, Nam SH, Chung J, Tarokh V (2003) Differential space time block codes using nonconstant modulus constellations. IEEE Trans Signal Process 51(11):2955–2964
10. Chen Z, Zhu G, Shen J, Liu Y (2003) Differential space-time block codes from amicable orthogonal designs. In: Wireless communications and networking conference, vol 2, pp 768–772
11. Chen ZL, Zhu GX, Chen J, Zhu DM (2004) Differential space-time grouping codes based on mutual full orthogonal design. Acta Electron Sin 32(4):548–551
12. Tao M, Cheng RS (2003) Trellis-codes differential unitary space-time modulation over flat fading channels. IEEE Trans Com 51(4):587–596
13. Li Y, Vucetic B (2007) Design of differential space-time trellis codes. IEEE Trans Inf Theory 6(5):1631–1637
14. Schlegel C, Grant A (2003) Differential space-time turbo codes. IEEE Trans Inf Theory 49(9):2298–2306
15. Zhu Y, Jafarkhani H (2006) Differential super-orthogonal space-time trellis coded modulation. IEEE Trans Wireless Com 5(12):747–751
16. Vucetie B, Yuan J, Wang XH et al (2004) Space-time coding technology. Machinery Industry Press, Beijing, p 88
17. Abou-Rjeily C (2009) Orthogonal space-time block codes for binary pulse position modulation. IEEE Trans Com 57(3):602–605
18. Atomic Energy Commission (2009) Differential space-time coding method. China Patent:200910134688.1
19. SG Wilson, M Brandt-Pearce, Q Cao, JH Leveque (2005) Free-space optical MIMO transmission with Q-ary PPM. IEEE Trans Commun 52:1402–1412
20. M Uysal, M Yu (2006) Error rate performance analysis of coded free-space optical links over gamma-gamma atmospheric turbulence channels. IEEE Trans Wirel Commun 5:1229–1233
21. Wang Z, Zhong W-D, Lin C (2009) Performance comparison of different modulation formats over free-space optical (FSO) turbulence links with space diversity reception technique. IEEE Photonics J 1(6):277–285
22. Simon MK, Vilnrotter V (2005) Alamouti-type Space-time coding for free-space optical communication with direct detection. IEEE Trans Wirel Commun 4(1):35–39

Chapter 9
Unitary Space–Time Coding

Channel-state information (CSI) must be known at the receiver for space–time code decoding to function. It is very difficult, sometimes even unrealistic, to estimate the channel effectively in a mobile environment with rapidly changing channel characteristics. Therefore, it is of great practical value to study space–time coding without estimating the channel at the receiver.

Hughes, Hochwald, Marzetta, and other scholars [1–3] studied blind space–time block codes without channel-parameter estimation at the receiver. Hochwald et al. [4] derived the unitary space–time code (USTC) according to the channel capacity of a Rayleigh fading channel in information theory. If the sending code block is a unitary matrix, the receiver can perform maximum-likelihood decoding without knowing the channel state.

Unitary space–time coding satisfies the orthogonal design criteria of space–time block codes; thus, the properties and decoding methods of many space–time block codes are also applicable to unitary space–time codes. The decoding complexity of unitary space–time codes varies exponentially with the coding dimension. Constructing simple and efficient unitary space–time codes is a difficult problem.

Unitary space–time modulation is the best modulation method to achieve the channel capacity under a high signal-to-noise ratio. Hochwald further extended the unitary space–time code to differential unitary space–time modulation (DUSTM), also known as differential unitary space–time coding (DUSTC), which is suitable for continuous fading channels. This chapter discusses unitary space–time coding in wireless optical communications.

X. Ke, *Space-Time Coding Theory in Optical Wireless MIMO Systems*,
Optical Wireless Communication Theory and Technology,
https://doi.org/10.1007/978-981-97-6449-5_9

9.1 Coding Principle of USTCs and Differential USTC Modulation

Hochwald [4, 5] constructed a unitary space–time code without channel estimation from the perspective of information theory. Unitary space–time codes are similar to space–time block codes in form; however, the transmission-code matrix must be a unitary matrix φ. It completes the coding by mapping certain bits of information into a unitary matrix.

The difference between the design of unitary space–time codes and other space–time codes (such as space–time block codes, space–time trellis codes, and layered space–time codes) is that a unitary space–time code does not optimize the Euclidean distance, but optimizes the norm of different symbol-correlation F matrices $\left(\left\|\varphi_l^H \varphi_{l'}\right\|_F\right)$. The smaller the correlation of different unitary space–time symbols, the better the performance of unitary space–time codes $\left\|\varphi_l^H \varphi_{l'}\right\|_F$. Because the unitary matrix φ has good orthogonality and $\phi\phi^H = I$, it can ensure that the signals transmitted by each antenna are orthogonal to each other and the correlation of the transmitted signals is reduced to the greatest extent.

$L\, N_T \times T$ matrix elements, $X_l, l = 1, 2, \ldots, L, L = 2^{\tau N_T}$, and τ in the modulation constellation show the carrier-frequency utilization. $X_l = \sqrt{T}\varphi_l$ and φ_l are unitary matrices, satisfying

$$\varphi_l \varphi_l^H = I, X_l X_l^H = TI, l = 1, 2, \ldots, L, \tag{9.1}$$

where I is an $N_T \times N_T$ dimensional unit matrix. The Chernoff upper bound of the pairwise error probability that the receiving end misjudges the transmitted signal X_l to $\hat{X}_l$ with $P(X_l \to \hat{X}_l)$ is [7, 8]

$$\begin{aligned} P(X_l \to \hat{X}_l) &\le \left| I + \frac{\rho_{N_T} T^2}{4(1 + \rho_{N_T} T)} \left[I - \frac{1}{T^2} \hat{X}_l X_l^H X_l \hat{X}_l^H \right] \right|^{-N_R} \\ &= \frac{1}{2} \prod_{i=1}^{N_T} \left[I + \frac{\rho_{N_T} T^2}{4(1 + \rho_{N_T} T)} \left(1 - d_i^2\right) \right]^{-N_R}, \end{aligned} \tag{9.2}$$

where ρ is the signal-to-noise ratio of the receiving antenna and $\rho_{N_T} = \rho / N_T$. d_i is the i th singular value $1 \ge d_1 \ge \cdots \ge d_{N_T} \ge 0$ of the correlation matrix $\varphi_l \varphi_l^H$. It can be seen that when ρ_{N_T} is large, the diversity gain v and coding gain μ that can be achieved by Eq. (9.2) are given by the following equation:

$$v = rank\left(I - \frac{1}{T^2} \hat{X}_l X_l^H X_l \hat{X}_l^H \right). \tag{9.3}$$

$$\mu = \lambda \left\langle TI - \frac{1}{T^2} \hat{X}_l X_l^H X_l \hat{X}_l^H \right\rangle^{1/v}, \tag{9.4}$$

where $rank(A)$ is the rank of matrix A and $\lambda\langle A\rangle$ is the product of its non-zero eigenvalues. In this case, the pairwise error probability can be expressed as

$$P\left(X_l \to \hat{X}_l\right) \leq \left|I + \frac{\mu^v \rho_{N_T}}{4}\right|^{-N_R}, \tag{9.5}$$

From Eq. (9.5), it can be seen that the upper bound of the bit-error rate is mainly determined by the diversity gain v, while the coding gain μ has little effect. When ρ_{N_T} is large, the selected signal constellation should maximize the diversity gain $v = N_T$ and minimize the coding gain $\mu_{\min}$.

When the channel state is unknown, differential space–time modulation is another effective transmission method. The use of unitary space–time codes in differential modulated multiple-input multiple-output (MIMO) systems is called differential unitary space–time modulation (DUSTM) [9–12]. In a DUSTM system, T is generally set equal to M; that is, the transmitted signal matrix X is the unitary matrix of $M \times M$, and the system equation is

$$Y_\varsigma = \sqrt{\rho} X_\varsigma H_\varsigma + N_\varsigma, \tag{9.6}$$

where ς is the time index, indicating the (ς) code block in transmission, X_ς is the $M \times M$ transmission-code matrix, Y_ς is the received signal matrix, H_ς is the channel-state information matrix, and N_ς is the noise matrix. The transmission-code block information in the ς time period is equal to the product of the unitary space–time code selected in the constellation set and the transmission code block in the $(\varsigma - 1)$ time period.

The transmission signal is obtained by the following formula:

$$X_\varsigma = \Phi_{Z_\varsigma} X_{\varsigma-1}, \tag{9.7}$$

where $\Phi_{Z_\varsigma} \in \Omega \equiv \{\Phi_1, \Phi_2, \ldots, \Phi_L\}, L = 2^{RM}$, Φ_{Z_ς} is the coded set of unitary matrices, L is the number of matrices in the set, R is the transmission rate, and $X_0 = I_M$. Assuming that the channel-state information remains unchanged for two consecutive code blocks, i.e., $H_\varsigma = H_{\varsigma-1}$,

$$X_\varsigma = \Phi_{Z_\varsigma} X_{\varsigma-1} + W'_\varsigma, \tag{9.8}$$

where

$$W'_\varsigma = W_\varsigma - \Phi_{Z_\varsigma} W_{\varsigma-1}. \tag{9.9}$$

It can be seen that the channel-state matrix does not appear in Eq. (9.8), so as long as the channel-state information is consistent within the time of two adjacent transmission code blocks, the DUSTM system receiver can perform differential decoding without knowing the channel-state information. Because Φ_{Z_ς} is a unitary matrix and

W'_{ς} is an independent additive complex Gaussian distribution, the system can use the following maximum-likelihood decoding [13]:

$$\hat{z}_{\varsigma} = \arg \max_{i=0,\ldots L-1} \left\| X_{\varsigma} - \Phi_{Z_{\varsigma}} X_{\varsigma-1} \right\|_F. \tag{9.10}$$

In the case of a high signal-to-noise ratio, the upper bound of the average pairwise error probability (PEP) that Φ_i is misjudged as Φ_j is [14]

$$P(\Phi_i \to \Phi_j) \leq \frac{1}{2}\left(\frac{8}{\rho}\right)^{MN} \frac{1}{\left|\det(\Phi_i \to \Phi_j)\right|^{2N}}, \tag{9.11}$$

where M is the number of transmitting antennas and N is the number of receiving antennas. It can be seen that the upper bound is inversely proportional to $\left|\det(\Phi_i \to \Phi_j)\right|^{2N}$; therefore, we can use

$$\xi = \frac{1}{2} \min_{1 \leq i \leq j \leq L-1} \left|\det(\Phi_i - \Phi_j)\right|^{1/M} \tag{9.12}$$

to represent the diversity product of the DUSTM system.

It can be seen that the diversity product is non-negative. If the diversity product is not zero, the unitary space–time code is said to be full diversity. The coefficient $\frac{1}{2}$ and index $\frac{1}{M}$ in the formula are used for normalization. After normalization, the diversity product of the unitary space–time code constellation set is between 0 and 1.

9.2 Unitary Space–Time Coding Based on a Rotation Factor

9.2.1 Unitary Space–Time Coding Based on Phase-Shift Keying in RF

The key to unitary space–time coding is the construction of a unitary matrix. To simplify this construction, one researcher [13] constructed a unitary matrix by introducing an adaptive rotation factor into the quasi-orthogonal space–time block codes of the Tirkkonen–Boariu–Hottinen (TBH) and Jafarkhani coding schemes. TBH and Jafarkhani codes are the two most commonly used quasi-orthogonal space–time block codes, which are extended from the Alamouti scheme.

9.2.1.1 Alamouti Space–Time Code Design

Alamouti [16] proposed a space–time block code suitable for dual-transmit antennas in 1998. The key to this coding is to obtain the diversity gain using a simple maximum-likelihood decoding algorithm. Because of the orthogonality between the columns of the coding matrix, the signal transmitted on the antenna is also orthogonal; hence, the receiver can use maximum-likelihood decoding, which greatly reduces the decoding complexity.

Alamouti coding encodes a vector $x = [x_1, x_2]^T$ composed of every two modulation symbols into a space–time coding matrix:

$$X = \begin{bmatrix} x_1 & -x_2^* \\ x_2 & x_1^* \end{bmatrix}. \tag{9.13}$$

That is, the output of the encoder transmits the transmission space–time codeword matrix X from two antennas using two symbol periods. In the first transmission cycle, antennas 1 and 2 send signals x_1 and x_2, respectively. In the second transmission cycle, antennas 1 and 2 send signals $-x_2^*$ and x_1^*, respectively, where x_1^* represents the complex conjugate of x_1.

If x^1 represents the transmission sequence of antenna 1 and x^2 represents the transmission sequence of antenna 2, then

$$\begin{aligned} x^1 &= \left[x_1, -x_2^*\right] \\ x^2 &= \left[x_2, x_1^*\right]. \end{aligned} \tag{9.14}$$

The main feature of Alamouti coding is that the transmission-signal vectors of the two transmission antennas are orthogonal to each other, namely,

$$x^1 \cdot x^2 = x_1 x_2^* - x_2^* x_1 = 0. \tag{9.15}$$

The coding matrix has the following characteristics:

$$X \cdot X^H = \begin{bmatrix} |x_1|^2 + |x_2|^2 & 0 \\ 0 & |x_1|^2 + |x_2|^2 \end{bmatrix} = \left(|x_1|^2 + |x_2|^2\right) I_2, \tag{9.16}$$

where X^H represents the conjugate transpose matrix of X and I_2 represents a 2×2 identity matrix.

9.2.1.2 Quasi-orthogonal Space–Time Block Code Design

Orthogonal space–time block codes have orthogonality between the rows of the transmission matrix, which makes it have full diversity gain. The receiver can make a separate decision for each signal, and the maximum-likelihood decoding algorithm can be used to simplify the decoding complexity. Unfortunately, orthogonal space–time block codes inevitably have some defects; for example, they cannot achieve a full-rate transmission except when using the Alamouti scheme.

When the number of transmit antennas is greater than two, the code rate is less than 1, which leads to a loss of transmission rate. Therefore, a reasonable compromise must be made between the rate and diversity. The full rate is obtained by sacrificing partial orthogonality, which is similar to the orthogonal matrix in form, and the orthogonality of only some rows of the transmission matrix is maintained. Based on this idea, Jafarkhani et al. proposed the quasi-orthogonal space–time block code.

Based on the Alamouti scheme, Jafarkhani [15] designed a quasi-orthogonal space–time block code with four transmitting antennas and a code rate of 1:

$$X_4 = \begin{bmatrix} X_2(x_1, x_2) & -X_2^*(x_3, x_4) \\ X_2(x_3, x_4) & X_1^*(x_1, x_2) \end{bmatrix} = \begin{bmatrix} x_1 & -x_2^* & -x_3^* & x_4 \\ x_2 & x_1^* & -x_4^* & -x_3 \\ x_3 & -x_4^* & x_1^* & -x_2 \\ x_4 & x_3^* & x_2^* & x_1 \end{bmatrix}. \tag{9.17}$$

It can be seen that the rows of the matrix in (9.17) have a certain orthogonality; however, no two rows are orthogonal to each other. The main idea is to expand the 2×2 matrix to a 4×4 matrix so that its emission rate is 1. The inner product of each row in the transmission matrix of a quasi-orthogonal space–time block code with another specific row is not zero; however, the inner product with other rows is zero; that is,

$$XX^H = \rho_x\big(|x_1|^2 + |x_2|^2 + \cdots |x_k|^2\big)I_{N_T} + Q, \tag{9.18}$$

where Q is a quasi-orthogonal matrix. There are other quasi-orthogonal matrices with similar structures, such as,

$$X_4^a = \begin{bmatrix} X_2(x_1, x_2) & X_2(x_3, x_4) \\ X_2(x_3, x_4) & X_1(x_1, x_2) \end{bmatrix} = \begin{bmatrix} x_1 & -x_2^* & x_3 & -x_4^* \\ x_2 & x_1^* & x_4 & x_3^* \\ x_3 & -x_4^* & x_1 & -x_2^* \\ x_4 & x_3^* & x_2 & x_1^* \end{bmatrix} \tag{9.19}$$

$$X_4^b = \begin{bmatrix} X_2(x_1, x_2) & -X(x_3, x_4) \\ X_2(x_3, x_4) & X_1(x_1, x_2) \end{bmatrix} = \begin{bmatrix} x_1 & -x_2^* & -x_3 & x_4^* \\ x_2 & x_1^* & -x_4 & -x_3^* \\ x_3 & -x_4^* & x_1 & -x_2^* \\ x_4 & x_3^* & x_2 & x_1^* \end{bmatrix} \tag{9.20}$$

$$X_4^\tau = \begin{bmatrix} X_2(x_1, x_2) & X_2(x_3, x_4) \\ -X_2(x_3, x_4) & X_1(x_1, x_2) \end{bmatrix} = \begin{bmatrix} x_1 & -x_2^* & x_3 & -x_4^* \\ x_2 & x_1^* & x_4 & x_3^* \\ -x_3 & x_4^* & x_1 & -x_2^* \\ -x_4 & -x_3^* & x_2 & x_1^* \end{bmatrix}. \tag{9.21}$$

According to the coding rules of quasi-orthogonal space–time block codes, many different coding and decoding methods have been proposed. The two most commonly used codes are as follows.

(1) TBH code [14]

$$\Omega_{TBH} = \begin{bmatrix} \phi(x_1, x_2) & \phi(x_3, x_4) \\ \phi(x_3, x_4) & \phi(x_1, x_2) \end{bmatrix} = \begin{bmatrix} x_1 & -x_2^* & x_3 & -x_4^* \\ x_2 & x_1^* & x_4 & x_3^* \\ x_3 & -x_4^* & x_1 & -x_2^* \\ x_4 & x_3^* & x_2 & x_1^* \end{bmatrix}. \tag{9.22}$$

In the formula, $\phi(x_1, x_2)$ and $\phi(x_3, x_4)$ adopt an Alamouti structure. Through calculation, we can obtain

$$XX^H = \rho_x\left(|x_1|^2 + |x_2|^2 + |x_3|^2 + |x_4|^2\right)I_{N_r} + Q_{TBH}; \tag{9.23}$$

of which,

$$Q_{TBH} = 2\mathrm{Re}\left(x_1x_3^* + x_2x_4^*\right)\begin{bmatrix} 0 & 0 & 1 & 0 \\ 0 & 0 & 0 & 1 \\ 1 & 0 & 0 & 0 \\ 0 & 1 & 0 & 0 \end{bmatrix}. \tag{9.24}$$

(2) Jafarkhani code [15]

$$\Omega_J = \begin{bmatrix} \phi(x_1, x_2) & \phi(x_3, x_4) \\ \phi(x_3, x_4) & \phi(x_1, x_2) \end{bmatrix} = \begin{bmatrix} x_1 & -x_2^* & x_3 & -x_4^* \\ x_2 & x_1^* & x_4 & x_3^* \\ -x_3^* & x_4 & x_1^* & -x_2 \\ -x_4^* & -x_3 & x_2^* & x_1 \end{bmatrix}, \tag{9.25}$$

where $\phi(x_1, x_2)^H$ and $\phi(x_3, x_4)^H$ respectively represent the conjugation of the matrix. Through calculation, we can obtain

$$XX^H = \rho_x\left(|x_1|^2 + |x_2|^2 + |x_3|^2 + |x_4|^2\right)I_{N_r} + Q_J; \tag{9.26}$$

of which,

$$Q_J = 2\mathrm{Re}\left(x_1x_4^* - x_2x_3^*\right)\begin{bmatrix} 0 & 0 & -1 & 0 \\ 0 & 0 & -1 & 0 \\ 0 & -1 & 0 & 0 \\ 1 & 0 & 0 & 0 \end{bmatrix}. \tag{9.27}$$

Reference [13] adds a rotation factor $e^{j\alpha}$ to an Alamouti [16] structure to construct a matrix in the following form:

$$S_{TBH}: \begin{bmatrix} x_1 & -e^{j\alpha}x_2^* & x_3 & -x_4^* \\ e^{j\alpha}x_2 & x_1^* & x_4 & x_3^* \\ e^{j\alpha}x_3^* & -x_4^* & x_1 & -x_2^* \\ x_4^* & e^{j\alpha}x_3 & x_2 & x_1^* \end{bmatrix}, \begin{bmatrix} x_1 & -x_2^* & e^{j\alpha}x_3 & -x_4^* \\ x_2 & x_1^* & x_4 & e^{j\alpha}x_3^* \\ x_3 & -x_4 & x_1 & -e^{j\alpha}x_2^* \\ x_4 & x_3^* & e^{j\alpha}x_2 & x_1^* \end{bmatrix}. \tag{9.28}$$

$$S_J: \begin{bmatrix} x_1 & -e^{j\alpha}x_2^* & x_3 & -e^{j\alpha}x_4^* \\ e^{j\alpha}x_2 & x_1^* & e^{j\alpha}x_4 & x_3^* \\ -x_3^* & x_4^* & x_1^* & -x_2 \\ -x_4^* & -x_3 & x_2^* & x_1 \end{bmatrix}, \begin{bmatrix} x_1 & -x_2^* & x_3 & -x_4^* \\ x_2 & x_1^* & x_4 & e^{j\alpha}x_3^* \\ -x_3^* & e^{j\alpha}x_4 & x_1^* & -e^{j\alpha}x_2^* \\ -e^{j\alpha}x_4^* & -x_3 & -e^{j\alpha}x_2^* & x_1^* \end{bmatrix}. \tag{9.29}$$

We take the following matrix as an example:

$$S = \begin{bmatrix} x_1 & -e^{j\alpha}x_2^* & x_3 & -x_4^* \\ e^{j\alpha}x_2 & x_1^* & x_4 & x_3^* \\ e^{j\alpha}x_3^* & -x_4^* & x_1 & -x_2^* \\ x_4 & e^{j\alpha}x_3^* & x_2 & x_1^* \end{bmatrix}. \tag{9.30}$$

To meet the definition of a unitary matrix, the following unitary space–time coding matrices are developed:

$$SS^H = \begin{bmatrix} a & 0 & b & 0 \\ 0 & a & 0 & b \\ b & 0 & a & 0 \\ 0 & b & 0 & a \end{bmatrix} \text{ or } SS^H = \begin{bmatrix} a & 0 & 0 & b \\ 0 & a & b & 0 \\ 0 & b & a & 0 \\ b & 0 & 0 & a \end{bmatrix},$$

where $a = \sum_{i-1}^{4} |x_i|^2 = 1$ represents its conjugate transpose matrix. If (4) is satisfied and $b = 0$ is satisfied:

$$e^{j\alpha} = -\frac{x_1x_3^* + x_2^*x_4}{x_1^*x_3 + x_2x_4^*}. \tag{9.31}$$

$x_i (i = 1, 2, 3, 4)$ represents the equal-gain modulation symbols sent at a certain time; based on quadrature phase-shift keying (QPSK) mapping, the mapping set can be obtained: $u = \left\{\frac{1}{2}, \frac{j}{2}, -\frac{j}{2}, -\frac{1}{2}\right\}$. It is found that when $x_1x_3^* + x_2^*x_4 = 0$, $x_1^*x_3 + x_2x_4^*$ must be zero; therefore, $e^{j\alpha} = 1$. When $x_1^*x_3 + x_2x_4^* \neq 0$, there must be $x_1^*x_3 + x_2x_4^* \neq 0$, so $e^{j\alpha} = -\frac{x_1x_3^* + x_2^*x_4}{x_1^*x_3 + x_2x_4^*}$.

Phase-shift keying (PSK) is used for constellation mapping in unitary space–time coding in the radio-frequency (RF) field. The modulated signal is the negative or complex form of the signal itself and the signal. However, in intensity modulation/direct detection (IM/DD) free-space optical (FSO) communication, the information can only be transmitted in the "*on*" and "*off*" states of the optical signal; the negative or complex form of the signal cannot be expressed. This is a key problem to be solved in this area.

9.2.2 Unitary Space–Time Coding Based on PPM Modulation

Suppose there is a 4 × 1 bit stream to be encoded. It first adopts QPSK mapping, and then quaternary pulse-position modulation (4-PPM) modulates the mapped constellation points. For example, the transmitted data sequence is 00011011, and the mapping relationship is shown in Fig. 9.1.

As shown in Fig. 9.1, we assume $x_1 \rightarrow 00$, $x_2 \rightarrow 01$, $x_3 \rightarrow 10$, and $x_4 \rightarrow 11$. In the RF field, QPSK is used to map information bits to constellation points in complex form. After constellation mapping $x_1x_2x_3x_4$, four effective results can be obtained:

$$\frac{1}{2}, \frac{j}{2}, -\frac{1}{2}, \text{ and } -\frac{j}{2}.$$

That is, the unitary matrix $\begin{bmatrix} x_1 & -e^{j\alpha}x_2^* & x_3 & -x_4^* \\ e^{j\alpha}x_2 & x_1^* & x_4 & x_3^* \\ e^{j\alpha}x_3 & -x_4^* & x_1 & -x_2^* \\ x_4 & e^{j\alpha}x_3^* & x_2 & x_1^* \end{bmatrix}$ can be expressed as

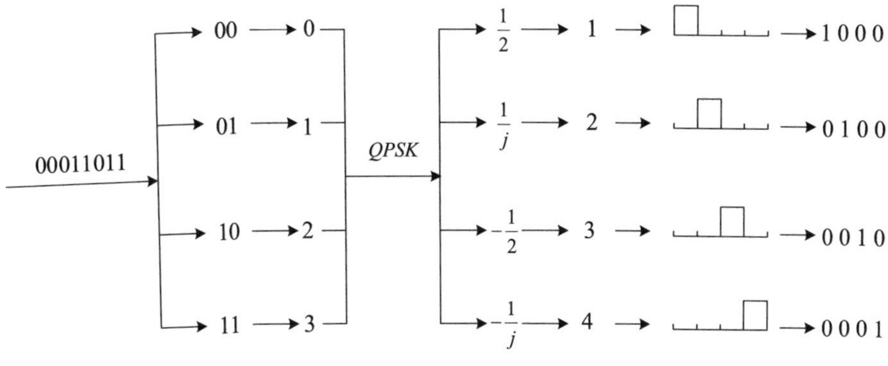

Fig. 9.1 Process diagram of bit-stream mapping in free-space optical communication

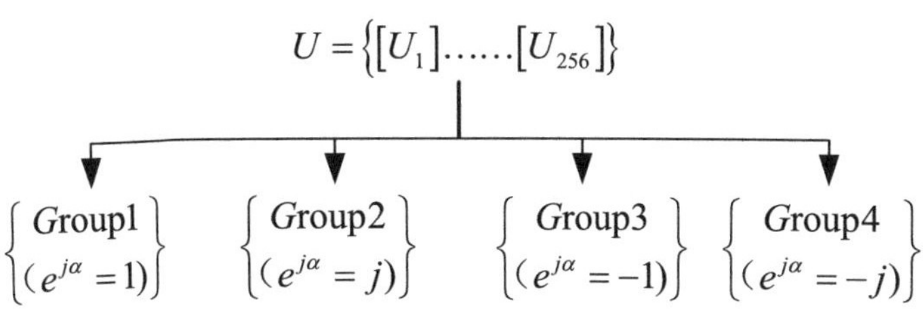

Fig. 9.2 Grouping diagram of estimation matrix set

$$\begin{bmatrix} \frac{1}{2} & -\frac{j}{2} & -\frac{1}{2} & -\frac{j}{2} \\ -\frac{j}{2} & \frac{1}{2} & -\frac{j}{2} & -\frac{1}{2} \\ \frac{1}{2} & -\frac{j}{2} & \frac{1}{2} & \frac{j}{2} \\ -\frac{j}{2} & \frac{1}{2} & \frac{j}{2} & \frac{1}{2} \end{bmatrix}.$$

Because the matrix contains complex and negative numbers, we use 4-PPM modulation to make it suitable for an IM/DD FSO communication system. Values $\frac{1}{2}$, $\frac{j}{2}$, $-\frac{1}{2}$, and $-\frac{j}{2}$ correspond to (1000), (0100), (0010), and (001), respectively, and the unitary space–time coding matrix is converted into a real matrix expressed by the pulse position.

That is, matrix $\begin{bmatrix} \frac{1}{2} & -\frac{j}{2} & -\frac{1}{2} & -\frac{j}{2} \\ -\frac{j}{2} & \frac{1}{2} & -\frac{j}{2} & -\frac{1}{2} \\ \frac{1}{2} & -\frac{j}{2} & \frac{1}{2} & \frac{j}{2} \\ -\frac{j}{2} & \frac{1}{2} & \frac{j}{2} & \frac{1}{2} \end{bmatrix}$ is transformed into a matrix such as $\begin{bmatrix} (1000) & (0001) & (0010) & (0001) \\ (0001) & (1000) & (0001) & (0010) \\ (1000) & (0001) & (1000) & (0100) \\ (0001) & (1000) & (0100) & (1000) \end{bmatrix}.$

After the above conversion, the final emission matrix is expressed as

$$A = \begin{bmatrix} a_{11} & a_{12} & a_{13} & a_{14} \\ a_{21} & a_{22} & a_{23} & a_{24} \\ a_{31} & a_{32} & a_{33} & a_{34} \\ a_{41} & a_{42} & a_{43} & a_{44} \end{bmatrix},$$

and each element in A is a vector with one row and four columns; that is, it represents the pulse positions after 4-PPM modulation. In this manner, FSO communication can be realized using intensity modulation / direct detection. It solves the problem of transplanting code from the RF field to the optical-communication field.

9.3 Decoding Principle of Unitary Space–Time Codes

9.3.1 Known Channel-State Information

For a system with four transmitting antennas, as given in Sect. 9.2.2, in the first transmission cycle, signals a_{11}, a_{21}, a_{31}, and a_{41} are simultaneously transmitted from antennas 1, 2, 3, and 4, respectively. In the second transmission cycle, signals a_{12}, a_{22}, a_{32}, and a_{42} are simultaneously transmitted from antennas 1, 2, 3, and 4, respectively. By analogy, each column element in transmission matrix A is simultaneously transmitted from four antennas in the same transmission cycle. The channel-state equation of the system can be expressed as:

$$R = AH + N, \tag{9.32}$$

where R represents the receiving matrix at the receiving end, H represents the light-intensity fading coefficient, and N represents the noise of the detector and its auxiliary circuit, which follows a Gaussian distribution.

r_1, r_2, r_3, and r_4 are the information received by the receiving antenna in cycles 1, 2, 3, and 4, respectively; $R = [r_1\ r_2\ r_3\ r_4]^T$ and T represents the transpose of the matrix. h_1, h_2, h_3, and h_4 are the light-intensity fading coefficients from transmitting antennas 1, 2, 3, and 4 to the receiving antenna, respectively; $H = [h_1\ h_2\ h_3\ h_4]^T$. n_1, n_2, n_3, and n_4 are the additive noise from transmitting antennas 1, 2, 3, and 4 to the receiving antenna, respectively; $N = [n_1\ n_2\ n_3\ n_4]^T$.

The information received by the receiving end can be expressed as

$$\begin{aligned} r_1 &= a_{11} \cdot h_1 + a_{21} \cdot h_2 + a_{31} \cdot h_3 + a_{41} \cdot h_4 + n_1 \\ r_2 &= a_{12} \cdot h_1 + a_{22} \cdot h_2 + a_{32} \cdot h_3 + a_{42} \cdot h_4 + n_2 \\ r_3 &= a_{13} \cdot h_1 + a_{23} \cdot h_2 + a_{33} \cdot h_3 + a_{43} \cdot h_4 + n_3 \\ r_4 &= a_{14} \cdot h_1 + a_{24} \cdot h_2 + a_{34} \cdot h_3 + a_{44} \cdot h_4 + n_4 \end{aligned} \tag{9.33}$$

We use an improved maximum-likelihood decoding algorithm. After the current four cycles are sent, the value of rotation factor $e^{j\alpha}$ of this group of the information matrix will be sent in the fifth cycle.

Let the set of all estimated matrices be $U = \{U_1 \ldots U_{256}\}$. At the receiving end, all matrices with $e^{j\alpha} = 1$ are summarized in [group 1]; the matrix of $e^{j\alpha} = j$ is summarized in [group 2]; the matrix of $e^{j\alpha} = -1$ is summarized in [group 3]; the $e^{j\alpha} = -j$ matrix is summarized in [group 4]; then, the estimated matrix set can be divided into four groups, according to the value of $e^{j\alpha}$.

The maximum-likelihood decoding decision criteria are [16–18]

$$\sum_t \sum_{j=1}^{n_R} |r_t^j - \sum_{i=1}^{n_T} h_{j,i}^t u_t^i|^2. \tag{9.34}$$

The decoder selects the codeword with the minimum decision metric as the output, where n_R and n_T are the number of receiving and transmitting antennas, respectively.

During decoding, first, information r_3 and r_4 are received in cycles 3 and 4. The information without rotation factors in r_3 and r_4 is used as reference information and the information in the estimated matrix is set for a maximum-likelihood decision (that is, the minimum measure of the maximum-likelihood decoding decision at 3 and 4). Then, the rotation factor of the output code matrix is compared with the received rotation factor. If the two are consistent, the output is determined as the final output. If not, the best output is sought in the set of estimation matrices divided according to the received rotation factor.

9.3.2 Unknown Channel-State Information

Let A_t be the transmission-code matrix at time t. The reception matrix R_t can be expressed as

$$R_t = A_t H_t + N_t, \tag{9.35}$$

where N_t is additive white Gaussian noise with a mean value of 0 and variance of σ^2, which is the noise generated by the detector and its auxiliary circuit. The system first sends an $N_T \times N_T$ unit array A_0. Through differential coding, the coding matrix (5) at a subsequent time t can be obtained as follows:

$$A_t = S_t A_{t-1}, \tag{9.36}$$

where S_t is a unitary matrix that contains the transmitting data at time t. Assuming that the channel is constant for two consecutive transmission cycles, R_t can be expressed as

$$R_t = S_t A_{t-1} H_t + N_t = S_t R_{t-1} + N', \tag{9.37}$$

where $N' = -S_t N_{t-1} + N_t$ and N' are also additive Gaussian random noise. At the receiving end, the maximum-likelihood criterion is used to decide:

$$\tilde{x}_t = \arg\min ||R_t - S_t R_{t-1}||^2, \tag{9.38}$$

where $\tilde{x}_t$ is the estimated value of the codeword transmitted in period T.

After the 4-PPM conversion, matrix multiplication will lead to coefficient accumulation; this is reflected in the continuous accumulation of pulse amplitudes in PPM modulation, which cannot be realized in reality. Therefore, in this section, the transmission matrix is modified under a differential-transmission framework, and the unit matrix is uniformly transmitted before each transmission matrix; namely,

$$A_t = S_t I. \tag{9.39}$$

The maximum-likelihood decision is expressed as

$$\tilde{x}_t = \arg\min ||R_t - S_t R_I||^2, \tag{9.40}$$

where R_I represents the information of the unit matrix received by the receiving end after channel transmission.

9.4 Simulation Results

The simulation conditions assume that the total transmit power E_S remains unchanged and the normalized value $E_S = 1$ is used. According to the frozen-field theory, assuming that the channel parameters remain unchanged for two consecutive transmit-code matrix blocks and the photoelectric-conversion efficiency $\eta = 0.5$, the number of binary data-stream samples is 100,000. Using the unitary space–time coding method in FSO communications, the bit-error rate and system bit-error rate are simulated. The intensity of the atmospheric turbulence is determined by the scintillation index (SI). The simulation results are shown in Figs. 9.3, 9.4, 9.5, 9.6, 9.7 and 9.8.

Figure 9.3 shows the system bit-error rate and bit-error rate of the unitary space–time coding of four transmitting antennas when the channel-state information is known, using the lognormal-distribution turbulence model and scintillation index SI = 0.4 (weak turbulence). As can be seen from the figure, the overall bit-error performance of the 4 × 2 system is better than that of the 4 × 1. When the signal-to-noise ratio is 25 dB, the bit-error rate of the 4 × 2 system is below 10^{-4}, and the bit-error rate of the 4 × 1 system is between 10^{-3} and 10^{-4}. It is shown that the performance of the unitary space–time coding scheme is greatly improved with the increase in the number of receiver apertures.

Figure 9.4 shows the bit-error rate and bit-error rate of the unitary space–time coding of four transmitting antennas when the channel-state information is known, using the lognormal-distribution turbulence model and scintillation index SI = 0.8 (medium turbulence). It can be seen that the BER performance curves in Figs. 9.3 and 9.4 are basically consistent. The system performance improves with the increase in the number of receiving apertures.

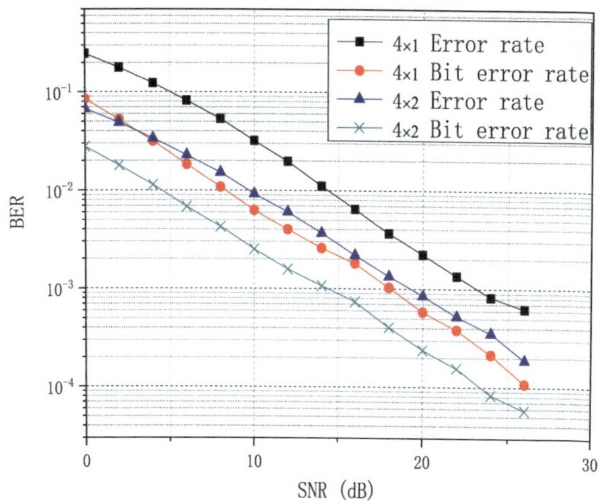

Fig. 9.3 Bit-error rate of unitary space–time coding with known channel-state information when SI = 0.4

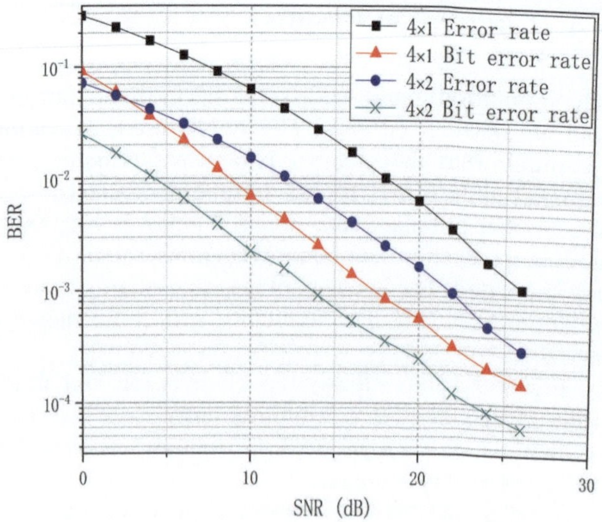

Fig. 9.4 Bit-error rate of unitary space–time coding with known channel-state information when SI = 0.8

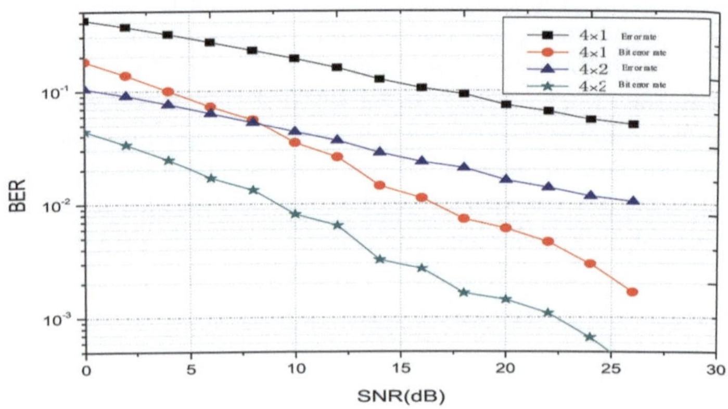

Fig. 9.5 Bit-error rate of unitary space–time coding with known channel-state information when SI = 1.12

When the signal-to-noise ratio is 25 dB, the bit-error rate of the 4 × 2 system is below 10^{-4}, and the bit-error rate of the 4 × 1 system is between 10^{-3} and 10^{-4}; however, when SI = 0.8 (medium turbulence) is compared with SI = 0.4 (weak turbulence), the lowest bit-error rate of the 4 × 1 system cannot reach 10^{-4}, and the bit-error rate of the 4 × 2 system can still reach below 10^{-4}. The turbulence suppression ability of the 4 × 2 system is better than that of the 4 × 1 system.

As shown in Figs. 9.3 and 9.4, when SI = 0.4 and the bit-error rate is 10^{-3}, the system signal-to-noise ratio of the 4 × 1 system is almost maintained at 17 dB. The signal-to-noise ratio of the 4 × 2 system is 14 dB. When SI = 0.8 (medium turbulence) and the bit-error rate is 10^{-3}, the system signal-to-noise ratio of the 4 × 1 system is around 18 dB. The signal-to-noise ratio of the 4 × 2 system is around

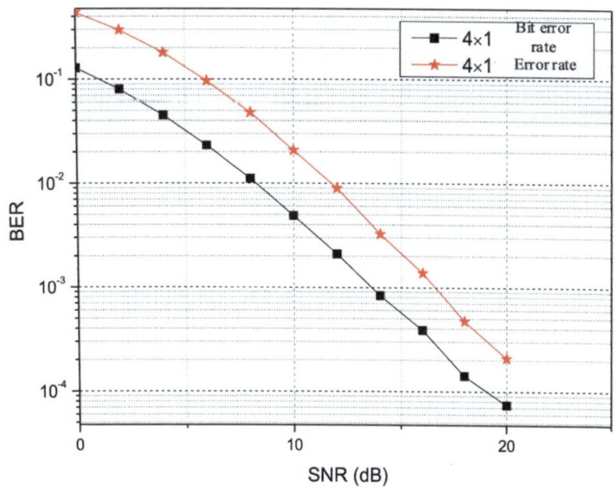

Fig. 9.6 Bit-error rate of unitary space–time coding with unknown channel-state information when SI = 0.4

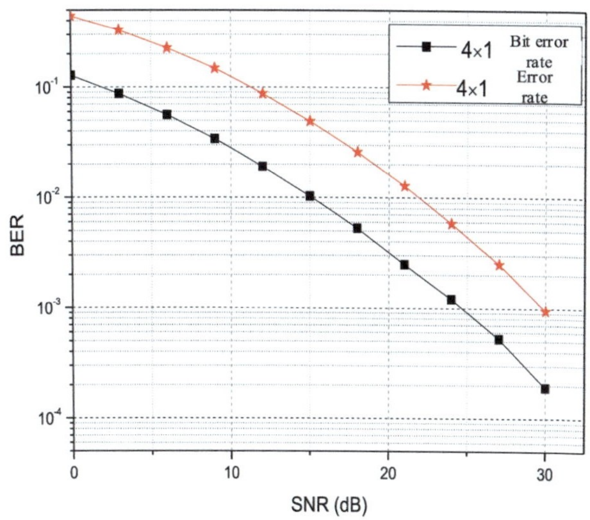

Fig. 9.7 Bit-error rate of unitary space–time coding with unknown channel-state information when SI = 0.8

13 dB. However, in the case of SI = 0.8 (medium turbulence), the bit-error rate of the 4 × 1 system is unable to reach, and the bit-error rate of the 4 × 2 system can still reach 10^{-4} to 10^{-5}. It can be seen that, in terms of suppressing the influence of turbulence, the 4 × 2 system is better than the 4 × 1 system.

Comparing Figs. 9.3 and 9.4, it can be seen that when the signal-to-noise ratio is 25 dB, whether the turbulence is weak (SI = 0.4) or medium (SI = 0.8), the bit-error rate of the 4 × 1 system is between 10^{-3} and 10^{-4}. The bit-error rate of the 4 × 2 system is between 10^{-4} and 10^{-5}. The bit-error rates of the system under the two turbulent conditions are similar.

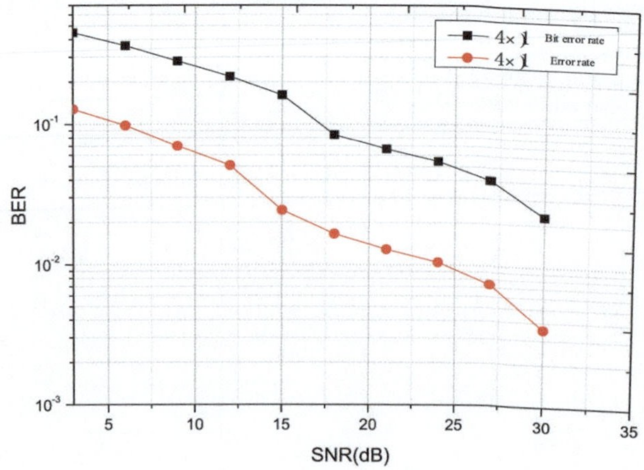

Fig. 9.8 Bit-error rate of unitary space–time coding with unknown channel-state information when SI = 1.12

Figure 9.5 shows the bit-error rate and bit-error rate of unitary space–time coding of four transmitting antennas when the turbulence model adopts the distributed turbulence model, the scintillation index SI = 1.12 (strong turbulence), and the channel-state information is known. The results are similar to the simulation results under the lognormal distribution model. As can be seen from the figure, the overall bit-error performance of the 4 × 2 system is better than that of the 4 × 1. The performance of the system under the influence of strong turbulence is different from that under the conditions of weak and medium turbulence. The bit-error rate of the 4 × 1 system is close 10^{-3}. The bit-error rate of the 4 × 2 system is between 10^{-3} and 10^{-4}.

Comparing Figs. 9.3, 9.4 and 9.5, we observe the 4 × 2 system, when the number of binary data-stream samples is 100,000. If the signal-to-noise ratio is 20 dB, the bit-error rate under weak turbulence is between 10^{-3} and 10^{-4} close to 10^{-4}. In the case of medium turbulence, the bit-error rate is between and. In the case of strong turbulence, the bit-error rate is between 10^{-2} and 10^{-3}; close to 10^{-3}. It can be seen that atmospheric turbulence has a great impact on the BER performance of the system.

Figure 9.6 shows the bit-error rate and system bit-error rate of the unitary space–time coding with four transmitting antennas when the lognormal-distribution turbulence model is used, scintillation index SI = 0.4 (weak turbulence) and the channel-state information is unknown. As can be seen from the figure, the bit-error rate of the 4 × 1 system is between 10^{-3} and 10^{-4}, and the bit-error rate is between 10^{-4} and 10^{-5}.

Figure 9.7 shows the bit-error rate and bit-error rate of unitary space–time coding of the system when the lognormal-distribution turbulence model is adopted, the scintillation index SI = 0.8 (medium turbulence), and the channel-state information is unknown. As can be seen from the figure, the bit-error rate of the 4 × 1 system under medium turbulence is around 10^{-3}, and the bit-error rate is between 10^{-3} and 10^{-4}.

Figure 9.8 shows the bit-error rate and bit-error rate of unitary space–time coding of the system when a distributed turbulence model is adopted, the scintillation index SI = 1.12 (strong turbulence), and the channel-state information is unknown. As can be seen from the figure, the performance of the 4 × 1 system in suppressing channel turbulence is inferior to that of the 4 × 2 system, such that, in the case of strong turbulence, the bit-error rate of the 4 × 1 system is only between 10^{-2} and 10^{-3}.

Comparing the system-simulation diagrams for unknown channel-state information in Figs. 9.6, 9.7 and 9.8, it can be seen that under the conditions of a 25-dB signal-to-noise ratio and SI = 0.4 (weak turbulence), the bit-error rate of the 4 × 1 system is better than 10^{-4}. When SI = 0.8, the bit-error rate of the 4 × 1 system is between 10^{-3} and 10^{-4}. When SI = 1.12 (strong turbulence), the bit-error rate of the 4 × 1 system is between (4) and (5), which is obviously inferior to the bit-error performance of the known channel-state information. It shows that the system with known channel-state information can suppress the turbulence better than the system with unknown channel-state information.

Atmospheric turbulence and other factors have a great impact on the bit-error performance of a free-space optical communication system. Therefore, we investigate the effect of disturbing the channel state in a communication system with known channel-state information; the other conditions remain unchanged. When SI = 0.4, simulations were carried out while disturbing the channel state; the channel-estimation deviations were taken as 2 and 4%. The simulation results are shown in Figs. 9.9 and 9.10.

As shown in Fig. 9.9, the channel-estimation deviation is 2%. When SI = 0.4, the bit-error rate of the 4 × 1 system is between 10^{-2} and 10^{-3}; close to 10^{-3}. The bit-error rate of the 4 × 2 system can reach below 10^{-3}, maintaining the trend of decreasing with the increase of the signal-to-noise ratio. When the channel-estimation deviation

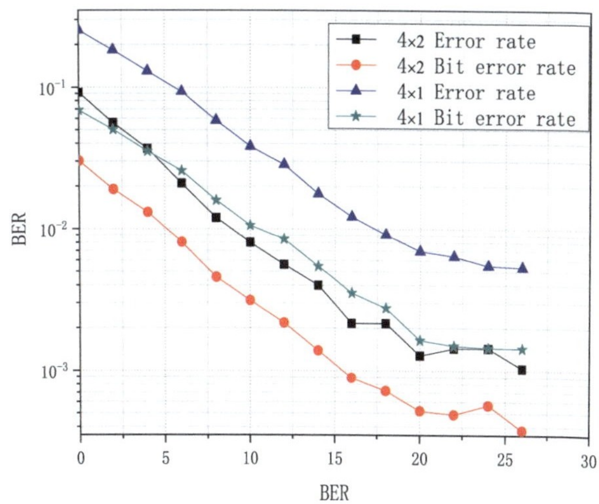

Fig. 9.9 Bit-error rate of unitary space–time coding when the channel-estimation deviation is 2% (SI = 0.4)

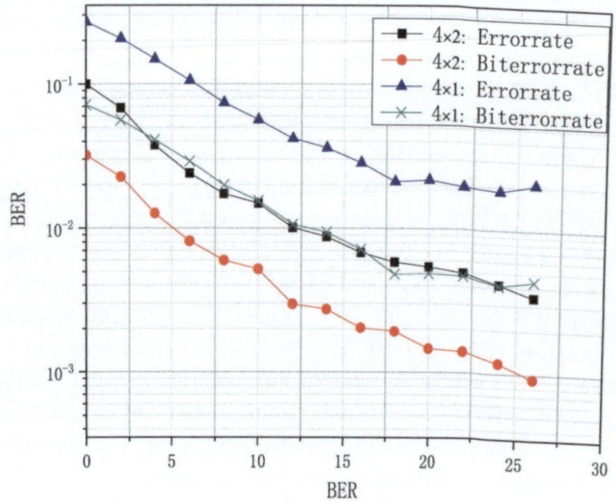

Fig. 9.10 Bit-error rate of unitary space–time coding when the channel-estimation deviation is 4% (SI = 0.4)

is 4%, as shown in Fig. 9.10, the bit-error rate of the 4×1 system drops to 10^{-2}–10^{-3}; close to 10^{-2}. The bit-error rate of the 4×2 system just reaches 10^{-3}. With the increase of the signal-to-noise ratio, the bit-error rate has no obvious downward trend, and the bit-error performance worsens. The results show that the channel-state estimation plays an important role. The more accurate the channel estimation is, the better the BER performance is.

9.5 Summary

In this chapter, unitary space–time coding and differential unitary space–time modulation in the RF field were discussed in detail, and a rotation-factor-based unitary space–time coding and decoding principle in the RF field was introduced. Using the orthogonality of pulse-position modulation (PPM), unitary space–time coding was applied to a free-space optical MIMO system to convert from RF communication to free-space optical communication. Finally, unitary space–time codes with known channel-state information and unknown channel-state information were simulated using MATLAB, and the system was simulated when the channel-state estimation was biased.

The results showed that the BER performance of unitary space–time coding improved with the increase in the number of receiving antennas in a free-space optical communication system. In addition, it was verified that the channel-state estimation plays an important role in the BER performance of the system.

References

1. Hughes BL (2000–11). Differential space-time modulation. IEEE Trans Inform Theory 46(7):2567–2578
2. Bertrand MH et al (2000–09) Systematic design of unitary space-time constellations. IEEE Trans Inform Theory 46(6):1962–1973
3. Hochwald BM, Marzetta TL (2000–03) Unitary space-time modulation for multiple-antenna communications in Rayleigh flat fading. IEEE Trans Inform Theory 46(2):543–564
4. Hochwald BM (2000) Systematic design of unitary space-time constellations. IEEE Trans Inform Theory 46(6):1962–1973
5. Hochwald BM, Marzetta TL (2000) Unitary space–time modulation for multiple-antenna communications in Rayleigh flat fading. IEEE Trans Inform Theory 46(2):543–564
6. Chen X, Zhou K, Aravena JL (2006) A new family of unitary space-time codes with a fast parallel sphere decoder algorithm. IEEE Trans Inform Theory 52(1):115–140
7. Tong H (2008) Study on the construction and bounds of unitary space-time codes. University of Science and Technology of China
8. Zhang D (2009) Research on unitary space-time code modulation technology in wireless MIMO communication system. Shandong University
9. Liang X, Xia X (2000) Some unitary signal constellations for differential space-time modulation. In: 2000.conference record of the thirty-fourth Asilomsr conference on signals, systems and computers, vol 1, pp 716–720
10. Liang X, Xia X (2002) Unitary signal constellations for differential space-time modulation with two transmit antennas: parametric codes, optimal designs, and bounds. IEEE Trans Inform Theory 48:2291–2322
11. Lv S, Wei G, Zhu J et al (2004) Differential unitary space-time modulation in fast fading channel. IEEE Veh Technol Conf Los Angeles CA 4:2374–2378
12. Hassibi B, Hochwald BM (2002) Cayley differential unitary space-time codes. IEEE Trans Inform Theory 48(9):1485–1503
13. Jiang C (2010) Design of differential unitary space-time code and its application in MIMO-OFDM. Suzhou University
14. Tirronen O, Boariu A, Hottinen A (2000) Minimal non-orthogonally rate 1 space-time block code for 3+Tx antennas. In: Spread spectrum techniques and applications, 2000 IEEE sixth international symposium, vol 2, pp 429–432
15. Jafarkhani H (2001) A quasi-orthogonal space-time block code. IEEE Trans Commun 49(1):1-4
16. Alamouti SM (1998) A simple transmit diversity technique for wireless communication. IEEE J Selected Areas Commun 16(8):1451–1458
17. Hochwald B, Sweldens W (2000) Differential unitary space—time modulation. IEEE Trans Commun 48(12):2041–2052
18. Li G, Hu Y, Zhu Z (2001) Decoding algorithm of space–time block coding. J Natl Univ Defense Technol 23(6):70–73

Chapter 10
Adaptive Layered Space–Time Coding

To improve the spectrum utilization of a system with fading channels, researchers have designed adaptive coding and modulation (ACM) technology. This technology can decide the channel-coding and modulation modes according to the traffic volume, average signal-to-noise ratio, average delay, interruption probability, and data rate. Adaptive coding and modulation technology can effectively improve the spectrum utilization, increase the system throughput, and inhibit the impact of atmospheric turbulence on system transmission performance [1–4].

Combining the characteristics of orthogonal space–time block codes (OSTBC) and layered space–time codes, this chapter discusses a 4×4 multiple-input multiple-output (MIMO) transmission scheme with adaptive layered space–time coding by referring to the layered space–time block coding method proposed by Tarokh. According to the received signal-to-noise ratio (SNR), the coding-modulation mode is adaptively adjusted to optimize the bit-error rate and transmission rate under different channel characteristics.

At the receiving end, a packet-interference suppression decoding method is adopted. The symbols sent by other packet-sending antennas are eliminated as interference, and then the symbols of the sent packet are decoded. This method can greatly reduce the decoding complexity of the system.

10.1 ACM Technology Research Progress

Adaptive coding and modulation (ACM) technology can be traced back to trellis-coded modulation, which was proposed in 1982 by Ungerboeck et al. [5]. It departs from the traditional mode that coding and modulation are independent of each other. For the first time, coding and modulation are considered as a whole. This technology can obtain a coding gain of 3–6 dB in systems with limited power and bandwidth.

X. Ke, *Space-Time Coding Theory in Optical Wireless MIMO Systems*,
Optical Wireless Communication Theory and Technology,
https://doi.org/10.1007/978-981-97-6449-5_10

In 1991, Vucetic proposed changing the coding rate and coding scheme to adapt the communication link. When the channel conditions were good, the coding scheme with a high bit rate was adopted, while when the channel conditions were bad, the coding scheme with a low bit rate was adopted to improve the error-correction performance of the system [6].

In 1997, Goldsmith and Alouini proposed an adaptive *m*-ary quadrature amplitude modulation (MQAM) scheme in fading channels [7] whose data-transmission rate and power were variable. By discretizing the modulation order and combining the optimal and suboptimal schemes, the spectrum utilization in fading channels was derived. When 56 different signal constellations were used, the gain was as high as 20 dB, compared with non-adaptive modulation.

In 2000, Vishwanath and Goldsmith studied the use of turbo codes in channel coding, different adaptive-modulation techniques in non-frequency selective fading channels [8], and used the optimal adaptive turbo-coded modulation (TuCM) scheme under various system constraints. The results showed that, compared with the traditional TuCM, this scheme adaptively adjusted the turbo encoder and transmission power, and achieved a gain of 3 dB.

In 2004, Maaref and Aissa studied the application of automatic repeat request (ARQ) adaptive-modulation technology in a multi-antenna system [9]. In this document, a cross-layer method combining ARQ (truncated ARQ) in the data-link layer and cooperative diversity in the physical layer was proposed. The source node and the relay node used orthogonal space–time block codes for retransmission. Cooperative diversity was only carried out when the destination node received the signal error of the source node, and the relay node selectively forwarded the information with cyclic redundancy checks (CRC). The analysis results showed that the proposed scheme could greatly improve the system throughput.

In 2006, Ko and Tepedelenlioglu combined OSTBC with variable-rate adaptive M-QAM modulation [10] to improve the STBC throughput.

In 2005, Li and Bai proposed an adaptive system scheme based on a combination of hybrid automatic repeat request (HARQ) technology and a concatenated two-state trellis-coded modulation (CT-TCM) scheme, and its performance was improved to some extent [11].

In 2007, Li and Shen proposed to apply low-density parity-check (LDPC) codes to an adaptive coding and modulation system. This scheme overcame the time-varying characteristics of fading channels and enabled the system to achieve better performance at low SNRs [12].

In 2011, Zhang and Ji proposed to apply adaptive bit and power allocation (ABPA) to a space–time block code (STBC) MIMO-orthogonal frequency-division multiplexing (OFDM) system, and to introduce the Fischer algorithm with the goal of minimizing the bit-error rate in adaptive modulation technology. Compared with the general MIMO-OFDM system, the proposed system model obtained better bit-error performance [13].

10.2 Basic Adaptive Coding and Modulation Principle

At the end of the 1960s, Hayes proposed the adaptive coding concept [11]: "An effective way to combat the fading effect is to adaptively adjust the modulation and channel-coding methods and other system parameters according to the instantaneous channel-quality information obtained by the receiver, and then the channel-quality information is fed back to the sender through the feedback channel." Adaptive coding and modulation (ACM) technology is an effective method to improve the communication performance of a system. The basic principle of ACM is to obtain instantaneous channel-state information (CSI) by channel estimation at the receiving end. The channel-state information is returned to the sender through a feedback link, and the sender adjusts the transmission parameters in real time, according to the change of the current channel state, to select the optimal modulation mode.

The channel-state information (CSI) refers to time-varying channel-fading information, which can be the bit-error rate, frame-error rate, signal-to-noise ratio, etc. CSI can be directly characterized by the channel-state information fed back from the previous transmission (for slow-changing channels), or the current channel state can be estimated according to the channel-state information fed back from the previous n transmissions (for fast-changing channels with continuous changes). When the channel condition is good, a high-order coding and modulation method is adopted to improve the transmission rate and system capacity. When the channel condition deteriorates, a low-order coding and modulation method is adopted to ensure the required system bit-error rate.

The adaptive coding and modulation principle is shown in Fig. 10.1. At the receiving end, the wireless fading channel characteristics are estimated, the received signal is compensated for fading, the coding and modulation mode adopted by the transmitting end is extracted, and then the received data are demodulated, decoded, and output. Simultaneously, according to the channel-estimation result of the receiver, an optimal coding and modulation mode is selected, according to a given adaptive algorithm, and the coding and modulation mode of the transmitter is adjusted through the feedback link.

10.3 Adaptive Layered Space–Time Coding of a 4 × 4 FSO-MIMO System

Multiple antennas at the sending end are divided into multiple groups, and send information symbols, after independently processing the codes in the groups. At the receiving end, the influence of the transmission symbols on the other packet-transmission antennas is eliminated by packet-interference suppression, and the transmission symbols in this packet are space–time decoded [12].

Based on this idea, combined with the characteristics of orthogonal space–time block codes and layered space–time codes (LSTC), a scheme of adaptively adjusting

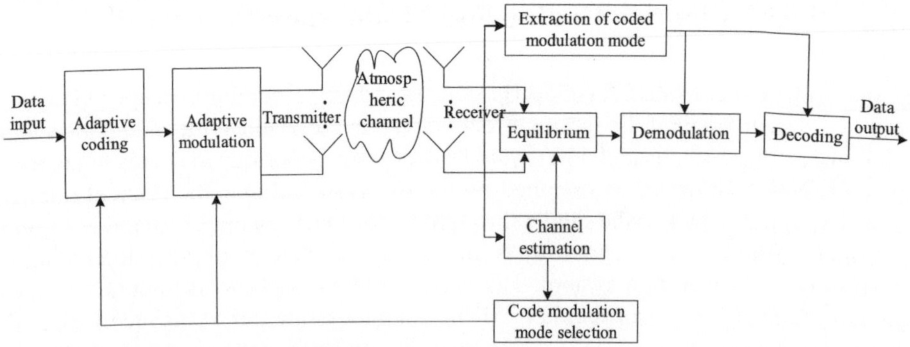

Fig. 10.1 Principle of adaptive coding and modulation

the modulation mode of the space–time codes, according to the received signal-to-noise ratio in free-space optical (FSO) communication is proposed. This should ensure that the data-transmission rate and system capacity can be increased as much as possible while satisfying a certain bit-error rate.

10.3.1 Principle of Adaptive Layered Space–Time Coding of a 4 × 4 System in FSO-MIMO

Orthogonal space–time block coding (OSTBC) has good noise-suppression performance at low signal-to-noise ratios because of its full diversity; however, its data-transmission rate is low. Layered space–time coding (LSTC) has receive diversity, but no transmit diversity. Although it can achieve a higher rate and larger capacity at high SNRs, its error performance cannot meet the requirements at low SNRs. Therefore, combining with the advantages and disadvantages of OSTBC and LSTC, based on the layered space–time coding principle of multi-antenna grouping, a four transmit-antenna and four receive-antenna (4 × 4) system is shown in Fig. 10.2.

Figure 10.2 shows the schematic diagram of a 4 × 4 adaptive layered space–time coding system in FSO-MIMO. After a serial–parallel conversion, the binary data stream sent by the source is divided into two data-stream layers. The data streams of each layer are coded by the corresponding space–time encoder, and then transmitted by the corresponding antenna after pulse-position modulation (PPM).

At the receiving end, four optical receiving antennas receive data. The received signal-to-noise ratios of these two layers are detected and compared. Then, the modulation-mode decider selects which decoding method these two groups of space–time decoders adopt. The modulation control information is transmitted to the space–time coding-modulation mode controller through the feedback channel. It determines whether the two groups of encoders adopt bit-complement orthogonal space–time block coding (BCOSTBC) or vertical layered space–time coding (V-BLAST).

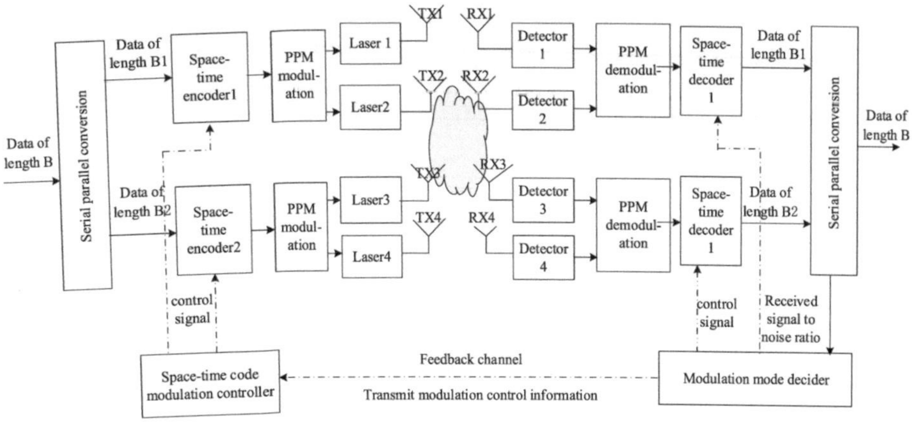

Fig. 10.2 Principle of adaptive layered space–time coding for 4 × 4 antennas in FSO-MIMO

While the same noise-suppression performance is achieved, OSTBC and LSTC have different SNR ranges. In adaptive layered space–time coding, two thresholds, SNR0 and SNR1, need to be set. The modulation-mode decider will determine the decoding mode of the two groups of space–time decoders at the receiving end, according to the relationship between the detected SNR value and the set threshold. In addition, the control information is transmitted to the transmitter through the feedback channel to determine which encoding and modulation mode the two groups of space–time encoders adopt. This ensures that the encoding and modulation mode at the transmitter is synchronized with the demodulation and decoding mode at the receiver. The rules for setting the coding modulation mode are as follows:

(1) When SNR < SNR0, the two groups of space–time encoders adopt BCOSTBC, which is called transmission-mode one.
(2) When SNR0 < SNR < SNR1, one group of encoders adopts BCOSTBC, and the other group adopts V-BLAST, which is called transmission-mode two.
(3) When SNR > SNR1, the two groups of encoders adopt V-BLAST, which is called transmission-mode three.

10.3.2 Orthogonal Space–time Block Codes and Vertical Layered Space–time Codes in Optical Communication

Using traditional Alamouti space–time coding [12], the transmission-code matrix in two adjacent transmission periods can be expressed as

$$X = \begin{bmatrix} x_1 & -x_2^* \\ x_2 & x_1^* \end{bmatrix}. \tag{10.1}$$

According to Eq. (10.1), at time t of the first transmission cycle, signals x_1 and x_2 are simultaneously transmitted from antennas 1 and 2, respectively. At the second transmission cycle time, $t + T$, the signals $-\mathrm{x}_2{}^*$ and $\mathrm{x}_1{}^*$ are simultaneously transmitted from antennas 1 and 2, respectively, where $\mathrm{x}_1{}^*$ is the complex conjugate of x_1 and $-\mathrm{x}_2{}^*$ is the negative conjugate of x_2.

Because of the difference between atmospheric laser communication and radio-frequency (RF) communication, traditional space–time coding cannot be directly applied to atmospheric laser communication. In radio-frequency communications, Alamouti codes can represent negative and complex numbers; however, only real numbers can be transmitted in atmospheric laser communications because complex conjugate and negative numbers cannot be represented by "*on*" and "off".

At present, atmospheric laser communication usually adopts intensity modulation, so researchers [13, 14] improved the traditional Alamouti code, and proposed a bit-complement orthogonal space–time code (BCOSTBC) suitable for an intensity modulation/direct detection (IM/DD) atmospheric-turbulence channel. It replaces the negative and complex signals existing in the traditional Alamouti code by seeking a symbolic complement. The improved coding matrix can be expressed as follows:

$$X = \begin{bmatrix} x_1 & \overline{x}_2 \\ x_2 & x_1 \end{bmatrix}, \tag{10.2}$$

where $\overline{x}_2$ represents the complement of symbol x_2. Using 4-PPM modulation, when $x = [1, 0, 0, 0]$, $\overline{x} = [0, 0, 0, 1]$; when $x = [0, 1, 0, 0]$, $\overline{x} = [0, 0, 1, 0]$; when $x = [0, 0, 1, 0]$, $\overline{x} = [0, 1, 0, 0]$; and when $x = [0, 0, 0, 1]$, $\overline{x} = [1, 0, 0, 0]$ $[1, 0, 0, 0]$. A scalar 0 indicates that no light pulse exists in the corresponding position, and a scalar 1 indicates that a light pulse exists in the corresponding position.

The vertical layered space–time coder (V-BLAST) receives the output from the parallel channel coder and performs spatial coding in the vertical direction. In two consecutive transmission periods, its transmission code matrix is

$$X = \begin{bmatrix} x_1 & x_3 \\ x_2 & x_4 \end{bmatrix}. \tag{10.3}$$

10.3.3 Encoding Adaptive Layered Space–Time Codes

According to the three modes proposed in the above scheme, we divide these four antennas into two groups with two antennas each, and design a transmission-code matrix for these three transmission modes.

10.3.3.1 Transmission-Mode One

When the transmission mode is one, both groups of space–time encoders adopt BCOSTBC codes. The transmission-code matrix in two symbol periods is

$$X = \begin{bmatrix} x_1 & \overline{x}_2 \\ x_2 & x_1 \\ x_3 & \overline{x}_4 \\ x_4 & x_3 \end{bmatrix}. \tag{10.4}$$

The encoded symbols are transmitted from the two antennas, respectively. In the transmission cycle of the first symbol, symbols x_1, x_2, x_3, and x_4 are simultaneously transmitted from transmission antennas 1, 2, 3, and 4, respectively. In the second symbol-transmission cycle, symbols $\overline{x}_2, x_1, \overline{x}_4$, and x_3 are simultaneously transmitted from transmission antennas 1, 2, 3, and 4, respectively, where $\overline{x}_2$ and $\overline{x}_4$ are the complements of x_2 and x_4, respectively.

10.3.3.2 Transmission-Mode Two

When the transmission mode is two, one group adopts BCOSTBC and the other adopts V-BLAST. The transmission-code matrix in two symbol periods is

$$X = \begin{bmatrix} x_1 & \overline{x}_2 \\ x_2 & x_1 \\ x_3 & x_5 \\ x_4 & x_6 \end{bmatrix}. \tag{10.5}$$

The coded symbols are transmitted from two antennas. In the first symbol-transmission cycle, symbols x_1, x_2, x_3, and x_4 are simultaneously transmitted from transmitting antennas 1, 2, 3, and 4, respectively. In the second symbol-transmission cycle, symbols $\overline{x}_2$, x_1, x_5, and x_6 are simultaneously transmitted from transmitting antennas 1, 2, 3, and 4, respectively. Among them, $\overline{x}_2$ is the complement of x_2.

10.3.3.3 Transmission-Mode Three

When the transmission mode is three, both groups adopt V-BLAST, and the transmission-code matrix in two symbol periods is

$$X = \begin{bmatrix} x_1 & x_5 \\ x_2 & x_6 \\ x_3 & x_7 \\ x_4 & x_8 \end{bmatrix}. \tag{10.6}$$

The coded symbols are transmitted by two antennas. In the first symbol-transmission cycle, symbols x_1, x_2, x_3, and x_4 are sent out simultaneously from sending antennas 1, 2, 3, and 4. In the second symbol-transmission cycle, symbols x_5, x_6, x_7, and x_8 are simultaneously transmitted from transmitting antennas 1, 2, 3, and 4, respectively.

In mode 1, BCOSTBC coding is adopted for both groups of antennas, and the diversity gain of each code group is 8. Four transmit antennas transmit a total of four different real signals at two symbol intervals, with a symbol rate of $R = K/P = 4/2 = 2$ (symbol/time slice). In mode 2, one group of antennas adopts BCOSTBC coding with a diversity gain of 8, and the other group of antennas adopts V-BLAST coding with a diversity gain of 4. The four transmit antennas transmit a total of six different real signals at two symbol intervals at a symbol rate of $R = K/P = 6/2 = 3$ (symbol/time slice). In mode 3, V-BLAST is used, and its diversity gain is 4; four transmit antennas transmit a total of eight different real signals at two symbol intervals, with a symbol rate of $R = K/P = 8/2 = 4$ (symbol /time slice).

It can be seen that mode one adopts full BCOSTBC coding, which has the largest diversity gain but the slowest transmission rate. The third mode adopts full V-BLAST coding, with the maximum transmission rate but the minimum diversity gain. The second mode adopts a mix of BCOSTBC and V-BLAST, and the diversity gain and transmission rate are between those of the other modes.

10.3.4 Decoding of Adaptive-Layered Space–time Codes

For transmission modes one and three, full BCOSTBC coding and full V-BLAST coding are adopted, respectively. See Refs. [14, 16] for their decoding methods. In the second mode, the performance advantages of BCOSTBC and V-BLAST are combined, and it can be seen from its transmission-code matrix (10.5) that the first data channel sends a set of 2 × 2 BCOSTBC symbols, and the last two data channels send a set of 2 × 2 V-BLAST symbols. The equations are as follows:

$$X_{BCOSTBC} = \begin{bmatrix} x_1 & \bar{x}_2 \\ x_2 & x_1 \end{bmatrix} \tag{10.7}$$

$$X_{VBLAST} = \begin{bmatrix} x_3 & x_5 \\ x_4 & x_6 \end{bmatrix}. \tag{10.8}$$

In the second mode, a mixed space–time coding scheme is adopted. First, we decode with BCOSTBC coding, eliminate its influence on the first layer, and perform V-BLAST decoding on the rest.

Linear-decoding and nonlinear-decoding algorithms exist. Linear-decoding algorithms include the zero forcing (ZF) and minimum mean-squared error (MMSE) algorithms, while nonlinear algorithms include the successive interference-cancellation (SIC) and ordered successive interference-cancellation algorithms (OSIC).

Consider that the channel model is a quasi-static flat fading channel; that is, the elements of the channel matrix are random, but the channel is constant in a certain discrete time interval T. When changing to another time interval t, the channel changes to another constant accordingly. In addition, it is assumed that the channel-state information (CSI) in the received signal is

$$y = \eta HX + n. \tag{10.9}$$

η—photoelectric-conversion efficiency, which is a constant related to the light intensity;

$y = \left[y_i\right]^T (i = 1, 2, \ldots, N)$—signals received on n receiving antennas;

$X = [x_i]^T (i = 1, 2, \ldots M)$—signals transmitted on m transmitting antennas;

H—channel-gain matrix, whose element h_{ij} represents the channel-gain coefficient from transmitting antenna j to receiving antenna i, expressed as

$$H = \begin{bmatrix} h_{11} & h_{12} & h_{13} & h_{14} \\ h_{21} & h_{22} & h_{23} & h_{24} \\ h_{31} & h_{32} & h_{33} & h_{34} \\ h_{41} & h_{42} & h_{43} & h_{44} \end{bmatrix}. \tag{10.10}$$

n—noise matrix. Because we are studying laser communication, n is in the real-number domain, in which each component of n is independent of the others. The additive noise can be considered to be the noise generated by the electronic circuit and detector, and follows a Gaussian distribution with a mean value of 0 and variance of σ^2. It is expressed as

$$n = \begin{bmatrix} n_{11} & n_{12} \\ n_{21} & n_{22} \\ n_{31} & n_{32} \\ n_{41} & n_{42} \end{bmatrix}. \tag{10.11}$$

To solve the two symbols of the first layer in 4 × 4 hybrid coding, the concept of packet interference suppression is introduced here. The core idea of packet interference suppression is to perform a separate decoding operation on the output $C_i(k)(1 \le i \le p)$ of the space–time encoder and simultaneously suppress the signals from other groups [17, 18]. The advantage of this method is low complexity; however, the disadvantage is that full diversity *MN* cannot be obtained. (In the 4×4 system in this example, the packet interference suppression algorithm is used to suppress the signals transmitted by the second group of transmitting antennas; thus, the diversity gain obtained at this time is $2 \times 4 = 8$, instead of the full diversity gain of $4 \times 4 = 16$.)

In the above layered space–time coding method, based on multi-antenna grouping, the first two transmit antennas are divided into a group represented by $G1$, the last two transmit antennas are a group represented by $G2$, and the corresponding transmit symbol vector of the group is represented by Gi ($i = 1, 2$), which is a 2×1 matrix. Without loss of generality, the transmission symbols corresponding to the first group of transmission antennas are selected as the desired signal, and the packet interference suppression method is used for decoding. At this time, there is only one interference signal.

Assuming that the receiver knows the channel information (that is, the known channel-transmission matrix H), the definition matrix will be divided into two parts, H_1, H_2, as follows:

$$H_1 = \begin{bmatrix} h_{11} & h_{12} \\ h_{21} & h_{22} \\ h_{31} & h_{32} \\ h_{41} & h_{42} \end{bmatrix} \tag{10.12}$$

$$H_2 = \begin{bmatrix} h_{13} & h_{14} \\ h_{23} & h_{24} \\ h_{33} & h_{34} \\ h_{43} & h_{44} \end{bmatrix}. \tag{10.13}$$

It can be seen from Eqs. (10.12) and (10.13) that $H1$ and $H2$ are the first two columns and the last two columns of the channel-gain matrix, respectively. Its rank $rank[H_2] \leq 2$ defines P as the zero space of the matrix; P is formed by all row-vector matrices u satisfying the following conditions: $uH_2 = 0$, where 0 is a null vector, which is defined by the null space of the matrix. Let $A \in C^{m\times n}$, $N(A) = \{x|Ax = 0, x \in C^n\}$ be called the null space of the matrix A.

In addition, $\dim N(A) = m - rankA$ and $\dim(u) + rank(H_2) = 4$. Because of the dimension of $rank(H_2) \leq 2$, the dimension of u is $\dim(u) \geq 2$. Therefore, we can find a set of orthogonal normalization vectors in P: $\{u_1, u_2\}$. u_1, u_2 is a 1×4 row vector, respectively. u is defined as a 2×4 matrix composed of u_i, and its ith state is u_i, $u = [u_1, u_2]^T$. Because u_i is an orthogonal normalized vector,

$$u_j u_i^H = \begin{cases} 1, j = i \\ 0, j \neq i \end{cases}. \tag{10.14}$$

Then, $u \cdot u^H = I_2$, where I_2 is the identity matrix of order two. Both sides of the equation of the received signal are simultaneously multiplied by the orthogonal normalization vector u; because H is composed of $H1$ and $H2$, $uH_2 = 0$; therefore, $uy = \eta uHX + un = \eta uH_1 X_{BCOSTBC} + un$, where η is the photoelectric-conversion efficiency, and the values of $X_{BCOSTBC}$, H_1, and n are shown in Eqs. (10.7), (10.12), and (10.1), respectively. It can be seen that the signals on the third and fourth antennas are suppressed. Therefore, a 4 × 4 system becomes a 2 × 2 system after packet interference suppression. In this manner, the BCOSTBC code of the first layer is isolated through packet interference suppression, and the first two symbols can be solved using the BCOSTBC decoding method by itself.

10.3.4.1 Joint V-BLAST and BCOSTBC Decoding

The first step is decoding the BCOSTBC code.

To apply the above-mentioned packet-interference suppression method to the BCOSTBC of the first layer, we must first find a matrix u that can make $uH_2 = 0$ work. Here, the singular-value decomposition (SVD) method is adopted, namely,

$$H_2 = U \begin{pmatrix} \sum & 0 \\ 0 & 0 \end{pmatrix} V^H, \tag{10.15}$$

where U is a unitary matrix of order four, V is a unitary matrix of order two, $\sum = diag(\sigma_1, \sigma_2, \ldots \sigma_r)$, and $\sigma_i (i = 1, 2, \ldots r)$ is a non-zero singular value of H_2.

Let $U = [U_1 U_2]$, where U_1 is the first two columns and U_2 is the last two columns of unitary matrix U. $U_1^H U_1 = I$, $U_2^H U_2 = I$, and I is a 2 × 2 unit array. $U_2^H U_1 = 0$. That is, the two columns of U_1 and U_2 are pairwise orthogonal unit vectors:

$$U^H H_2 V = \begin{bmatrix} U_1^H \\ U_2^H \end{bmatrix} H_2 V = \begin{bmatrix} U_1^H H_2 V \\ U_2^H H_2 V \end{bmatrix} = \begin{bmatrix} \sum \\ 0 \end{bmatrix}$$

$$\therefore U_2^H H_2 V = 0 \text{ and } \because V \neq 0, \therefore U_2^H H_2 = 0.$$

As can be seen from the above formula, U_2^H is the row-vector matrix constituting the zero space u of P:

$$u = U_2^H = \begin{bmatrix} u_{11} & u_{12} & u_{13} & u_{14} \\ u_{21} & u_{22} & u_{23} & u_{24} \end{bmatrix}. \tag{10.16}$$

After multiplying both sides of Eq. (10.9) by u left, the 2 × 2 BCOSTBC coding matrix of the first layer will be separated from the right side of the formula, except that the noise matrix composed of noise terms is correspondingly left-multiplied by a 2 × 4 normalized vector. According to the existing theoretical analysis [19], this does not change the noise distribution, and correspondingly, does not degrade the received signal. Similarly, the original received signal on the left side of the formula can obtain a new received signal y after packet interference suppression, which is expressed as follows:

$$
\begin{aligned}
r = uy = \eta uHX + un &= \eta uH_1X_{BCOSTBC} + un \\
&= \eta \begin{bmatrix} u_{11} & u_{12} & u_{13} & u_{14} \\ u_{21} & u_{22} & u_{23} & u_{24} \end{bmatrix} \begin{bmatrix} h_{11} & h_{12} \\ h_{21} & h_{22} \\ h_{31} & h_{32} \\ h_{41} & h_{42} \end{bmatrix} \begin{bmatrix} x_1 & \overline{x}_2 \\ x_2 & x_1 \end{bmatrix} \\
&+ \begin{bmatrix} u_{11} & u_{12} & u_{13} & u_{14} \\ u_{21} & u_{22} & u_{23} & u_{24} \end{bmatrix} \begin{bmatrix} n_{11} & n_{12} \\ n_{21} & n_{22} \\ n_{31} & n_{32} \\ n_{41} & n_{42} \end{bmatrix}.
\end{aligned} \tag{10.17}
$$

Next, the BCOSTBC coding matrix of the first layer, separated from the original BCOSTBC and V-BLAST mixed coding matrix after packet interference suppression, is decoded. Set

$$
\begin{pmatrix} a & b \\ c & d \end{pmatrix} = \begin{bmatrix} u_{11} & u_{12} & u_{13} & u_{14} \\ u_{21} & u_{22} & u_{23} & u_{24} \end{bmatrix} \begin{bmatrix} h_{11} & h_{12} \\ h_{21} & h_{22} \\ h_{31} & h_{32} \\ h_{41} & h_{42} \end{bmatrix}. \tag{10.18}
$$

The change of noise after packet interference suppression is represented by the following formula:

$$
\begin{bmatrix} N_{11} & N_{12} \\ N_{21} & N_{22} \end{bmatrix} = \begin{bmatrix} u_{11} & u_{12} & u_{13} & u_{14} \\ u_{21} & u_{22} & u_{23} & u_{24} \end{bmatrix} \begin{bmatrix} n_{11} & n_{12} \\ n_{21} & n_{22} \\ n_{31} & n_{32} \\ n_{41} & n_{42} \end{bmatrix}. \tag{10.19}
$$

Therefore, the received signals in two consecutive time slots, that is, one BCOSTBC symbol interval, are as follows:

$$
\begin{bmatrix} r_{11} & r_{12} \\ r_{21} & r_{22} \end{bmatrix} = \begin{bmatrix} \eta a x_1 + \eta b x_2 + N_{11} & \eta a \overline{x}_2 + \eta b x_1 + N_{12} \\ \eta c x_1 + \eta d x_2 + N_{21} & \eta c \overline{x}_2 + \eta d x_1 + N_{22} \end{bmatrix}. \tag{10.20}
$$

Combining the received signal and the channel-state information structure generates two decision statistics, and the statistical results can be expressed as

$$
\tilde{x}_1 = \eta(ar_{11} + cr_{21} + br_{12} + dr_{22}) \tag{10.21}
$$

$$\tilde{x}_2 = \eta(br_{11} + dr_{21} + a\overline{r}_{12} + c\overline{r}_{22}). \tag{10.22}$$

Next, we find the point with the smallest distance measurement from $\tilde{x}_1, \tilde{x}_2$, and η is the photoelectric-conversion efficiency, which is a constant; therefore,

$$\hat{x}_1 = \arg\min_{\hat{x}_1 \in s} |\tilde{x}_1 - \hat{x}_1|^2 \tag{10.23}$$

$$\hat{x}_2 = \arg\min_{\hat{x}_2 \in s} |\tilde{x}_2 - \hat{x}_2|^2, \tag{10.24}$$

where s is all possible sets of modulation symbols. In this manner, two estimated values $\hat{x}_1, \hat{x}_2$ corresponding to x_1, x_2 are obtained.

The second step is to eliminate the interference of the first two symbols.

After the subsystem is decoded by decoding the space–time block code in the first step, it is returned to the original system, as shown in Eq. (10.5). The matrix obtained after eliminating the influence of the symbol of the subsystem, as shown in Eq. (10.7), is

$$r_1 = r - \eta H_1 \begin{bmatrix} x_1 & \overline{x}_2 \\ x_2 & x_1 \end{bmatrix} = \eta H_2 X_{VBLAST} + n. \tag{10.25}$$

In the equation, the value of H_2 is shown in Eq. (10.13), and the value of n is shown in Eq. (10.11). If x_1, x_2 is completely correctly translated in the first step, some values of X_{VBLAST} are shown in Eq. (10.8).

In the third step, the remaining symbols after the BCOSTBC code is eliminated are decoded using V-BLAST. Among the decoding methods of V-BLAST, the commonly used linear decoding methods are the minimum mean-squared error (MMSE) detection algorithm and the zero-forcing algorithm (ZF). See Chap. 12 or Refs. [15, 16] for the specific algorithm processes.

10.4 Calculation of the Receiving SNR

The 4×4 adaptive layered space–time coding scheme adaptively selects the modulation mode of the transmitter, according to the received signal-to-noise ratio. Assuming that the transmitting antenna is controllable, that is, only one antenna is allowed to transmit at a time, four subsystems are formed. Each subsystem consists of one transmitting antenna and four receiving antennas, which are referred to as a transmitting-antenna subset.

The received signal-to-noise ratio (SNR) values corresponding to the four transmit-antenna subsets are calculated, and then the modulation modes of the four transmit-antenna subsets are selected, according to the comparison between the SNR values and the set threshold. According to the frozen-field theory, the attenuation of optical signals caused by atmospheric turbulence is a slow-changing process relative to the communication rate, and the atmospheric channel can be regarded as a time-varying channel with memoryless quasi-static flat fading.

The equivalent mathematical model of an atmospheric channel is [19, 20]

$$y(t) = sx(t) + n(t) = \eta Ix(t) + n(t), \tag{10.26}$$

where $y(t)$ is the signal received by the receiver; $x(t)$ is the signal sent by the transmitter; η is the photoelectric-conversion efficiency; $s = \eta I$ is the instantaneous light-intensity gain, and $n(t)$ is the channel state. It is a statistically independent additive white Gaussian noise with a mean value of 0 and a variance of $\sigma_v^2 = N_0$ (N_0 is the average power of the noise) in the real number field.

The received signal-to-noise ratio (SNR) of the optical communication system is defined as [21]

$$SNR = \frac{P_r}{P_v} = \frac{\eta^2 I^2}{N_0} = \frac{s^2}{N_0}, \tag{10.27}$$

where P_r is the total received power, P_v is the noise power, η is the photoelectric-conversion efficiency, I is the light intensity, $s = \eta I$ is the instantaneous light-intensity gain, and N_0 is the variance of the Gaussian distribution, that is, the average power of noise.

In the adaptive layered space–time coding scheme in laser communication, the noise matrix $n(t) = [n_1(t), n_2(t), n_3(t), n_4(t)]^T$ is in the real-number domain, and the components of $n(t)$ are independent of each other. It is Gaussian white noise with a mean value of 0 and variance of $\sigma_v^2 = N_0$; thus, the noise power of each receiving antenna at the receiving end is equal.

Because of this, the received signal-to-noise ratio of each antenna subset only depends on the total received power P_r of each antenna subset, according to Eq. (10.27). In addition, because $s = \eta I$ is the instantaneous light-intensity gain (i.e., channel state) and the channel state is determined by the channel-gain matrix $H(t)$, the total received power of each antenna subset at the receiving end only depends on the sum of the squares of the channel gains corresponding to each antenna subset.

For the channel-gain matrix, the sum of squares of the elements of the 4×4 channel-gain matrix in each column is as follows:

$$s_j = \|H_j\|_F^2, \tag{10.28}$$

where H_j represents the channel-gain matrix corresponding to the jth antenna subset, $j = 1, 2, 3, 4$. $\|\cdot\|_F^2$ represents the square of the F-norm. If A is a matrix, the F-norm of matrix A can be expressed as $\|A\|_F = [tr(A^H A)]^{1/2} = [tr(AA^H)]^{1/2}$.

The received signal-to-noise ratio (SNR) corresponding to each antenna subset is

$$SNR = \frac{\eta^2 I^2}{N_0} = \frac{\|H_j\|_F^2}{N_0}. \quad (10.29)$$

From Eq. (10.29), the values of the received signal-to-noise ratio corresponding to the four transmit-antenna subsets can be calculated. The two antennas with a small signal-to-noise ratio are taken as a group, and the remaining two are taken as a group. Then, the modulation mode of the transmit antenna is selected by comparing the calculated SNR of the two groups with the set thresholds $SNRO$ and $SNR1$. In the above equation, N_0 is a constant. Knowing the value of H_j ($j = 1, 2, 3, 4$), we can calculate the value of SNR.

In practice, the channel-state information is unknown and must be estimated. Channel estimation is the process of estimating the parameters of the channel model from the received data, which is a key technology in wireless communication. In the following, two commonly used channel-estimation algorithms are introduced.

10.4.1 Channel-Estimation Algorithms

10.4.1.1 Least-Squares (LS) Estimation

Least-squares (LS) estimation is a widely used method that is suitable for linear-observation models and does not need any description of the probability or statistical characteristics of the data to be estimated and observed. For consistency in expression, the estimation model containing the parameters to be estimated is shown in Eq. (10.9). LS channel estimation is obtained from the sense of least squares. According to least-squares theory [22–25], to least-squares estimate parameter h, the cost function of the least-squares estimation is defined by the channel model of laser communication as

$$C(h) = (Y - \eta\hat{H}X)^H (Y - \eta\hat{H}X), \quad (10.30)$$

where Y, $\hat{H}$, and X represent the received sequence of pilot positions, and the estimated values of the channel and pilot sequence, respectively. The $\hat{H}$ that minimizes the cost function is the least-squares estimation of H; that is, the partial derivative of H in the cost function is obtained and set equal to 0. From the least-squares criterion, the channel response H_{LS} is

$$H_{LS} = \min\{(Y - \eta\hat{H}X)^H (Y - \eta\hat{H}X)\} \quad (10.31)$$

$$\frac{\partial\left\{(Y - \eta\hat{H}X)^H (Y - \eta\hat{H}X)\right\}}{\partial H_{LS}} = 0. \quad (10.32)$$

Then, the estimated value in the time domain is

$$\hat{h}_{LS} = \frac{1}{\eta}\left(x^T x\right)^{-1} x^T y, \tag{10.33}$$

where η is the photoelectric-conversion efficiency, x is the transmitted pilot information vector, and y is the received vector.

The estimated value in the frequency domain is

$$\hat{H}_{LS} = \frac{1}{\eta} X^{-1} Y \tag{10.34}$$

with order

$$\hat{H}_{LS} = \frac{1}{\eta}[H(0)H(1)\ldots H(N-1)]^T.$$

Then,

$$\hat{H}_{LS} = \frac{1}{\eta} X^{-1} Y = \frac{1}{\eta}\left[\frac{Y(0)}{X(0)}\,\frac{Y(1)}{X(1)} \cdots \frac{Y(N-1)}{X(N-1)}\right]. \tag{10.35}$$

Here,

$$X = \begin{bmatrix} X(0) & & 0 \\ & \ddots & \\ 0 & & X(N-1) \end{bmatrix}. \tag{10.36}$$

X is a diagonal matrix of the transmission-sequence vector, and the elements on the diagonal are the information symbols transmitted on the corresponding subchannels. When X is not a square matrix, or X is a square matrix but its determinant is 0 ($\det(X) = 0$), the inverse matrix of X does not exist. To solve the problem that the above matrix cannot be directly inverted, the LS algorithm replaces the direct inversion of matrix X with a generalized inverse matrix $(X^H X)^{-1} X^H$. The expression of such a channel-estimation result is

$$\begin{aligned} \hat{H}_{LS} &= \frac{1}{\eta} X^{+} Y = \frac{1}{\eta}(X^H X)^{-1} X^H Y \\ &= \frac{1}{\eta} Y X^{+} = \frac{1}{\eta} Y X^H (X X^H)^{-1} \end{aligned} \tag{10.37}$$

where $X^{+} = (X^H X)^{-1} X^H$ is the generalized inverse matrix of X. To ensure the existence of the generalized inverse matrix of X, the pilot sequence must satisfy the row full rank. In a flat fading channel, the Hadamard sequence, Gold sequence, and other orthogonal sequences can be used as pilot sequences.

The channel-estimation algorithm based on the LS criterion has a simple structure, only one division operation, and the amount of calculation is small. However, the

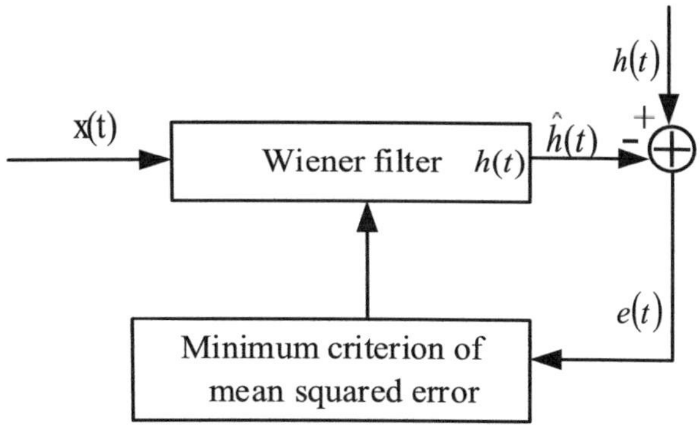

Fig. 10.3 Block diagram of Wiener filtering system

correlation characteristics of the frequency and time domains of the channel are not used in LS estimation, and the influence of noise is ignored; hence, the channel-estimation value is sensitive to the influence of noise. When the signal-to-noise ratio is small, the estimation accuracy is not high, which affects the parameter estimation of the data subchannels.

10.4.1.2 Channel Estimation Using Wiener Filtering Based on the MMSE Criterion

If the signal $x(n)$ or $X(n)$ and the observed signal $y(n)$ or $Y(n)$ is wide-sense stationary, and its autocorrelation function or power-spectral density is known or available, the process of estimating $x(n)$ or $X(n)$ using the minimum mean-squared error (MMSE) criterion is called Wiener filtering. The idea of Wiener filtering is to find the best impact function or transfer function of a linear filter to minimize the mean-squared error between the output waveform and the input waveform. The system block diagram of a Wiener filter is as follows (Fig. 10.3).

According to the MMSE criterion, the cost function is defined as

$$C(h) = E[(H_k - \hat{H}_k)(H_k - \hat{H}_k)^H]. \tag{10.38}$$

According to Wiener filter theory, it is assumed that the estimated value of the channel is

$$\hat{H} = \sum_{K=1}^{M} w_k y_k = W_k^H Y_k = Y_k^H W_k, \tag{10.39}$$

Where $Y_k = [Y_1, Y_2, \ldots, Y_M]^T$ is the received signal vector at the pilot position, $W_k = [w_1, w_2, \ldots, w_M]^T$ is the tap coefficient of the filter, and M is the number of pilots. The estimation error is defined as

$$e_k = H_k - \hat{H}_k = H_k - W_k^H Y_k = H_k - Y_k^H W_k \tag{10.40}$$

$$\begin{aligned} C(h) &= E[e_k e_k^H] = E[(H_K - \hat{H}_K)(H_K - \hat{H}_K)^H] \\ &= E[H_K H_K] - 2E\big[H_K Y_K^H\big] W_K \\ &+ W_K^H E[Y_K Y_K^H] W_K \end{aligned} \tag{10.41}$$

Order it; then, $R_{YY} = E[Y_k Y_k^H]$, $R_{HY} = E[H_k Y_k^H]$ and

$$C(h) = E\big[|e_k|^2\big] = E\big[|H_K|^2\big] - 2R_{HY} W_k + W_k^H R_{YY} W_k. \tag{10.42}$$

To minimize $C(h)$, we take the partial derivative of W_k and set it equal to zero:

$$\frac{\partial C(h)}{\partial W_k} = 2R_{YY} W_k - 2R_{HY} = 0. \tag{10.43}$$

Then,

$$W_k = R_{YY}^{-1} R_{HY} = R_{HY} R_{YY}^{-1}. \tag{10.44}$$

Therefore,

$$\hat{H} = W^H Y = R_{HY} R_{YY}^{-1} Y, \tag{10.45}$$

where

$$R_{HY} = E[HY^H] = E[H(\eta XH + V)^H] = \eta R_{HH} X^H \tag{10.46}$$

$$\begin{aligned} R_{YY} &= E[YY^H] = E[(\eta XH + V)(\eta XH + V)^H] \\ &= \eta^2 X R_{HH} X^H + \sigma_V^2 I, \end{aligned} \tag{10.47}$$

where $X = diag(X_1, X_2, \ldots, X_M)^T$ is a diagonal matrix composed of M pilots. I is the identity matrix and σ_V^2 is the noise variance. The Wiener-filtering channel-estimation algorithm based on the MMSE criterion is expressed as follows:

$$\begin{aligned} \hat{H} &= R_{HY} R_{YY}^{-1} Y = \eta R_{HH} X^H (\eta^2 X R_{HH} X^H + \sigma_V^2 I)^{-1} Y \\ &= \eta R_{HH} [X^{-1} (\eta^2 X R_{HH} X^H + \sigma_V^2 I)(X^H)^{-1}]^{-1} X^{-1} Y \\ &= \eta R_{HH} [\eta^2 R_{HH} + \sigma_V^2 (X^H X)^{-1}]^{-1} X^{-1} Y \\ &= R_{HH} [\eta^2 R_{HH} + \sigma_V^2 (X^H X)^{-1}]^{-1} \hat{H}_{LS}. \end{aligned} \tag{10.48}$$

10.5 Simulation Analysis

To analyze the performance of the adaptive multi-layer space–time code in three transmission modes, the Monte Carlo method is used to simulate the bit-error performance, and the influence of the different transmission modes on the bit-error rate of the system, from weak to strong atmospheric turbulence, is analyzed. The simulation conditions are as follows:

(1) Assume that the total transmission power Es is constant and take the normalized value $Es = 1$;
(2) The channel environment t is a flat fading channel; the receiving end already knows the channel-state information and accurately estimates the noise;
(3) The photoelectric-conversion efficiency $\eta = 0.5$;
(4) We adopt multi-level pulse-position modulation, where $Q = 4$;
(5) The total number of symbols sent by the transmitter is $N = 500{,}000$;
(6) Under weak turbulence, the light-intensity flicker follows a lognormal distribution, while under moderate turbulence, the light-intensity flicker follows a gamma–gamma distribution;
(7) It is assumed that the signals transmitted by the four antennas arrive at the receiving antenna synchronously and simultaneously.

10.5.1 *Decoding Using Zero-Forcing*

With a 4×4 system, 4PPM modulation, and weak and strong turbulence, three modes of adaptive multilayer space–time codes are simulated, when ZF is adopted in the V-BLAST decoding algorithm.

Figure 10.4 shows the bit-error rate curve of the V-BLAST code using the ZF algorithm under 4PPM modulation with weak turbulence (scintillation index (SI) = 0.6). The three curves in the figure respectively show the error rate of the system when it adopts one of the three modes (full BCOSTBC code, full V-BLAST code, and mixed BCOSTBC–V-BLAST code). The following can be seen:

(1) After using the packet-interference suppression algorithm, the bit-error rate of the system gradually decreases with the increase of the signal-to-noise ratio, which shows that it is feasible to apply the packet-interference suppression algorithm to the decoding of atmospheric laser communication.
(2) The BCOSTBC system has the best performance, followed by the BCOSTBC–V-BLAST mixed coding system, and finally the V-BLAST system. This shows that the BCOSTBC system obtains a large diversity gain at the loss of some multiplexing gain, and its anti-fading ability is the best. The V-BLAST system has receive diversity but no transmit diversity, so it has the smallest diversity gain and the worst anti-fading ability. The bit-error performance of the BCOSTBC–V-BLAST hybrid code system is between the other two systems, which indicates

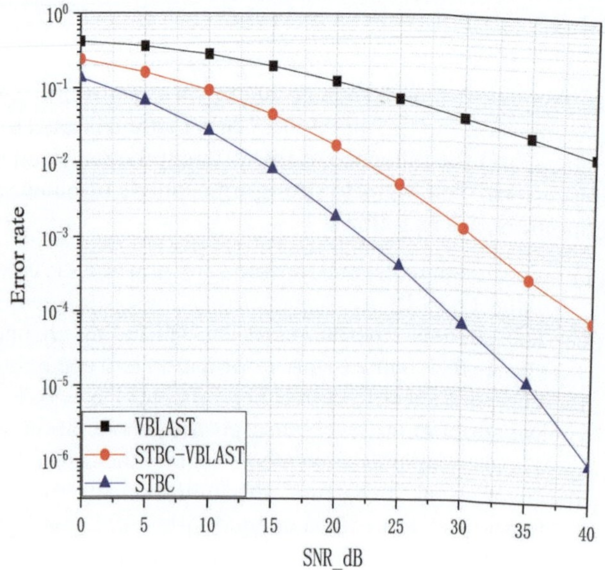

Fig. 10.4 Bit-error rate curve under weak turbulence, with 4PPM modulation

that the 4 × 4 hybrid space–time code system has both receive and transmit diversity gain; the obtained diversity gain is in the middle.

(3) The error performance of the 4 × 4 hybrid space–time coding system is between that of the STBC and V-BLAST systems, which shows that the error performance of hybrid space–time coding is better than that of V-BLAST code. When the bit-error rate is 10^{-2}, the performance of the 4 × 4 hybrid space–time coding system is improved by about 18 dB, compared with that of the V-BLAST system. It shows that the BCOSTBC code should be used in the first layer of hybrid space–time coding, which increases the diversity gain of the first-layer symbols and improves the system performance.

Figure 10.5 shows the bit-error rate curve of V-BLAST code when the ZF algorithm is adopted under 4PPM modulation with moderate and strong turbulence (SI = 1.2). The rule reflected in this figure is the same as that in Fig. 4.8; however, the improvement of the bit-error rate is different. This shows that atmospheric turbulence will affect the performance of the system to a certain extent; that is, with the increase of the scintillation index, the anti-fading performance of the system will decline to some extent.

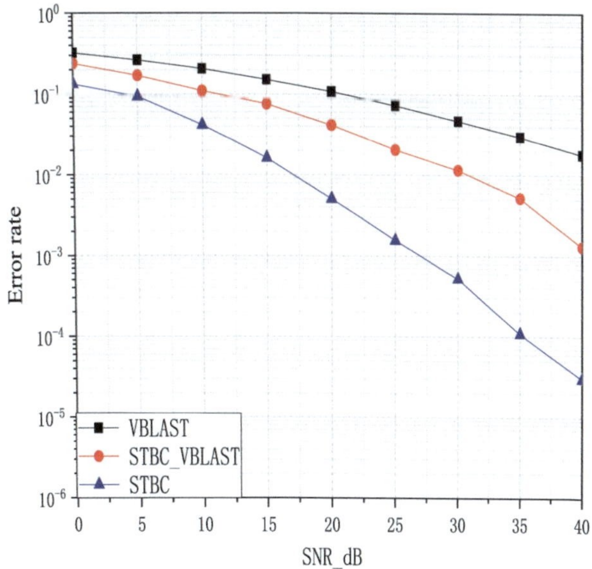

Fig. 10.5 Bit error rate curve under strong turbulence, with 4PPM modulation

10.5.2 Decoding Using MMSE

With a 4 × 4 system modulated by 4PPM, the V-BLAST decoding algorithm adopts MMSE to simulate three modes of adaptive multilayer space–time codes under weak and strong turbulence.

Figure 10.6 shows the BER curve of V-BLAST using the MMSE algorithm under weak turbulence (SI = 0.6) with 4PPM modulation. The following can be seen from the figure:

(1) When the bit-error rate is at the same level (10^{-2}), the performance of the BCOSTBC–V-BLAST hybrid system is improved by about 7 dB compared with the BCOSTBC system; compared with the V-BLAST system, the performance of the hybrid system is improved by about 21 dB. The hybrid space–time coding system obtains both receive and transmit diversity gain.

(2) Two decision thresholds, SNR0 and SNR1, can be set according to the required performance. For 4PPM modulation, if the bit-error rate is required to be 10^{-2}, SNR0 can be set to 14 dB and SNR1 to 35 dB.

(3) If the SNR is less than 14 dB, the two groups of antennas all adopt BCOSTBC coding. At this time, the transmission rate is low, but the diversity gain is large, which ensures reliable and stable communication when the channel characteristics are poor. If the SNR is greater than 35 dB, both groups of antennas adopt V-BLAST coding. At this time, although the diversity gain is the smallest, the throughput of the system with the largest data-transmission rate is the largest,

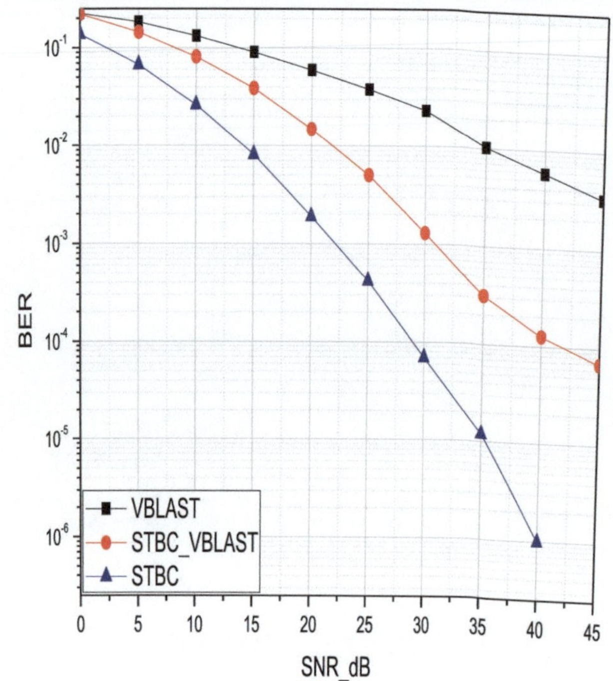

Fig. 10.6 BER curves under weak turbulence with 4PPM modulation

which makes full use of the spectrum resources. If the SNR is detected to be 14–35 dB, one group uses BCOSTBC coding and the other group uses V-BLAST coding, which compromises between diversity gain and multiplexing gain.

Comparing Figs. 10.4 and 10.6, when the bit-error rate is 10^{-2}, the signal-to-noise ratio is improved by approximately 5 dB when the MMSE algorithm is adopted, compared with the ZF algorithm in the V-BLAST system. This is because the MMSE algorithm considers the influence of noise on the signal. The performance of the MMSE algorithm is obviously better than that of the ZF algorithm. However, in the BCOSTBC–V-BLAST hybrid space–time coding system, the signal-to-noise ratios of the two algorithms are different. This shows that after the correctness of the V-BLAST first-layer decoding is ensured, noise has little influence on the signal, which further shows that the error in the V-BLAST system is mainly caused by the first-layer error.

Figure 10.7 shows the BER curve of V-BLAST using the MMSE algorithm under moderate and strong turbulence (SI = 1.2) with 4PPM modulation. The rule reflected in this figure is the same as that in Fig. 4.10; however, the improvement of the bit-error rate is different.

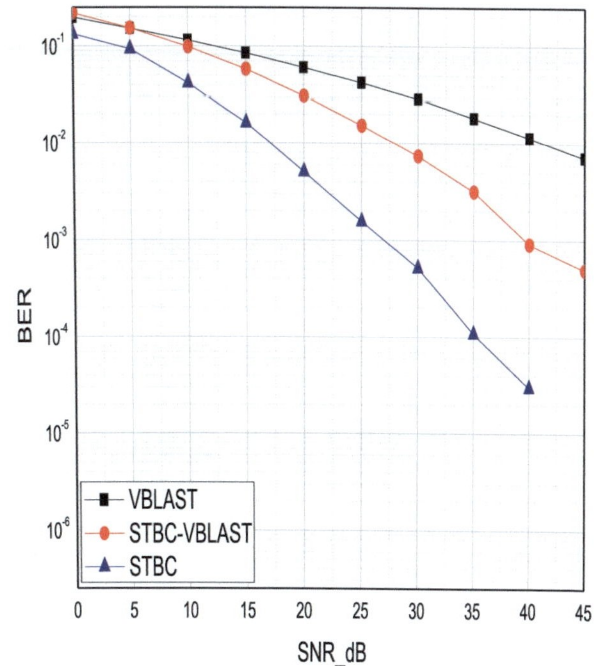

Fig. 10.7 BER curves under strong turbulence with 4PPM modulation

10.5.3 Comparison of the Three Modes

The bit-error rates and data-symbol rates that can be achieved by the three modes under weak turbulence (SI = 0.6) with 4PPM modulation are represented by a histogram, as follows.

It can be seen from Fig. 10.8 that the full BCOSTBC code has the lowest transmission rate, but the lowest bit-error rate; thus, it can reliably and stably communicate. Using full V-BLAST coding, the bit-error rate is the highest, but the transmission rate is the fastest. Using this mode can make full use of spectrum resources to maximize the throughput.

When BCOSTBC–V-BLAST coding is adopted, its bit-error rate and symbol rate are between those of the other two codes. Compared with full V-BLAST, the transmission diversity is added, which improves the bit-error rate performance of the system. Compared with full BCOSTBC, because multiplexing technology is used to improve the transmission rate, this coding achieves a compromise between diversity and multiplexing gain.

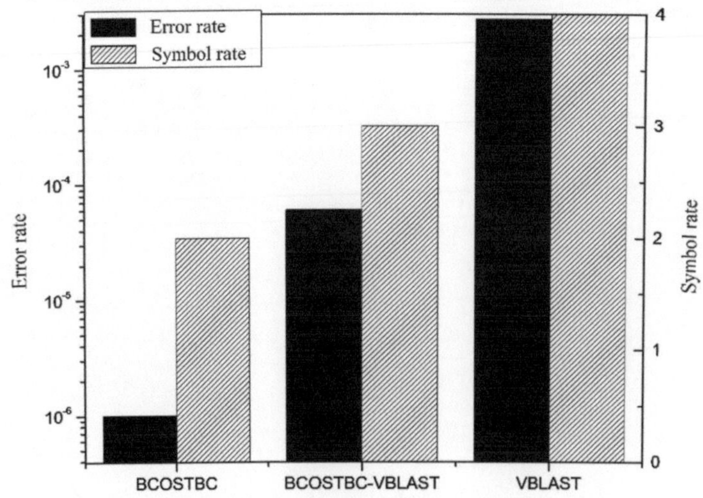

Fig. 10.8 Comparison of BER and symbol rate of three modes under S.I = 0.6 with 4PPM modulation

It can be seen that when the channel characteristics are poor, that is, when the signal-to-noise ratio is low, a low-order coding method, such as BCOSTBC coding, should be adopted to ensure reliable and stable communication. When the channel characteristics are good, that is, in the case of a high SNR, a high-order coding method, such as V-BLAST coding, should be adopted to ensure that the transmission rate of the system can be increased when the required bit-error rate is reached. When the channel characteristics are in the middle (that is, the signal-to-noise ratio is in the middle), BCOSTBC–V-BLAST coding is adopted to obtain a reasonable compromise between diversity gain and multiplexing gain. In this manner, adaptive coding modulation is realized, which allows the system to optimize the bit-error rate and transmission rate under different channel characteristics.

This coding scheme is more suitable for an actual channel with time-varying characteristics. It also eliminates the traditional fixed space–time coding mode, giving full play to the advantages of a MIMO system. This transmission scheme has a certain application prospect.

Reference [26] proposed a MIMO scheme for RF adaptive coding modulation mode. This scheme and the scheme proposed in this paper can adaptively select the coding modulation mode. They have the following differences.

Reference [26] uses phase-shift keying (PSK) instead of pulse-position modulation, and orthogonal space–time block code (OSTBC) coding instead of BCOSTBC. The simulation parameters of Ref. [26] are as follows: The channel is a quasi-static flat attenuated Rayleigh channel, the noise is an independent complex Gaussian random variable, and the modulation mode is 4PSK modulation.

It can be seen from the simulation diagram that when the error rate is 10^{-2}, SNR0 can be set to 7 dB and SNR1 to 19 dB. If the SNR value is less than 7 dB, the two groups of antennas all adopt OSTBC coding. If the SNR value is greater than 19 dB, both groups of antennas adopt V-BLAST coding. If the SNR value is 7–19 dB, one group uses OSTBC coding and the other group uses V-BLAST coding. In this manner, a given error rate can be satisfied. The transmission rate can be maximized at the bit-error rate.

10.6 Summary

In this chapter, 4 × 4 adaptive multi-layer space–time coding in free-space optical communication was described in detail. Combining the characteristics of bit-complement orthogonal space–time block coding (BCOSTBC) and vertical layered space–time coding (V-BLAST), the method was based on the principle of antenna-grouping multi-layer space–time coding. This scheme adaptively adjusted the coding modulation mode of the sender, according to the received signal-to-noise ratio (SNR), to meet the requirements of different channels.

At the receiving end, the packet interference suppression method was adopted for decoding. The concept was to eliminate the symbols transmitted by other packet-transmitting antennas as interference, and then decode the symbols of the transmitted packet. In this method, singular value decomposition and matrix multiplication were performed once, and each cycle in the set of modulation symbols only needed to be searched 16 times, which greatly reduced the decoding complexity of the system.

The simulation parameters were as follows: The channel was a quasi-static flat fading channel; the noises were independent Gaussian random variables, and all of them were in the real-number domain. The modulation mode was 4PPM modulation.

It can be seen from the simulation results in Fig. 4.11 that under weak turbulence, when the bit-error rate was 10-2, SNR0 could be set to 14 dB and SNR1 to 35 dB. If the SNR value was less than 14 dB, the two groups of antennas all adopted BCOSTBC coding. If the SNR value was greater than 35 dB, both groups of antennas adopted V-BLAST coding. If the SNR was detected to be 14–35 dB, one group used BCOSTBC coding and the other group used V-BLAST coding. Thus, the mode was changed to optimize the bit-error rate and transmission rate under different channel characteristics.

Different space–time modulation modes can be processed using similar methods, and this transmission scheme has a certain application prospect.

References

1. Shu F, Ao F, Sun L (2010) Research on the application of adaptive coding modulation technology in atmospheric laser communication. Microcomput Inform 15:148–149+147
2. Qing Y, Zhang Q (2010) Performance analysis of space-time block code in adaptive modulation and coding system. Dig Commun 05:45-47
3. Lin P, Zhao H (2005) Adaptive coding modulation technology in Rayleigh fading channel. J Equip Command Technol College 04:99–102
4. Zhou X, Min F, Zhou X (2008) A study of link adaptation technology based on LDPC. Commun Technol 04:31–32+35
5. Ungerboeck G (1982) Channel coding with multilevel/phase signals. IEEE Trans Inform Theory 28(1):55–67
6. Vucetic B (1991) An adaptive coding scheme for time-varing channels. IEEE Trans Commun 39(5):653–663
7. Goldsmith AJ, Chua SG (1997) Variable-rate variable-power MQAM for fading channels. IEEE Trans Commun 45(10):1218–1230
8. Vishwanath S, Goldsmith A (2000) Exploring adaptive turbo coded modulation for flat fading channels. In: Proceedings of VTC, vol 4, pp 1778–1783
9. Maaref A, Aissa S (2004) Combined adaptive modulation and truncated ARQ for packet data transmission in MIMO systems. IEEE Commun Soc Globecom 04(6):3818–3822
10. Ko Y, Tepedelenlioglu C (2006) Orthogonal space-time block coded rate adaptive modulation with outdata feedback. IEEE Trans Commun 5(2):290–295
11. Li W (2005) Research on adaptive coding modulation technology based on HARQ and CTTCM. Xidian University
12. Zhang S, Ji Z, Cui W (2011) Research on adaptive modulation technology in MIMO-OFDM system based on STBC. Comput Appl Softw 28(10):69–71
13. Li X (2007) Adaptive modulation and coding Technology Research Based on LDPC. Chongqing University of Posts and Telecommunications
14. Weistein SB, Ebert PM (1971) Data transmission by frequency-division multiplexing using the discrete Fourier transform. IEEE Transf Commun 19(5):628–634
15. Simon MK, Vilnrotter V (2005) Alamouti-type space-time coding for free-space optical communication with direct detection. IEEE Trans Wirel Commun 4(1):35–39
16. Wang H, Ke X (2009. Orthogonal space-time block codes in atmospheric laser communication. J Opt 36(S2):63–68
17. Tarokh V, Jafarkhani H, Calderbank AR (1999) Space-time block codes from orthogonal designs. IEEE Trans Inform Theory 45(5):1456–1467
18. Wang H, Ke X (2008) Free space optical communication based on vertical layered space-time code. China Laser 35(6):874–878
19. Hwang SK, Lyu DS, Chang KH (2003) 4G vision and technology development in Korea. Int Conf Commun Technol 1:26–27
20. Wang H, Ke X (2009) Hybrid space-time coding in free space optical communication. J Opt 29(1):132–137
21. Xu X, Liao G, Li Y, Feng W (2004) Space-time block codes based on hierarchical structure. J Radio Sci 19(2):129–134
22. Nistazakis HE, Tombras GS, Tsigopoulos AD et al (2008) Average capacity of optical communication systems over gamma-gamma atmospheric turbulence channels. IEEE MTT-S, Atlanta, pp 1561–1564
23. Huang G (2011) Research on the performance of wireless optical communication channel under atmospheric turbulence. J Xi'a Univ Technol 31(5):413–417
24. Kim J, Lee K-J, Sung CK, Lee I (2009) A simple SNR representation method for AMC schemes of MIMO systems with ML detector. IEEE Transf Commun 57(10):2971–2976

25. Qin S, Peide L, Xin Z, Laibo Z, Wang D (2007) Channel estimation with timing offset based on PSD & LS estimation for wireless OFDM systems. In: IEEE international symposium on intelligent signal processing and communication systems (ISPACS), pp 248–251
26. Liang Y, Yin J, Ding Y (2008) An adaptive multi-layer space-time coding MIMO transmission scheme. Commun Technol 12(41):84–86

Chapter 11
Indoor MIMO-VLC System Performance Analysis

This chapter analyzes the performance of visible-light communications (VLC) systems with multiple-input multiple-output (MIMO) configurations, focusing on the application of space–time coding in VLC systems. Two detection algorithms for MIMO-VLC systems, maximum-likelihood joint detection and equal-gain combined detection, were analyzed, and their bit-error rate (BER) performance and channel capacity were discussed. This chapter is an application of space–time coding.

11.1 Indoor MIMO-VLC System Model

Visible-light communication technology was first proposed by Masao Nakagawa of Keio University in Japan [1]. Because indoor visible-light communication can easily connect portable devices, such as cordless phones, personal digital assistants (PDAs), and laptops, to data-communication systems, the technology has also received strong support from Cyber Solutions Laboratories, the Advanced Long-Distance Communication Laboratory of the Post Office, Nippon Telegraph and Telephone Company, and the Sony Computer Science Laboratory [2]. In view of the good prospects of visible-light communications (VLC) technology, the IEEE has established a study group for VLC standards, namely the IEEE 802.15 Wireless Personal-Area Networks (WPAN) Interest Group on Visible-Light Communications.

A block diagram of the MIMO-VLC communication system is shown in Fig. 11.1 [3]. M white LEDs emit a light signal that is directed to a receiving array consisting of N photodetectors. An indoor multipath channel lies between each white LED and a photodetector; if the channel is unobstructed, the detector array can receive the light signal from all LEDs. Let the channel impact response from the m th white LED to the n th detector be $h_{nm}(t)$. Then, the n th photodetector output photocurrent can be expressed as

X. Ke, *Space-Time Coding Theory in Optical Wireless MIMO Systems*,
Optical Wireless Communication Theory and Technology,
https://doi.org/10.1007/978-981-97-6449-5_11

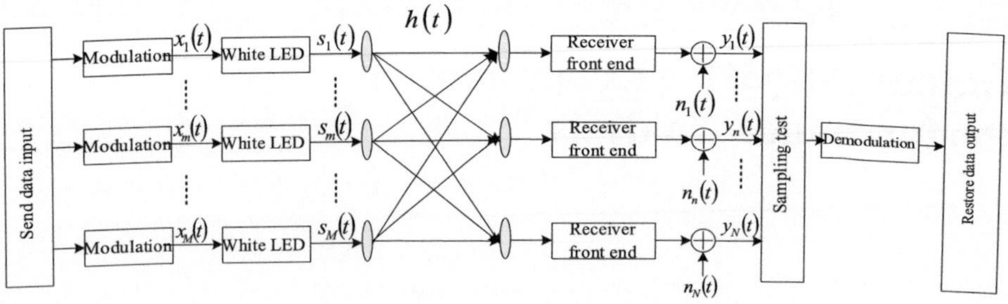

Fig. 11.1 Block diagram of a MIMO-VLC system composition

$$y_n(t) = \eta_n \sum_{m=1}^{M} s_m(t) \otimes h_{nm}(t) + n_n(t), \tag{11.1}$$

where η_n is the photoelectric-conversion efficiency of the n th photodetector, and $n_n(t)$ is the channel noise (white noise, mean μ_n, variance σ_n^2) relative to the n th photodetector, and $\otimes$ indicates a cyclic convolution. The N th receiver-branch output signal is sent to the detector for detection.

The main detection algorithms are joint and selective-merge detection. Joint detection samples the N-way input signal and then creates a verdict based on the sampling value; the maximum-likelihood joint-detection algorithm is typically used for the verdict. Selective-merge detection receives the input signal, according to a certain merging algorithm, and then uses a sampling decision to recover the original sent data.

11.1.1 Transmitter Design

Figure 11.2 shows a schematic diagram of the array light-source layout of a MIMO-VLC system [4]. When the light-source layout is basically the same (i.e., the same number of light sources, distribution symmetry, ring diameter, and rectangular-ring length), the rectangular ring-array antenna and the circular ring-array antenna in the MIMO-VLC system can receive light indoors. The effect of the power is essentially the same.

White LEDs were symmetrically distributed on a circular ring with a radius of d. Taking the symmetrical distributed-array antenna as the research object, let each plane ring perpendicular to the z-axis be defined as one layer. The LEDs are evenly distributed on the ring with a radius of $d_l (l = 0, 1, 2 \ldots L$, and $d_0 = 0)$. There are L layers in total, and the number of LEDs distributed in each layer is $N_l (N_0 = 1)$. The polar angle between the optical axis of the LED on the $l (l > 0)$ layer ring and the optical axis of the central LED (i.e., layer 0) is θ_l, and the horizontal angle between

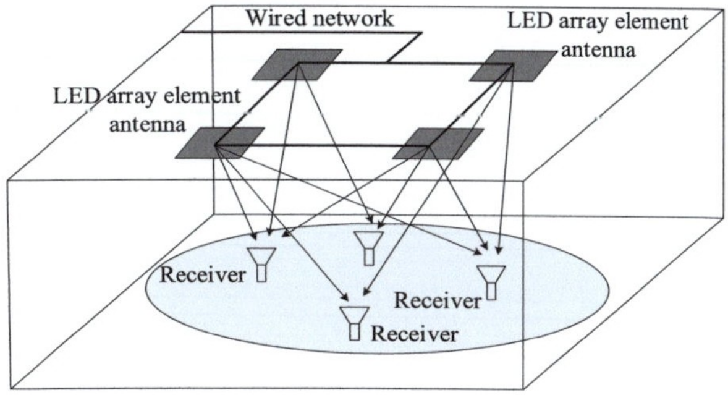

Fig. 11.2 Schematic diagram of a MIMO-VLC array light-source layout

the optical axes of adjacent LEDs is $\varphi_{n,l}$. The number of array light sources is U and the number of white LEDs in each array light source is V.

In a MIMO-VLC system, the intensity modulation/direct detection (IM/DD) method is used to transmit signals. On–off keying (OOK) and pulse-position modulation (PPM) are example modulation methods [5].

11.1.2 Receive-Array Design

As shown in Fig. 11.3, the receiving array adopts a hexagonal planar-array design [4]. The side length of the hexagon is d; the photodetector is located at the center of each hexagon, and the distance between adjacent photodetectors is $\sqrt{3}d$. They are numbered sequentially. To increase the diversity gain as much as possible, the distance between the detectors should satisfy certain conditions so that the channel characteristics of each detector are independent of each other. The choice of detector spacing is also related to the field of view (FOV) of the detector, as shown in Fig. 11.4.

11.1.3 Maximum-Likelihood Joint Detection

Consider a repeated-transmission MIMO-VLC system, where M transmitters transmit the same modulated signal simultaneously. After the indoor-channel transmission, the average output photocurrent of the nth receiver front-end, with and without optical pulses, is respectively [6]

$$\mu_{on,n} = \left(\sum_{m=1}^{M} \frac{\xi P_{s,m} T}{hf} h_{nm}(t-T) + \frac{\xi P_{b,n} T}{hf} \right) R_f q \tag{11.2}$$

$$\mu_{off,n} = \left(\frac{\xi P_{b,n} T}{hf} \right) R_f q, \tag{11.3}$$

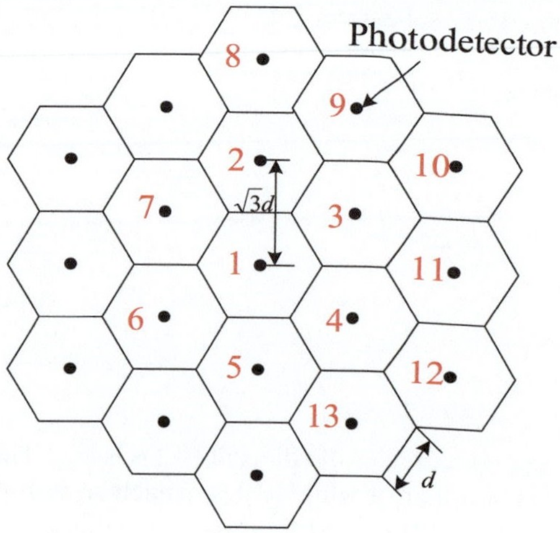

Fig. 11.3 Schematic diagram of a receiving-array design

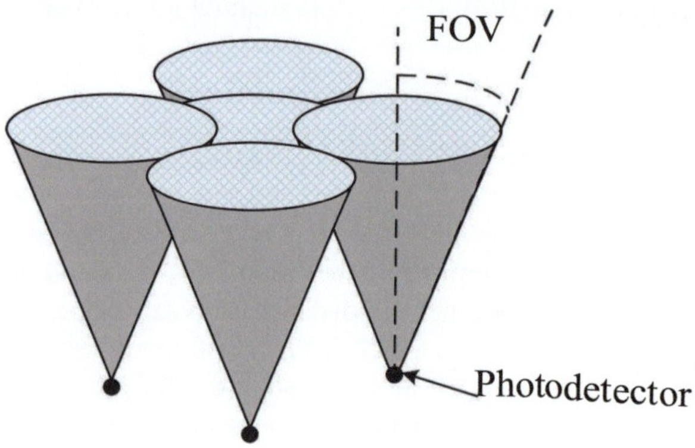

Fig. 11.4 Photodetector coverage area

where $P_{s,m}$ is the optical power of the mth transmitter, $P_{b,n}$ is the background light power received by the nth detector, and T is the period of the OOK signal. The variance of the output photocurrent of the nth receiver front-end, with and without the optical pulse, is respectively [6]

$$\sigma_{on,n}^2 = \left(\sum_{m=1}^{M} \frac{\xi P_{s,m} T}{hf} h_{nm}(t-T) + \frac{\xi P_{b,n} T}{hf}\right) R_f^2 q^2 + I_{bg,n} R_f^2 q + 2kT_K T R_f \quad (11.4)$$

$$\sigma_{off,n}^2 = \left(\frac{\xi P_{b,n} T}{hf}\right) R_f^2 q^2 + I_{bg,n} R_f^2 q + 2kT_K T R_f. \quad (11.5)$$

Assuming that the sampling signal-output vectors of the N receiver front-ends at time t are Y, $Y = [Y_1, Y_2, \ldots, Y_N]^T$, the maximum-likelihood joint-detection criterion for the input signal X_i is [7]

$$\widehat{X} = \arg\max_{X_i} f(Y|X_i), \tag{11.6}$$

where $f(Y|X_i) = \prod_{n=1}^{N} f(Y_n|X_i)$. The probability-density functions for $f(Y|on)$ and $f(Y|off)$ (probability-density functions judged as "*off*" (0) and "*on*" (1) at the receiving end) with and without light pulses, are respectively [7]

$$f(Y_n|on) = \frac{1}{\sqrt{2\pi\sigma_{on,n}^2}} \exp\left(-\frac{\left(Y_n - \mu_{on,n}\right)^2}{2\sigma_{on,n}^2}\right) \tag{11.7}$$

$$f(Y_n|off) = \frac{1}{\sqrt{2\pi\sigma_{off,n}^2}} \exp\left(-\frac{\left(Y_n - \mu_{off,n}\right)^2}{2\sigma_{off,n}^2}\right). \tag{11.8}$$

11.1.4 Choosing a Merge Algorithm

In the combined-detection method for MIMO-VLC system selection, each receive branch includes a detector, an amplifier, and a signal-to-noise ratio estimator. The detector receives optical signals from different transmitters and outputs photocurrents corresponding to the signals. The photocurrent is amplified by the amplifier and sent to the combination-selection circuit. The combination-selection circuit outputs the signal to be demodulated according to the different combination-selection techniques. The entire process is shown in Fig. 11.5. The algorithms for selection merging mainly include the optimal-selection algorithm, equal-gain merging algorithm, and maximum-ratio merging algorithm.

The so-called optimal selection (select best, SB) algorithm is used to compare the signal-to-noise ratios of the output signals of the N receiver front-ends, and the one with the largest signal-to-noise ratio is selected as the output signal; it can be expressed as [8]

$$y(t) = \arg\max_{y_n(t)}(SNR_n) = \arg\max_{n}\left(\frac{\xi^2 P_{R,n}^2}{\sigma_n^2}\right), 1 \leq n \leq N, \tag{11.9}$$

where $P_{R,n}$ is the optical power received by the nth detector, σ_n^2 is the variance of the output photocurrent at the front end of the nth receiver, and ξ is the quantum efficiency of the photodetector. Then, the output signal-to-noise ratio is

$$SNR_{SB} = \max_{n}\left(\frac{\xi^2 P_{R,n}^2}{\sigma_n^2}\right). \tag{11.10}$$

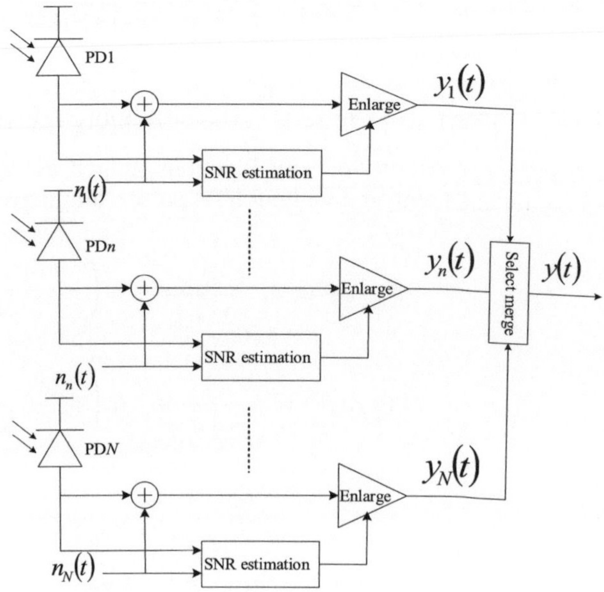

Fig. 11.5 Receive-array select–merge circuit

The equal-gain combining (EGC) algorithm means that the output of the signal is the equally weighted sum of all input signals; that is,

$$y(t) = \frac{1}{N}\sum_{n=1}^{N} y_n(t). \tag{11.11}$$

Then, the output signal-to-noise ratio is

$$SNR_{EGC} = \frac{1}{N}\sum_{n=1}^{N} \frac{\xi^2 P_{R,n}^2}{\sigma_n^2}. \tag{11.12}$$

In the maximum-ratio combining (MRC) algorithm, the output signal is the sum of the input signals with different weight ratios:

$$y(t) = \sum_{n=1}^{N} \omega_n y_n(t), \tag{11.13}$$

where ω_n is the weight coefficient of the nth channel signal, which is proportional to the ratio of the received optical power of the corresponding input signal to the noise variance:

$$\omega_n = \frac{\xi P_{R,n}}{\sigma_n^2}. \tag{11.14}$$

Then, the output signal-to-noise ratio is [8]

$$SNR_{MRC} = \frac{\left(\sum_{n=1}^{N} \xi P_{R,n} \frac{\xi P_{R,n}}{\sigma_n^2}\right)^2}{\sum_{n=1}^{N} \left(\frac{\xi P_{R,n}}{\sigma_n^2}\right)^2 \sigma_n^2}. \tag{11.15}$$

Both the SB selection algorithm and the MRC algorithm need to accurately determine the signal-to-noise ratio of each input signal, whereas the EGC algorithm is easy to implement because it does not need to obtain the signal-to-noise ratio of each signal in advance. The EGC algorithm is used as an example for analysis. After the EGC algorithm is combined, the mean value of the output signal is [8]

$$\mu_{on} = \frac{1}{N}\sum_{n=1}^{N}\mu_{on,n} = \frac{R_f q}{N}\left(\sum_{n=1}^{N}\sum_{m=1}^{M}\frac{\xi P_{s,m}T}{hf}h_{nm}(t-T) + \sum_{n=1}^{N}\frac{\xi P_{b,n}T}{hf}\right) \tag{11.16}$$

$$\mu_{off} = \frac{1}{N}\sum_{n=1}^{N}\mu_{off,n} = \frac{R_f q}{N}\sum_{n=1}^{N}\frac{\xi P_{b,n}T}{hf}. \tag{11.17}$$

The total variance of the total output photocurrent after combining the EGC algorithm is [8]

$$\begin{aligned}\sigma_{on}^2 = {} & \frac{R_f^2 q^2}{N}\left(\sum_{n=1}^{N}\sum_{m=1}^{M}\frac{\xi P_{s,m}T}{hf}h_{nm}(t-T) + \sum_{n=1}^{N}\frac{\xi P_{b,n}T}{hf}\right) \\ & + \frac{1}{N}\sum_{n=1}^{N} I_{bg,n}R_f^2 q + 2kT_K TR_f\end{aligned} \tag{11.18}$$

$$\sigma_{off}^2 = \frac{1}{N}\sum_{n=1}^{N}\left(R_f^2 q^2 \frac{\xi P_{b,n}T}{hf} + I_{bg,n}R_f^2 q\right) + 2kT_K TR_f. \tag{11.19}$$

11.2 Indoor MIMO-VLC System Performance

11.2.1 Bit-Error Performance

11.2.1.1 OOK System

(1) Maximum-likelihood joint detection

For a MIMO-VLC system using repeated OOK modulation, the maximum-likelihood joint-detection decision criterion is [7]

$$\widehat{X} = \arg\max_{X_i} \prod_{n=1}^{N} f(Y_n|X_i), \tag{11.20}$$

which is

$$\frac{\prod_{n=1}^{N} f(Y_n|on)}{\prod_{n=1}^{N} f(Y_n|off)} \mathop{\gtrless}_{off}^{on} \frac{P(off)}{P(on)}. \tag{11.21}$$

$P(off)$ and $P(on)$ are the probabilities of the sender sending "*off*" and "*on*," respectively, and it is assumed that the probabilities of the sender sending 0 and 1 are equal; that is, $P(off) = P(on) = 1/2$. Then, the likelihood criterion is [7]

$$\Lambda(Y) = \frac{\prod_{n=1}^{N} f(Y_n|on)}{\prod_{n=1}^{N} f(Y_n|off)}. \tag{11.22}$$

Substituting Eqs. (11.7) and (11.8) into Eq. (11.22), we obtain

$$\Lambda(Y) = \left(\prod_{n=1}^{N} \frac{\sigma_{off,n}}{\sigma_{on,n}}\right) \exp\left(\sum_{n=1}^{N} a_n Y_n^2 + b_n Y_n + c_n\right). \tag{11.23}$$

In the formula,

$$a_n = \frac{\sigma_{on,n}^2 - \sigma_{off,n}^2}{2\sigma_{on,n}^2 \sigma_{off,n}^2} \tag{11.24}$$

$$b_n = \frac{\mu_{on,n}\sigma_{off,n}^2 - \mu_{off,n}\sigma_{on,n}^2}{\sigma_{on,n}^2 \sigma_{off,n}^2} \tag{11.25}$$

$$c_n = \frac{\left(\mu_{off,n}\sigma_{on,n}\right)^2 - \left(\mu_{on,n}\sigma_{off,n}\right)^2}{2\sigma_{on,n}^2 \sigma_{off,n}^2} \tag{11.26}$$

$$d_n = \frac{\sigma_{off,n}}{\sigma_{on,n}}. \tag{11.27}$$

We take the logarithm to obtain the new likelihood decision criterion:

$$\Delta(Y) = \sum_{n=1}^{N} \ln(d_n) + \sum_{n=1}^{N} a_n Y_n^2 + b_n Y_n + c_n. \tag{11.28}$$

After each sampling calculation, when $\Delta(Y) > 0$, it is judged as 1, and when $\Delta(Y) < 0$, it is judged as 0.

(2) Equal-gain combined detection

For a MIMO-VLC system using repeated OOK modulation, after the EGC algorithm is combined, the probability-density functions that the receiver judges as "*off*" or "*on*" can be expressed as [8]

$$f(y|on) = \frac{1}{\sqrt{2\pi\sigma_{on}^2}} \exp\left(-\frac{(y-\mu_{on})^2}{2\sigma_{on}^2}\right) \tag{11.29}$$

$$f(y|off) = \frac{1}{\sqrt{2\pi\sigma_{off}^2}} \exp\left(-\frac{\left(y-\mu_{off}\right)^2}{2\sigma_{off}^2}\right). \tag{11.30}$$

The values of μ_{on}, μ_{off}, σ_{on}^2, and σ_{off}^2 are shown in Eqs. (11.16), (11.17), (11.18), and (11.19), respectively. Then, the maximum-likelihood decision criterion is [8]

$$\widehat{X} = \arg\max_{X_i} f(Y|X_i), \tag{11.31}$$

which is

$$\frac{f(Y|on)}{f(Y|off)} \underset{off}{\overset{on}{\gtrless}} \frac{P(off)}{P(on)}. \tag{11.32}$$

$P(off)$ and $P(on)$ are the probabilities of the sender sending "*off*" and "*on*," respectively, and it is assumed that the probabilities of the sender sending 0 and 1 are equal, that is, $P(off) = P(on) = 1/2$. Then, the likelihood criterion is [8]

$$\Lambda(Y) = \frac{f(Y|on)}{f(Y|off)}. \tag{11.33}$$

Substituting Eqs. (11.25) and (11.26) into Eq. (11.33) yields

$$\frac{\sigma_{off}}{\sigma_{on}} \exp\left(\frac{Y^2\left(\sigma_{on}^2 - \sigma_{off}^2\right) - 2Y\left(\mu_{off}\sigma_{on}^2 - \mu_{on}\sigma_{off}^2\right) + \left(\mu_{off}\sigma_{on}\right)^2 - \left(\mu_{on}\sigma_{off}\right)^2}{2\sigma_{on}^2\sigma_{off}^2}\right). \tag{11.34}$$

Taking the logarithm of the above formula, the new likelihood-judgment criterion can be obtained as [8]

$$\Delta(Y) = \frac{Y^2\left(\sigma_{on}^2 - \sigma_{off}^2\right) - 2Y\left(\mu_{off}\sigma_{on}^2 - \mu_{on}\sigma_{off}^2\right) + \left(\mu_{off}\sigma_{on}\right)^2 - \left(\mu_{on}\sigma_{off}\right)^2}{2\sigma_{on}^2\sigma_{off}^2} + \ln\left(\frac{\sigma_{off}}{\sigma_{on}}\right). \tag{11.35}$$

The symbols are ordered:

$$a = \frac{\sigma_{on}^2 - \sigma_{off}^2}{2\sigma_{on}^2\sigma_{off}^2} \tag{11.36}$$

$$b = \frac{\mu_{on}\sigma_{off}^2 - \mu_{off}\sigma_{on}^2}{\sigma_{on}^2\sigma_{off}^2} \tag{11.37}$$

$$c = \frac{(\mu_{off}\sigma_{on})^2 - (\mu_{on}\sigma_{off})^2}{2\sigma_{on}^2\sigma_{off}^2} \tag{11.38}$$

$$d = \frac{\sigma_{off}}{\sigma_{on}}. \tag{11.39}$$

Then,

$$\Delta(Y) = aY^2 + bY + c + \ln(d). \tag{11.40}$$

After each sampling calculation, if $\Delta(Y) > 0$, it is judged as 1, and if $\Delta(Y) < 0$, it is judged as 0.

(3) Error rate

The bit-error rate of the OOK system is

$$P_e = P(0)\ldots P(1|0) + P(1).P(0|1), \tag{11.41}$$

where $P(0)$ and $P(1)$ represent the probability of sending bits "0" and "1," respectively. We set $P(0) = P(1) = 1/2$; $P(1|0)$ represents the probability that the transmission bit "0" is judged as "1," and $P(0|1)$ represents the probability that the transmission bit "1" is judged as "0."

Then,

$$P(1|0) = \int_{\Delta(Y)>0} f(Y|on)dY \tag{11.42}$$

$$P(0|1) = \int_{\Delta(Y)<0} f(Y|off)dY. \tag{11.43}$$

The BER of OOK can be expressed as

$$P_e = \frac{1}{2}\int_{\Delta(Y)>0} f(Y|on)dY + \frac{1}{2}\int_{\Delta(Y)<0} f(Y|off)dY. \tag{11.44}$$

11.2.1.2 PPM System

(1) Maximum-likelihood joint detection

For a MIMO-VLC system using repeated PPM modulation, the transmitted PPM sequence is $X_i = \{X_q, q = 0, 1, \ldots, Q - 1\}$, and after an indoor-channel transmission, the output signal of the receiver array is a matrix of Y_i and Y_i to $Q \times N$:

$$Y_i = \begin{bmatrix} Y_{1,1} & Y_{1,2} & \cdots & Y_{1,N} \\ Y_{2,1} & Y_{2,2} & \cdots & Y_{2,N} \\ \vdots & \vdots & \ddots & \vdots \\ Y_{Q,1} & Y_{Q,2} & \cdots & Y_{Q,N} \end{bmatrix}, \tag{11.45}$$

where $Y_{q,n}$ represents the output signal of the nth method in the qth time slot. The conditional probability-density function of sending PPM sequence X_i and receiving decision X_i is

$$f(Y|X_i) = \prod_{n=1}^{N} \left[\frac{1}{\sqrt{2\pi\sigma_{on,n}^2}} \exp\left(-\frac{\left(Y_{Q_{on,n}^i} - \mu_{on,n}\right)^2}{2\sigma_{on,n}^2} \right) \times \prod_{q \in Q_{off}^i} \left(\frac{1}{\sqrt{2\pi\sigma_{off,n}^2}} \exp\left(-\frac{\left(Y_{q,n} - \mu_{off,n}\right)^2}{2\sigma_{off,n}^2} \right) \right) \right]. \tag{11.46}$$

The maximum-likelihood joint-detection decision criterion is

$$\widehat{X} = \arg\max_{X_i} f(Y|X_i). \tag{11.47}$$

We substitute (11.46) into (11.47) and remove the constant term to obtain

$$\hat{X} = \arg\max_{X_i} \prod_{n=1}^{N} \left[\left(\frac{1}{\sigma_{on,n}} \exp\left(-\frac{\left(Y_{Q_{on,n}^i} - \mu_{on,n}\right)^2}{2\sigma_{on,n}^2} \right) \right) \times \prod_{q \in Q_{off}^i} \left(\frac{1}{\sigma_{off,n}} \exp\left(-\frac{(Y_{q,n} - \mu_{off,n})^2}{2\sigma_{off,n}^2} \right) \right) \right]. \tag{11.48}$$

We take the logarithm to obtain

$$\hat{X} = \arg\max_{X_i}\left\{\sum_{n=1}^{N}\left[-\frac{\left(Y_{Q_{on,n}^i} - \mu_{on,n}\right)^2}{2\sigma_{on,n}^2} - \ln(\sigma_{on,n})\right]\right.$$
$$\left.+\sum_{n=1}^{N}\prod_{q\in Q_{off}^i}\left[-\frac{(Y_{q,n} - \mu_{off,n})^2}{2\sigma_{off,n}^2} - \ln(\sigma_{off,n})\right]\right\}. \quad (11.49)$$

After removing all constant terms,

$$\widehat{X} = \arg\max_{X_i}\sum_{n=1}^{N}\left(\left(\mu_{on,n}\sigma_{off,n}^2 - \mu_{off,n}\sigma_{on,n}^2\right)Y_{Q_{on,n}^i} + \left(\frac{\sigma_{on,n}^2 - \sigma_{off,n}^2}{2}\right)Y_{Q_{on,n}^i}^2\right). \quad (11.50)$$

(2) Equal-gain combined detection

For the MIMO-VLC system using repeated PPM modulation, the transmitted PPM sequence is $X_i = \{X_q, q = 0, 1, \ldots, Q-1\}$; after indoor-channel transmission and EGC-algorithm merging, the output signals of the receiver array are Y, $Y = \{Y_q, q = 0, 1, \ldots, Q-1\}$. The conditional probability-density function of the receiving decision of X_i is [8]

$$f(Y|X_i) = \frac{1}{\sqrt{2\pi\sigma_{on}^2}}\exp\left(-\frac{\left(Y_{Q_{on}^i} - \mu_{on}\right)^2}{2\sigma_{on}^2}\right)$$
$$\times\prod_{q\in Q_{off}^i}\left(\frac{1}{\sqrt{2\pi\sigma_{off}^2}}\exp\left(-\frac{(Y_q - \mu_{off})^2}{2\sigma_{off}^2}\right)\right). \quad (11.51)$$

The values of μ_{on}, μ_{off}, σ_{on}^2, and σ_{off}^2 are shown in Eqs. (11.16), (11.17), (11.18), and (11.19), respectively. The maximum-likelihood decision criterion is [8]

$$\widehat{X} = \arg\max_{X_i} f(Y|X_i). \quad (11.52)$$

Substituting Eq. (11.51) into Eq. (11.52) and removing the constant term, we obtain

$$\hat{X} = \arg\max_{X_i}\left[\left(\frac{1}{\sigma_{on}}\exp\left(-\frac{\left(Y_{Q_{on}^i} - \mu_{on}\right)^2}{2\sigma_{on}^2}\right)\right)\right.$$
$$\left.\times\prod_{q\in Q_{off}^i}\left(\frac{1}{\sigma_{off}}\exp\left(-\frac{\left(Y_q - \mu_{off}\right)^2}{2\sigma_{off}^2}\right)\right)\right]. \quad (11.53)$$

After taking the logarithm, we obtain

$$\hat{X} = \arg\max_{X_i}\left[\left(-\frac{\left(Y_{Q_{on}^i} - \mu_{on}\right)^2}{2\sigma_{on}^2} - \ln(\sigma_{on})\right) + \sum\nolimits_{q\in Q_{off}^i}\left(-\frac{\left(Y_q - \mu_{off}\right)^2}{2\sigma_{off}^2} - \ln(\sigma_{off})\right)\right], \tag{11.54}$$

After removing all constant terms,

$$\widehat{X} = \arg\max_{X_i}\left(\left(\mu_{on}\sigma_{off}^2 - \mu_{off}\sigma_{on}^2\right)Y_{Q_{on}^i} + \left(\frac{\sigma_{on}^2 - \sigma_{off}^2}{2}\right)Y_{Q_{on}^i}^2\right). \tag{11.55}$$

(3) Error rate

Let $P_q(q)$ denote the correct decision probability that the pulse is located in the qth time slot. Then, the average correct symbol rate of the PPM receiver can be expressed as [7]

$$P_{ave} = \frac{1}{Q}\sum_{q=1}^{Q} P_q(q). \tag{11.56}$$

Because $f(Y|X_i)$ is symmetric for different time slots,

$$P_{ave} = \left[\int\limits_{f(Y|X_i)\geq f(Y|X_1)} f(Y|X_i)dY\right]^{Q-1}. \tag{11.57}$$

Then, the symbol error rate of the PPM signal is

$$P_s = 1 - \left[\int\limits_{f(Y|X_i)\geq f(Y|X_1)} f(Y|X_i)dY\right]^{Q-1}. \tag{11.58}$$

The bit-error rate in binary is [7]

$$P_b = \frac{Q/2}{Q-1}P_s = \frac{Q/2}{Q-1}\left\{1 - \left[\int\limits_{f(Y|X_i)\geq f(Y|X_1)} f(Y|X_i)dY\right]^{Q-1}\right\}. \tag{11.59}$$

11.2.1.3 Simulation Results

The performance of a visible-light communication system under maximum-likelihood detection was simulated [9], and the simulation parameters are listed in Table 11.1. The parameters of the white LED and photodetector are shown in Table 11.2, and the communication rate was set to 100Mbps. The room parameters refer to the configuration data in Table 11.3.

The transmitting antenna is composed of multiple white LED arrays. We studied the bit-error performance of the OOK and PPM systems [10]. Figure 11.6 shows the simulation results for the bit-error performance of the OOK and PPM systems under different antenna configurations. The OOK system chooses two types of codes: non-return-to-zero (NRZ) and return-to-zero (RZ). The base numbers selected by the PPM system are four, eight, and 16, and the demodulation algorithm adopts maximum-likelihood joint detection.

Figure 11.7 shows the influence of the number of receive antennas on the system bit-error performance under the maximum-likelihood joint-detection and equal-gain combined-detection methods, when the number of transmit antennas is fixed. The BER of the system decreases with an increase in the number of receive antennas, thereby improving performance. However, using multiple transmit antennas restrains the fluctuation of the indoor received-optical power to a certain extent, thereby reducing the difficulty of transmitting optical-power control.

Figure 11.7 also reflects the degree to which maximum-likelihood joint detection and equal-gain combined detection affect the system performance. When the number

Table 11.1 Simulation parameters

Simulation parameters	Numerical value
ξ	0.75 A/W
P_b	− 10 dB mW
I_{nbf}	0.562
T_K	300 K
RR_f	10 kΩ
T	100 ns

Table 11.2 White LED and photodetector parameters

Transmitting optical power	20 (mW)
Reflection index	1.5
Transmit-power half angle	60°
Receiver FOV	60°
PD area	1.0 (cm^2)
Optical-filter gain	1.0
Photoelectric-conversion efficiency	0.53 (A/W)
Power-transmission half angle	60°

Table 11.3 Room configuration parameters

	Parameter	Config. 1	Config. 2	Config. 3	Config. 4
Room	Length (m)	5	7.5	7.5	7.5
	Width (m)	5	5.5	5.5	5.5
	Height (m)	3	3.5	3.5	3.5
Diffuse reflectance	East wall	0.8	0.30	0.30	0.58
	South window	0.8	0.56	0.001	0.001
	West wall	0.8	0.30	0.30	0.30
	North wall	0.8	0.12	0.72	0.72
	Ceiling	0.8	0.69	0.69	0.69
	Floor	0.3	0.09	0.09	0.09
Specular reflection coefficient	East wall	30	130	130	130
	South window	30	80	600	600
	West wall	30	130	130	130
	North wall	30	250	250	250
	Ceiling	30	60	60	60
	Floor	130	300	300	300

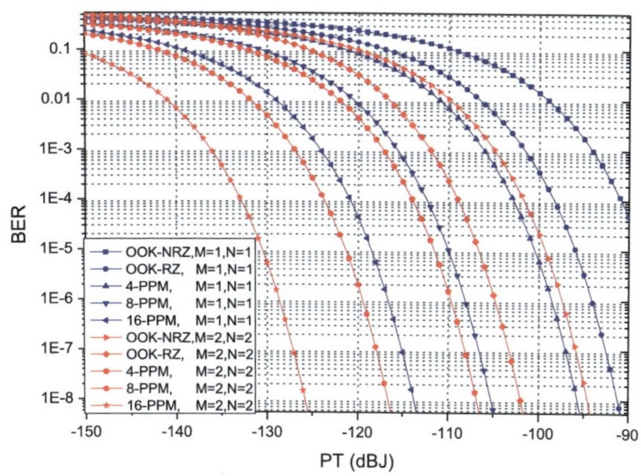

Fig. 11.6 Difference between OOK system and PPM system bit-error rates with varying numbers of antennas

of receiving antennas is one, the system bit-error rates of the two detection methods are the same; with an increase in the number of receiving antennas, the system bit-error rate of the maximum-likelihood joint detection is better than that of the equal-gain combined detection. The algorithm complexity of maximum-likelihood joint detection is greater than that of equal-gain combined detection, and the greater the number of receiving antennas, the higher the algorithm complexity.

Figure 11.8 shows the impact of the intersymbol interference (ISI) on the performance of the MIMO-VLC system. The root-mean-square (RMS) values were 30 ns, 20 ns, 10 ns, and 0 ns, respectively. As can be seen from the figure, the MIMO

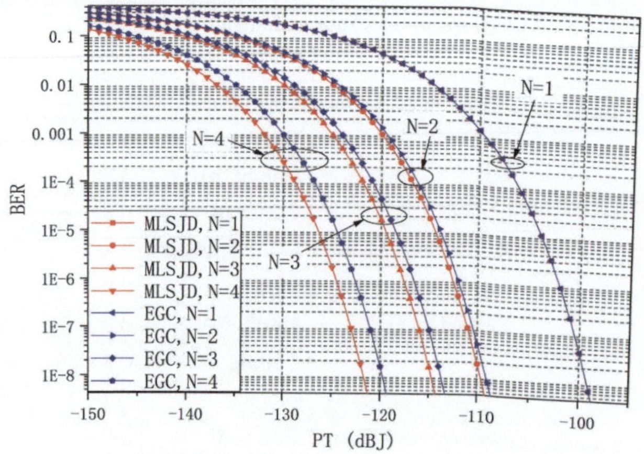

Fig. 11.7 Effect of the number of receive antennas on the bit-error performance; 4-PPM modulation, four transmit antennas

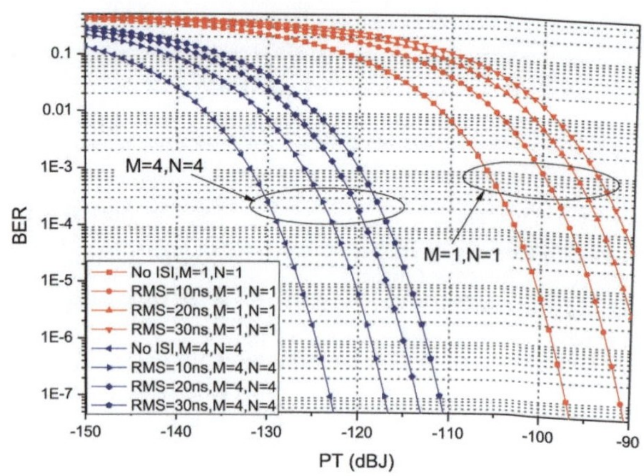

Fig. 11.8 Influence of ISI on the BER performance of a MIMO-VLC system with 4-PPM modulation

system has a stronger ability to suppress ISI than the single-input single-output (SISO) system.

11.2.2 Channel Capacity

11.2.2.1 OOK System

(1) Maximum-likelihood joint detection

In the MIMO-VLC system with repeated OOK modulation, the channel capacity under maximum-likelihood joint detection is [7]

$$C = \log 2 - E\left\{\log\left(1 + \frac{f(Y|on)}{f(Y|off)}\right)\right\} \quad (11.60)$$

$$C = \log 2 - E\left\{\log\left(1 + \frac{\prod_{n=1}^{N} f(Y_n|on)}{\prod_{n=1}^{N} f(Y_n|off)}\right)\right\} \quad (11.61)$$

$$C = \log 2 - E\left\{\log\left(1 + \left(\prod_{n=1}^{N} d_n\right)\exp\left[\sum_{n=1}^{N}\left(a_n Y^2 + b_n Y + c_n\right)\right]\right)\right\}. \quad (11.62)$$

(2) Equal-gain combined detection

The channel capacity under equal-gain combined detection is [7]

$$C = \log 2 - E\left[\log\left(1 + d\exp\left(aY^2 + bY + c\right)\right)\right]. \quad (11.63)$$

11.2.2.2 PPM System

(1) Maximum-likelihood joint detection

In a MIMO-VLC system with repeated PPM modulation, the channel capacity under maximum-likelihood joint detection is [7]

$$C = \log Q - E_{Y|X_0}\left[\log \frac{\sum_{i=0}^{Q-1} f\left(Y|\widehat{X}_i\right)}{f(Y|X_0)}\right], \quad (11.64)$$

where

$$f\left(Y|\widehat{X}_i\right) = \prod_{n=1}^{N} f\left(Y_n|\widehat{X}_i\right).$$

Therefore,

$$C = \log Q - E_{Y|X_0}\left[\log \frac{\sum_{i=0}^{Q-1}\prod_{n=1}^{N} f\left(Y_n|\widehat{X}_i\right)}{\prod_{n=1}^{N} f(Y_n|X_0)}\right]. \quad (11.65)$$

After simplification,

$$C = \log Q - E_{Y|X_0}\left\{\log\sum_{i=0}^{Q-1}\exp\left(\sum_{n=1}^{N}\left[a_n\left(Y^2_{Q^0_{on,n}} - Y^2_{Q^i_{on,n}}\right) - b\left(Y_{Q^0_{on,n}} - Y_{Q^i_{on,n}}\right)\right]\right)\right\}. \quad (11.66)$$

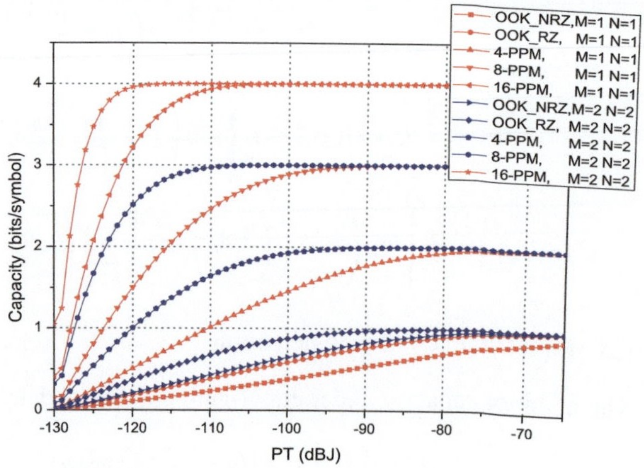

Fig. 11.9 MIMO-VLC system in different light-channel capacities at received pulse energy

(2) Equal-gain combined detection

The channel capacity under equal-gain combined detection is [7]

$$I(X;Y) = \log Q - E_{Y|X_0}\left[\log \sum_{i=0}^{Q-1} \exp\left(a\left(Y_{Q_{on}^0}^2 - Y_{Q_{on}^i}^2\right) - b\left(Y_{Q_{on}^0} - Y_{Q_{on}^i}\right)\right)\right]. \tag{11.67}$$

(3) Simulation results

Figure 11.9 shows the channel capacities of the MIMO-VLC system under different modulation schemes. Compared to the SISO-VLC system, the MIMO-VLC system has a higher channel capacity under the same received optical-pulse energy. Regardless of whether the system is MIMO or SISO, the channel capacity tends to a certain limit value as the received optical-pulse energy increases.

With an increase in the received optical-pulse energy, the MIMO system increases the channel capacity faster than the SISO system does. Therefore, as long as the optical transmit power is sufficiently large, the channel capacities of the SISO and MIMO systems will not differ significantly. The luminous power of white LEDs cannot be increased indefinitely, owing to limitations in device technology, human eye safety, and other factors. The MIMO system is more suitable for application to VLC systems.

Figure 11.10 shows the effect of the number of receiving antennas on the system channel capacity (the number of transmitting antennas remained unchanged). Two detection algorithms were used: maximum-likelihood joint detection and equal-gain combined detection. With an increase in the number of receiving antennas, the channel capacity also increased (with the same received pulse energy), and the performance of maximum-likelihood joint detection was better than that of equal-gain combined detection.

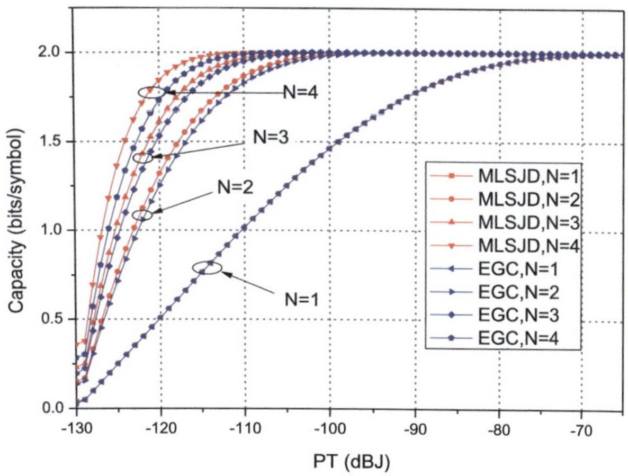

Fig. 11.10 Influence of the number of receiving antennas on the system channel capacity 4-PPM modulation

11.3 Space–Time Coding

In the analysis of the MIMO-VLC system, the signal is transmitted using a repeated-transmission method, and the advantages of multiple antennas are not fully utilized. This section combines space–time coding techniques with indoor MIMO-VLC systems [11].

11.3.1 Space–Time Coding Model of a MIMO-VLC System

Let the number of transmit antennas be M and the number of receive antennas be N, as shown in Fig. 11.11, for the MIMO-VLC space–time coding communication-system model. The transmitted data are sent to the space–time encoder for encoding. The space–time encoder maps the input block data into M parallel data $X^t = [X_1, X_2, \ldots, X_M]^T$, where $(.)^T$ means transposition, and modulates the white LED of the transmitting antenna at moment t. M parallel output signals are emitted simultaneously using M white LEDs. Among them, symbols X_m, $(m = 1, 2 \ldots, M)$, are emitted through the mth white LED, and all the emitted symbols have the same time width T_s. A vector of coded modulation symbols from different antennas is called a space–time symbol.

Without loss of generality, it can be assumed that the indoor VLC communication-link channel is memoryless. The channel between the transmit and receive antennas can be represented by the channel matrix H of $N \times M$:

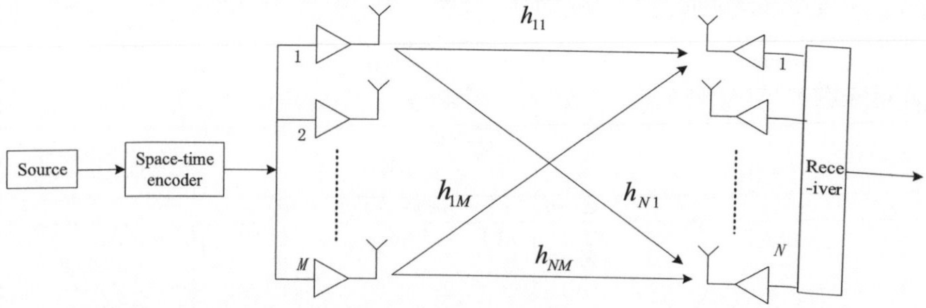

Fig. 11.11 MIMO-VLC space–time coding system model

$$H = \begin{bmatrix} h_{1,1} & h_{1,2} & \cdots & h_{1,M} \\ h_{2,1} & h_{2,2} & \cdots & h_{2,M} \\ \vdots & \vdots & \ddots & \vdots \\ h_{N,1} & h_{N,2} & \cdots & h_{N,M} \end{bmatrix}. \tag{11.68}$$

At the receiver end, the signals received by the Nth detector are the superposition of the optical signals emitted by the M white LEDs through the indoor multipath channel and the noise signal. At moment t, the photocurrent signal output by the front end of the nth branch receiver can be expressed as

$$Y_n^t = \sum_{m=1}^{M} h_{n,m} \cdot X_m^t + w_n^t, \tag{11.69}$$

where w_n^t is the noise component of the nth receiving branch, which is Gaussian noise. At the ith moment, the photocurrent output by the receiver-array antenna is

$$Y^t = \left[Y_1^t, Y_2^t, \ldots, Y_N^t\right]^T. \tag{11.70}$$

The noise at the receiver can be expressed as

$$n^t = \left[w_1^t, w_2^t, \ldots, w_N^t\right]^T. \tag{11.71}$$

Then, at moment t, the received signal vector can be expressed as

$$Y^t = HX^t + n^t. \tag{11.72}$$

The decoder of the receiver uses the maximum-likelihood algorithm to estimate the transmitted-signal sequence and calculates the decision criterion based on the squared Euclidean distance between the assumed received sequence and the sequence passing through the channel:

$$\sum_{t}\sum_{n=1}^{N}\left|Y_n^t - \sum_{m=1}^{M} h_{n,m}\cdot X_m^t\right|. \tag{11.73}$$

11.3.2 Dual-Transmit Antenna Space–Time Coding

In traditional Alamouti coding [12], the encoder divides two modulation symbols X_1 and X_2 into a group each time, and maps them to the transmit antenna through the coding matrix. The mapping matrix is

$$X = \begin{bmatrix} X_1 & -X_2^* \\ X_2 & X_1^* \end{bmatrix}, \tag{11.74}$$

where $*$ represents conjugation. The output of the encoder is transmitted from the two transmit antennas in two consecutive transmit cycles. In the first transmit cycle, antennas 1 and 2 transmit signals X_1 and X_2, respectively. In the second transmit cycle, antennas 1 and 2 transmit signals $-X_2^*$ and X_1^*. Because the signal emitted by a white LED can only be a positive signal, the Alamouti coding must be modified [13] (Fig. 11.12).

We consider adopting OOK and binary PPM modulation in an indoor VLC system. The transmitted codeword constellation of the OOK signal is $\mathfrak{X}_{OOK} = \{X_1, X_2\}$, where

$$\left.\begin{array}{l} X_1 = 0,\ \ 0 \le t \le T \\ X_2 = P_T,\ 0 \le t \le T \end{array}\right\}. \tag{11.75}$$

The transmitted codeword constellation of the binary PPM signal is $\mathfrak{X}_{PPM} = \{X_1, X_2\}$, where

$$\left.\begin{array}{l} X_1 = \begin{cases} 0, & \le t \le \frac{T}{2} \\ P_T, & \frac{T}{2} < \mathrm{t} \le T \end{cases} \\ X_2 = \begin{cases} P_T, & \le t \le T \\ 0, & \frac{T}{2} < \mathrm{t} \le T \end{cases} \end{array}\right\}, \tag{11.76}$$

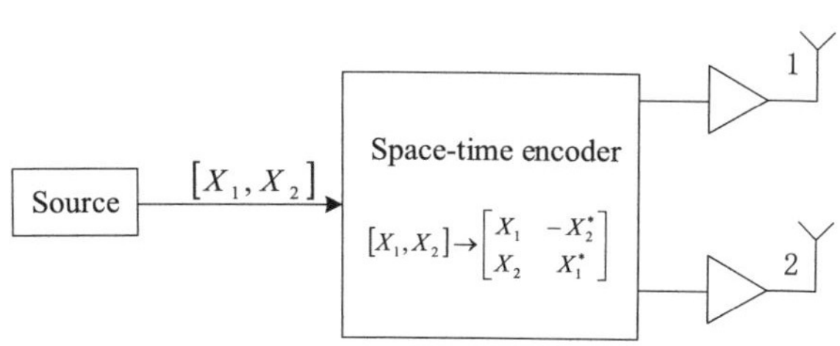

Fig. 11.12 Alamouti space–time encoder block diagram

where P_T denotes the emission power of the white LED. For these two modulation methods, any codeword can be represented by another codeword:

$$X_i = -X_j + P_T, i \neq j. \tag{11.77}$$

We define $\overline{X}_i$ as X_i's complement, i.e.,

$$\overline{X}_i = -X_i + P_T. \tag{11.78}$$

Then, the modified Alamouti-coding mapping matrix is [13]

$$X = \begin{bmatrix} X_1 & \overline{X}_2 \\ X_2 & X_1 \end{bmatrix}. \tag{11.79}$$

In the first transmission cycle, antenna 1 transmits signal X_1 and antenna 2 transmits X_2; whereas in the second transmission cycle, antenna 1 transmits signal $\overline{X}_2$, and antenna 2 transmits signal X_1. According to the modified Alamouti coding method [13], the current signal output by the nth receiving antenna at the receiver in the first symbol period is

$$Y_n^1 = h_{n,1}X_1 + h_{n,2}X_2 + w_n^1. \tag{11.80}$$

The current signal output by the nth receiving antenna in the second symbol period is

$$Y_n^2 = h_{n,1}\overline{X}_2 + h_{n,2}X_1 + w_n^2. \tag{11.81}$$

The equivalent outputs at two symbol periods are

$$Y_n^1 = h_{n,1}X_1 + h_{n,2}X_2 + w_n^1 \tag{11.82}$$

$$Y_n^2 = h_{n,1}P_T - h_{n,1}X_2 + h_{n,2}X_1 + w_n^2. \tag{11.83}$$

Using the channel-state information and transmit-power value of the white LED, the transmitted codeword can be reconstructed, as follows:

$$\widehat{X_1} = h_{n,1}Y_n^1 + h_{n,2}Y_n^2 - h_{n,1}h_{n,2}P_T \tag{11.84}$$

$$\widehat{X_2} = h_{n,2}Y_n^1 - h_{n,1}Y_n^2 + h_{n,1}^2 P_T. \tag{11.85}$$

We substitute Eq. (11.80) into Eq. (11.84) to obtain

$$\widetilde{X}_1 = h_{n,1}\big(h_{n,1}X_1 + h_{n,2}X_2 + w_n^1\big) + h_{n,2}\big(h_{n,1}\overline{X}_2 + h_{n,2}X_1 + w_n^2\big) - h_{n,1}h_{n,2}P_T \tag{11.86}$$

$$= \big(h_{n,1}^2 + h_{n,2}^2\big)X_1 + h_{n,1}h_{n,2}\big(\overline{X}_2 + X_2 - P_T\big) - h_{n,1}h_{n,2}P_T. \tag{11.87}$$

We substitute Eq. (11.81) into Eq. (11.84) to obtain

$$\widetilde{X}_2 = h_{n,2}\big(h_{n,1}X_1 + h_{n,2}X_2 + w_n^1\big) - h_{n,1}\big(h_{n,1}\overline{X}_2 + h_{n,2}X_1 + w_n^2\big) + h_{n,1}^2 P_T \tag{11.88}$$

$$= h_{n,1}^2\big(P_T - \overline{X}_2\big) + h_{n,2}^2 X_2 + h_{n,2}w_n^1 - h_{n,1}w_n^2. \tag{11.89}$$

Because $\overline{X}_2 = -X_2 + P_T$,

$$\widetilde{X}_1 = \big(h_{n,1}^2 + h_{n,2}^2\big)X_1 + h_{n,1}w_n^1 + h_{n,2}w_n^2 \tag{11.90}$$

$$\widetilde{X}_2 = \big(h_{n,1}^2 + h_{n,2}^2\big)X_2 + h_{n,2}w_n^1 + h_{n,1}w_n^2. \tag{11.91}$$

11.3.3 Multiple Transmit-Antenna Space–Time Coding

Abou-Rjeily proposed a multi-antenna space–time coding scheme based on free-space optical (FSO) communications [14]. We applied Abou-Rjeily's scheme to an indoor MIMO-VLC system to study space–time coding, based on multiple white LED transmit antennas. Considering the base PPM modulation, the constellation of the modulated signal can be expressed as

$$\mathfrak{X} = \{X_q;\, q = 0, \ldots, Q-1\}. \tag{11.92}$$

In this equation, X_q represents the transmitted PPM modulation signal, which is the $q+1$st column in a $Q \times Q$-dimensional unit matrix I_Q. Assuming that the transmitter sends a codeword at moment t, there are T discrete moments in total, the number of transmitters is M, and the number of receivers is N. The system model can be expressed as

$$Y = HX + W. \tag{11.93}$$

X is an $M \times QT$-dimensional transmitted codeword matrix, in which the codeword is represented at the qth slot position, the t symbol period, the codeword transmitted by the mth antenna, $q = 0, \ldots, Q-1$, $t = 1, \ldots, T$, and $m = 1, \ldots, M$. Y and W are both $N \times QT$ dimensional matrices; Y represents the output of the channel; and W represents additive white Gaussian noise, which is the noise generated by the

Table 11.4 Selection of the number of transmit antennas and the number of PPM modulation systems under full transmit diversity

	$Q = 2$	$Q = 4$	$Q = 8$
$M = 2$	√	√	√
$M = 3$	×	×	√
$M = 4$	×	×	√
$M \geq 5$	×	×	×

detector and its associated electronic circuits. H is an $N \times M$ channel matrix:

$$H = \begin{bmatrix} h_{1,1} & h_{1,2} & \cdots & h_{1,M} \\ h_{2,1} & h_{2,2} & \cdots & h_{2,M} \\ \vdots & \vdots & \ddots & \vdots \\ h_{N,1} & h_{N,2} & \cdots & h_{N,M} \end{bmatrix}. \tag{11.94}$$

For Q-ary PPM modulation with M transmit antennas, the minimum delay codeword can be expressed as

$$X(X_1, X_2, \cdots, X_M) = \begin{bmatrix} X_1 & X_2 & \cdots & X_M \\ \Omega X_M & X_1 & \ddots & \vdots \\ \vdots & \ddots & \ddots & X_2 \\ \Omega X_2 & \cdots & \Omega X_M & X_1 \end{bmatrix}, \tag{11.95}$$

$X_1, X_2, \cdots, X_M \in \mathfrak{X}$ and Ω is a cyclic permutation matrix of $Q \times Q$.

$$\Omega = \begin{bmatrix} 0_{1\times(Q-1)} & 1 \\ I_{Q-1} & 0_{(Q-1)\times 1} \end{bmatrix}, \tag{11.96}$$

where $0_{m\times n}$ represents the zero matrix of $m \times n$. Obviously, $\Omega X_m \in \mathfrak{X}$ and ΩX_m are the PPM waveform delay codes of X_m. This type of space–time coding is also called delayed coding. To obtain complete transmit diversity, the number of transmit antennas and the PPM modulation number should be selected reasonably, as shown in Table 11.4. "√" means that complete transmit diversity can be obtained, and "×" means that complete transmit diversity cannot be obtained.

11.3.4 Maximum-Likelihood Decoding

We assume that all transmitted codewords are equally probable, the receiver knows the ideal channel-state information, the decoder uses the Viterbi algorithm to perform maximum-likelihood decoding, and the decision criterion is the difference between the assumed and actual received symbols. The minimum norm is

$$\sum_{n=1}^{N}\left|Y_n^t - \sum_{m=1}^{M} h_{n,m}\cdot X_m^t\right|^2. \tag{11.97}$$

The maximum-likelihood decoding selects the path with the smallest path metric as the decoding sequence.

Figures 11.13 and 11.14 show the system error-performance simulation results using modified Alamouti coding and multi-transmit antenna space–time coding, respectively. Compared to the performance of the MIMO-VLC system that uses repetition coding without space–time coding, the resistance of the coded system to intersymbol interference (ISI) is enhanced.

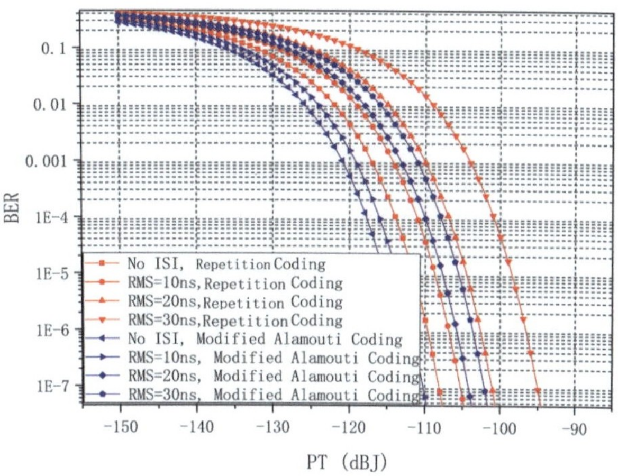

Fig. 11.13 Bit-error performance of a MIMO-VLC system using modified Alamouti coding, M = N = 2, 2-PPM modulation

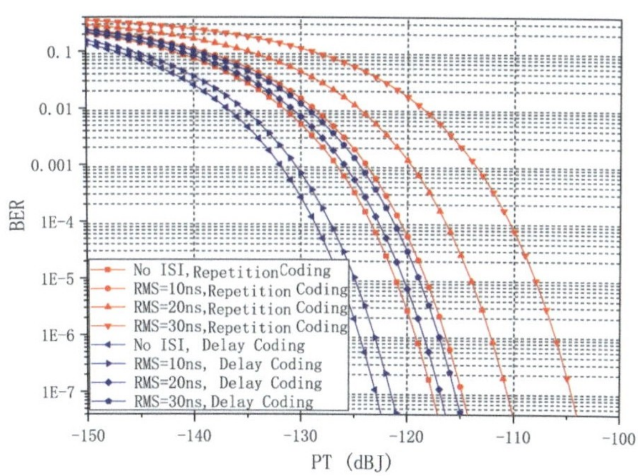

Fig. 11.14 Bit-error rate of a multiple-transmit antenna space–time coding system

11.4 Summary

This chapter analyzed the performance of a VLC system using multiple inputs and outputs and provided a design method and model for a MIMO-VLC system. The design and layout of a multi-array element white-LED transmitter and multi-array element receiver were studied. Two detection algorithms, maximum-likelihood joint detection and equal-gain combined detection, were provided for MIMO-VLC systems, and the calculation of the bit-error performance and channel capacity of MIMO-VLC systems using these two detection algorithms were studied.

Through simulation verification [11], when the number of receiving antennas increased, the maximum-likelihood joint detection algorithm was better than the equal-gain combined detection algorithm. From the simulation results of the space–time coding for the MIMO-VLC system with dual-antenna and multi-antenna arrays, the resistance of space–time coding to ISI is better than that of an uncoded system.

References

1. Komine T, Haruyama S, Nakagawa M (2003) Bidirectional visible-light communication using corner cube modulator. In: Proceedings of wireless and optical communication (WOC). Banff, Canada. IASTED
2. Nakamura S (1997) Present performance of InGaN basedblue/green/yellow LEDs. In: Proceedings of SPIE conference on light emitting diodes: research, manufacturing, and applications. SPIE, San Jose, CA, pp 26–35
3. Ding D, Ke X (2006) Research on visible light communication and its key technologies. Semicond Optoelectr 27(02):114–117
4. Ding D, Ke X, Li J (2011) Optimization design of multi-element transmitting antenna for indoor visible light communication. Chin J Appl Sci 29(03):238–242
5. Ding D, Ke X (2010) Research on a universal white LED mathematical luminescence model. Chin J Opt 30(09):2536–2540
6. Gagliardi RM, Karp S (1995) In: Proakis JG (ed) Optical communications. Wiley
7. Wilson SG, Brandt-Pearce M, Cao Q, Baedke M (2005) Optical repetition MIMO transmission with multipulse PPM. IEEE Selected Areas Commun 23(9):1901–1909
8. Nick L (2008) The Gaussian free space optical MIMO channel with Q-ary pulse position modulation. IEEE Trans Wirel Commun 7(5)
9. Ding D, Ke X (2007) Design and simulation of a wireless local area network system based on visible light communication. J Xi'an Univ Technol 23(01):29–32
10. Ding D, Ke X (2010) Performance of asymmetric retroreflection free space optical communication links. Chin J Appl Sci 28(04):337–341
11. Ding D Research on key technologies of indoor visible light communication. Doctoral Dissertation of Xi'an University of Technology, Xi'an
12. Lin Y, He F (2010) The principle and application of MIMO technology. People's Posts and Telecommunications Publishing House, Beijing
13. Simonand MK, Vilnrotter V (2003) Alamouti-type space-time coding for free-space optical communication with direct detection. IPN Progress Report, pp 42–155

14. Chadi A-R, Wissam F (2008) Space-time codes for MIMO ultra-wideband communications and MIMO free-space optical communications with PPM. IEEE J Selected Areas Commun 26(6):938–947

Chapter 12
MIMO-System Detection Algorithms

Multiple-input multiple-output (MIMO) signal detection is the key to improving the performance of an entire MIMO system [1]. MIMO-system detection algorithms can be divided into linear, nonlinear, and iterative detection algorithms. Linear detection algorithms include zero forcing (ZF) and minimum mean-squared error (MMSE). Nonlinear detection algorithms include a sphere decoding algorithm, semidefinite relaxation detector, intelligent methods, and iterative detection algorithms [2].

Most existing detection algorithms are limited to simple detector improvement. To approach the channel capacity as closely as possible, channel coding and interleaving are required to provide redundancy to suppress burst fading, interference, and additive noise at the receiver. In this chapter, the commonly used MIMO-system detection algorithms are analyzed, and their detection performance and computational complexity are discussed.

12.1 Basic MIMO-System Detection Algorithms

The structure of a basic MIMO-system detection algorithm is shown in Fig. 12.1. The basic concept is to separate the signals sent from the respective transmitting antennas, and to judge and estimate the transmitted signals. If channel coding is present, channel decoding is carried out after the corresponding space–time decoding is completed. De-interleaving, shown in Fig. 12.1, is the reverse process of interleaving at the transmitter. The purpose of interleaving is to improve the correlation of signals encoded by the channel.

Research results [3] show that the maximum-likelihood (ML) detection algorithm is the optimal hard-decision detection algorithm for MIMO systems. However, the computational complexity of the ML algorithm increases exponentially with the number of antennas and modulation order, and is difficult to implement. Researchers

X. Ke, *Space-Time Coding Theory in Optical Wireless MIMO Systems*,
Optical Wireless Communication Theory and Technology,
https://doi.org/10.1007/978-981-97-6449-5_12

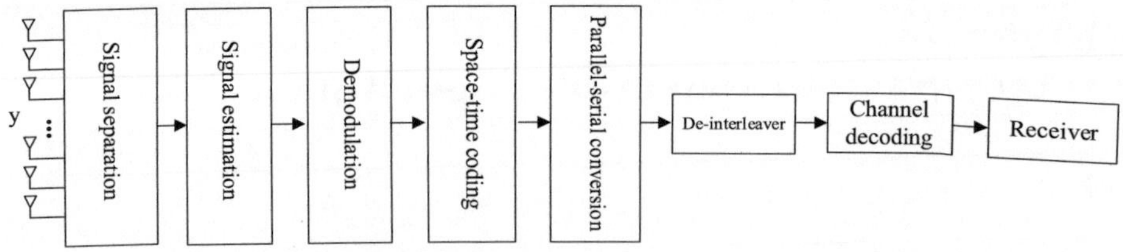

Fig. 12.1 Structure of a basic detection algorithm

often focus on the detection performance of ML detection algorithms, and turn to various suboptimal detection algorithms in space–time coding.

Representative suboptimal detection algorithms mainly include linear detection algorithms, based on the ZF and MMSE criteria, and nonlinear detection algorithms based on interference cancellation.

12.1.1 Maximum-Likelihood Detection Algorithm

The maximum-likelihood (ML) detection algorithm is the optimal decoding algorithm for layered space–time decoding with a minimum bit-error rate. Given a channel-characteristic matrix H and a known transmitted-signal vector, the probability-density function of the received signal at the receiver presents a Gaussian distribution [4]:

$$p(y|H, d) = \frac{1}{(2\pi\sigma_n^2)^{N/2}}\exp\left[-\frac{||y - Hd||^2}{2\sigma_n^2}\right], \tag{12.1}$$

where y represents the signal received by the receiving end, H represents the channel-characteristic matrix, and ${\sigma_n}^2$ represents the variance of additive noise n, which is the noise generated by photodetectors and their electronic circuits.

The basic idea of maximum-likelihood detection is to store all possible transmission-symbol vectors at the receiving end. After receiving a transmission signal, the receiving end calculates the Euclidean distance between the signal and the transmission-symbol vector stored at the receiving end in the receiving space. The signal corresponding to the minimum Euclidean distance can be used to estimate the received signal y [4]:

$$\begin{aligned} d_{ML} &= \arg \max_{d\in\Omega^M} (y|H, d) = \arg \max_{d\in\Omega^M} \frac{1}{\left(2\pi\sigma_n^2\right)^{N/2}} \exp\left(-\frac{\|y - Hd\|^2}{2\sigma_n^2}\right) \\ &= \arg \min_{d\in\Omega^M} \|y - Hd\|, \end{aligned} \tag{12.2}$$

where y represents the signal received by the receiving end; H is the channel characteristic matrix; σ_n^2 represents the variance of additive Gaussian noise n; $\arg\max_d(\cdot)$ represents the value of d corresponding to the maximum value of $(\cdot)$; and Ω^M is the set of constellation points of the modulated signal. For pulse-position modulation (PPM), Ω^M is the set of all modulation patterns.

Maximum-likelihood detection involves searching in set Ω^M to obtain the signal vector with the minimum Euclidean distance. The size of the search space increases exponentially with the number of transmit antennas and constellation points. When the modulation order or the number of transmit antennas increases, the computational complexity of the ML detection algorithm greatly increases. Theoretically, the minimum bit-error rate can be obtained on each antenna; however, it is difficult to achieve in practice because of its high computational complexity [5]. Various sub-optimal detection algorithms are based on the ML detection algorithm.

12.1.2 Linear Detection Algorithms

The linear detection algorithm is the simplest suboptimal detection algorithm, and its filter matrix can be obtained by inverting the channel matrix. To ensure a unique solution to matrix inversion, the number of transmitting antennas must be greater than the number of receiving antennas (i.e., the channel matrix H satisfies the column full rank). Linear detection algorithms include the ZF and MMSE criteria linear-detection algorithms. In a sense, the two criteria are similar.

12.1.2.1 ZF Criterion

The so-called zero-forcing (ZF) criterion sets the interference signal caused by channel matrix H to zero; hence, it is called "zero forcing." It is the simplest detection algorithm. The estimated signal obtained by the ZF criterion is [5]

$$\hat{d}_{ZF} = G_{ZF}y = H^{\&}y = d + H^{\&}n, \tag{12.3}$$

where $G_{ZF} = H^{\&} = \left(H^T H\right)^{-1} H^T$ is the pseudo inverse of channel matrix H; d is the transmission signal; and n is additive Gaussian white noise. The ZF-criterion estimation-error covariance matrix is [5]

$$\phi_{ZF} = E\{(\hat{d} - d)(\hat{d} - d)^H\} = \sigma_n^2 (H^T H)^{-1} = \sigma_n^2 H^{\&} H^{\&T}. \tag{12.4}$$

It can be seen from Eq. (12.4) that in the ZF detection algorithm, the noise at the receiver is amplified, and a small change of the $\Lambda(Y) = \frac{f(Y|on)}{f(Y|off)}$ eigenvalue will cause a significant estimation error, thereby affecting the performance of the entire system. Therefore, although the complexity of the zero-forcing criterion is low, its detection performance is not high. To improve the system-detection performance, researchers consider the influence of system noise when designing the filter matrix G; hence, a linear detection algorithm based on the minimum mean-squared error criterion appears.

12.1.2.2 MMSE Criterion

Although the ZF criterion eliminates the mutual interference between antennas, its detection performance is relatively low at the cost of increasing noise. The minimum mean-squared error (MMSE) criterion minimizes the mean-squared error between the actual transmitted signal and the estimated signal output by the detector [5]:

$$G_{MMSE} = \arg\min_{G} E\big((Gy - d)(Gy - d)^H\big). \tag{12.5}$$

Equation (12.5) takes the derivative of G and sets it to zero:

$$\frac{\partial G_{MMSE}}{\partial G} = 0. \tag{12.6}$$

Then, the filter matrix of the MMSE-criterion linear detection algorithm is [5]

$$\begin{aligned} G_{MMSE} =& H^T\left(H^T H + \sigma_n^2 I_N\right)^{-1} \\ =& \left(H^T H + \sigma_n^2 I_N\right)^{-1} H^T y. \end{aligned} \tag{12.7}$$

where H is the characteristic matrix of the channel; σ_n^2 represents the variance of additive noise n; and I_N represents the $N \times N$ dimensional identity matrix. Therefore, the estimated signal output by the detector can be obtained:

$$\hat{d}_{MMSE} = G_{MMSE} y = \left(H^T H + \sigma^2 I_N\right)^{-1} H^T y. \tag{12.8}$$

The estimation error covariance matrix is

$$\varphi_{MMSE} = E\Big\{\big(\hat{d} - d\big)\big(\hat{d} - d\big)^T\Big\} = \sigma_n^2\left(H^T H + \sigma^2 I_N\right)^{-1}. \tag{12.9}$$

Definitions:

$$\overline{H} = \begin{bmatrix} H \\ \sigma_n^2 I_N \end{bmatrix} \tag{12.10}$$

$$\overline{y} = \begin{bmatrix} y \\ 0_{N\times 1} \end{bmatrix}. \tag{12.11}$$

The estimation vector (12.8) can be rewritten as

$$\hat{d}_{MMSE} = \overline{H} + \overline{y} = \left(\overline{H}^T\overline{H}\right)^{-1}\overline{H}^T\overline{y}. \tag{12.12}$$

Then, the covariance matrix of the estimation error can be written as

$$\varphi_{MMSE} = \sigma^2\left(\overline{H}^T\overline{H}\right)^{-1} = \sigma^2\overline{H}^+\overline{H}^{+T}. \tag{12.13}$$

Comparing Eqs. (12.3), (12.12), (12.4), and (12.13), it can be seen that after introducing the extended matrix, the linear detection algorithms using the ZF and MMSE criteria are similar in expression. An extended model of the MMSE-criterion detection algorithm can be obtained by replacing H in Eqs. (12.8) and (12.9) with extended matrix $\overline{H}$ and the received-signal matrix y with extended matrix $\overline{y}$.

12.1.3 Nonlinear Detection Algorithms

The interference-cancellation algorithms include successive interference cancellation (SIC) and parallel interference cancellation (PIC). The successive interference-cancellation algorithm first detects a certain signal, treats the detected signal as an interference signal to other signals, and subtracts it from the remaining signals; it repeats this until all signals are detected. This detection algorithm can improve the accuracy of detecting residual signals. The following describes the unsorted and sorted successive interference-cancellation algorithms.

12.1.3.1 Successive Interference-Cancellation (SIC) Detection

In MIMO systems, the successive interference-cancellation (SIC) algorithm was first applied in vertical Bell Laboratories layered space–time (V-BLAST) systems; it is sometimes referred to as the V-BLAST detection algorithm or zero-cancellation algorithm. The algorithm can adopt either the ZF or MMSE criterion [6]. This method first detects the signal of an antenna through the ZF or MMSE criterion, cancels the interference it causes from the received signal, and iterates step-by-step until all signals are detected (that is, the transmitted data stream is detected and eliminated layer by layer).

The SIC detection algorithm detects the entire signal through M-level recursive processing, and detects a transmitted signal at each level. Each recursive-detection level is based on an equivalent signal model:

$$y_i = Hd_i + n,\ i = 1, 2, \ldots M, \tag{12.14}$$

where y_i is the signal after the detected signal is removed from the received signal, and d_i is the signal after the detected signal is removed from the transmitted signal. Two processes are required to execute the first-level processing: zeroing and interference elimination. The following describes the steps of each level of processing using the first-level processing as an example.

(1) Zeroing

We select a signal from the remaining undetected transmission signals for linear detection. Assuming that the selected signal is d_1, then,

$$\tilde{d}_1 = G_1 y \tag{12.15}$$

$$\hat{d}_1 = Q(\tilde{d}_1) \tag{12.16}$$

where G_1 is the first row of the filter matrix in linear detection, based on the ZF or MMSE criterion; $Q(\cdot)$ is a demodulation operation, which demodulates the estimated signal $\hat{d}_1$ detected linearly to obtain an estimated value of the transmitted signal.

(2) Interference cancellation

We eliminate the interference caused by signal $\hat{d}_2$, estimated in the previous step, from the signal at the receiving end to obtain the model of the remaining undetected signal, i.e.,

$$\begin{aligned} y_2 &= y - h_1 \hat{d}_1 \\ &= H_2 d_2 + n, \end{aligned} \tag{12.17}$$

where h_1 is the first column of the channel-characteristic matrix; y_2 is the N-dimensional equivalent received-signal vector obtained by canceling the interference of $\hat{d}_1$, from the received signal y; $H = [h_2, h_3, \ldots, h_M]$ represents the $N \times (M - 1)$-dimensional equivalent channel-characteristic matrix remaining after the first column is deleted from the channel-characteristic matrix H. $d_2 = [d_2, d_3, \ldots d_M]$ is the $(M - 1)$-dimensional equivalent transmitted-signal vector remaining after d_1 is deleted from d; this completes the first stage of processing. Then, we return to the first step, continue to detect the obtained equivalent model, detect the next transmission signal, and repeat the process until all the transmission signals have been detected.

The SIC detection algorithm reduces the dimension of the equivalent model after each detection stage; thus, it can obtain a higher diversity gain. The increase in diversity gain will improve the system performance. Therefore, the performance of the SIC detection algorithm is better than that of a linear detection algorithm. In practice, the error-propagation problem is inevitable; therefore, the performance of the entire system will be limited by the detection performance of the first detected signal.

12.1.3.2 Ordered Successive Interference-Cancellation (OSIC) Detection

In decoding, errors may propagate in the process of the successive interference-cancellation algorithm. If the signal with the highest reliability is offset first, the error-propagation probability will increase. To improve the detection performance, the signals are detected and deleted in a certain order. After each interference cancellation, the next detected signal is guaranteed to have the highest reliability. Compared with the ZF-criterion linear detection, MMSE-criterion linear detection, and unordered successive interference cancellation algorithms, this improved successive interference-cancellation algorithm has increased complexity, but can obtain good performance. This algorithm is called the ordered successive interference-cancellation (OSIC) algorithm.

The OSIC algorithm is an improvement on the unordered SIC algorithm. The signal-sequencing operation is carried out before the interference is eliminated from the SIC algorithm, which can effectively reduce the possibility of error propagation in the SIC algorithm and improve the detection performance of the system.

The emphasis of the OSIC algorithm is to perform the iterative interference-cancellation processing of multi-level optimal sequencing. The OSIC algorithm performs three basic operations in each processing level: sorting, zeroing, and interference cancellation. Sorting is the key operation in the OSIC algorithm, and the zeroing and interference cancellation are the same as those in the unsorted SIC algorithm. The purpose of sequencing is to improve the reliability of the first detected signal as much as possible, thereby reducing the error propagation, which improves the detection performance of the entire system.

The sorting principle is as follows: In each detection stage, the elimination and sorting criteria are used to select a signal with the best detection performance from the remaining undetected signals for detection and interference elimination. The sorting criterion was proposed by Golden et al. [7]: In each processing stage, one of the remaining signals with the largest signal-to-noise ratio is selected after passing through the linear filter for detection.

Because the global optimal ranking criterion is used, the detection performance of the SNR-OSIC detection algorithm is significantly improved compared with the SIC detection algorithm. The sorting criterion proposed by Golden et al. is implemented using the ZF criterion [6]. Theoretically, the OSIC algorithm based on the MMSE criterion (MMSE-OSIC) has better detection performance than the OSIC algorithm based on the ZF criterion (ZF-OSIC).

12.1.4 Simulation Results

In the following, the ZF-OSIC and MMSE-OSIC detection algorithms are simulated and analyzed. The simulation process of the basic detection algorithm is shown in Fig. 12.2. The channel coding adopts a low-density parity check (LDPC) code; the

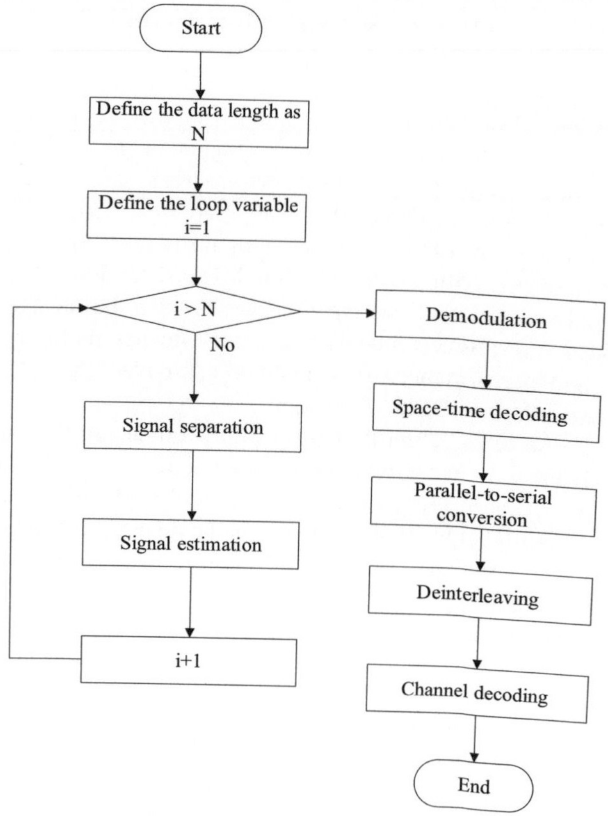

Fig. 12.2 Flowchart of the basic detection algorithm

coding length of each frame is 2048, the code rate is 1/2, the number of frames is 1000, and the modulation mode is pulse-position modulation (PPM).

For weak turbulence, the channel-attenuation coefficient is assumed to follow a lognormal distribution and the scintillation index SI = 0.3. For strong turbulence, the channel-attenuation coefficient is assumed to follow a gamma–gamma distribution and the SI = 0.9. The numbers of simulated antennas are 4 transmit × 4 receive and 4 transmit × 6 receive. The performance of each suboptimal detection algorithm under different atmospheric-turbulence intensities is shown in Figs. 12.3, 12.4, 12.5, 12.6, 12.7 and 12.8.

Figures 12.3 and 12.4 show that when the MMSE-criterion linear detection algorithm is adopted, the performance of the 4 × 6 system is about 6 dB higher than that of the 4 × 4 system. At a low signal-to-noise ratio (SNR), the detection performance of the MMSE algorithm is about 5 dB higher than that of the ZF algorithm; when the SNR is 30 dB, the performance of the MMSE algorithm converges with that of the ZF algorithm. Different atmospheric-turbulence intensities have different effects on the performance of a MIMO system. Strong turbulence interferes more with the signal, and the system performance is relatively poor.

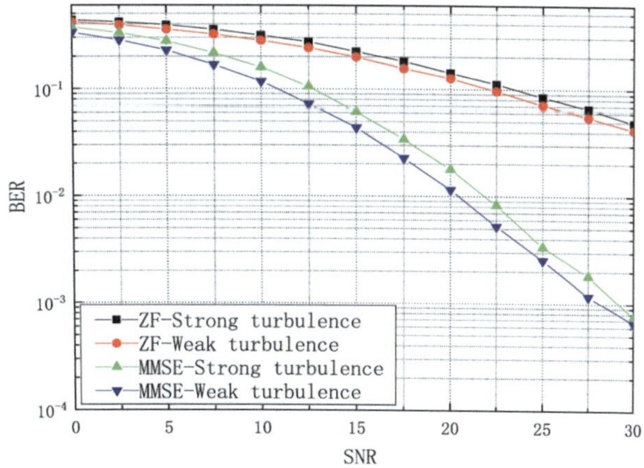

Fig. 12.3 Results for 4 × 4 antennas

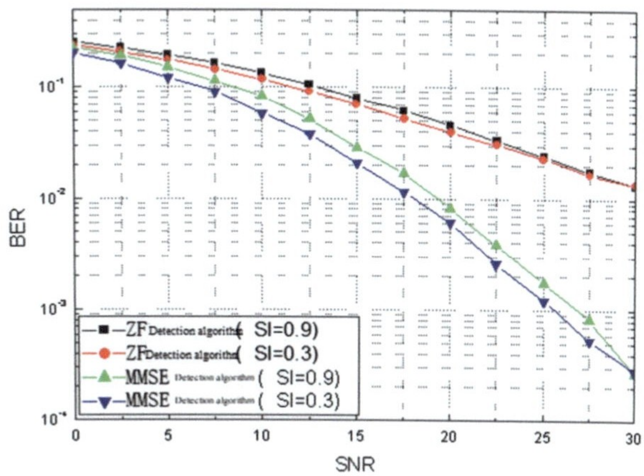

Fig. 12.4 Results for 4 × 6 antennas

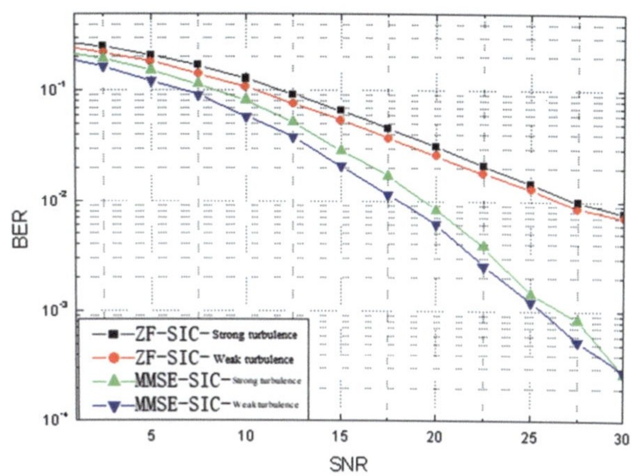

Fig. 12.5 Comparison of the SIC algorithm detection performance under different turbulence conditions, with 4 × 4 antennas

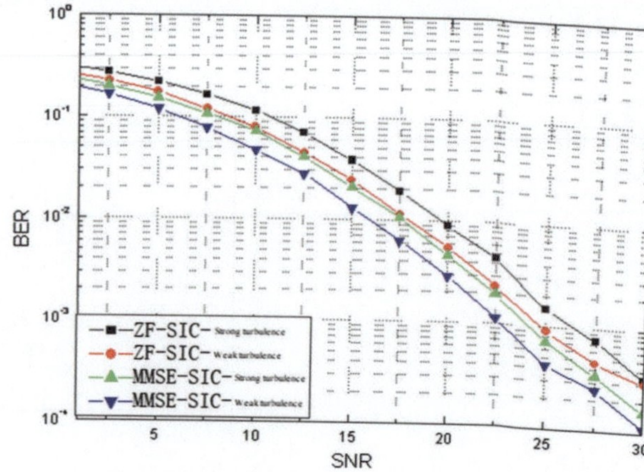

Fig. 12.6 Comparison of the SIC algorithm detection performance under different turbulence conditions with 4 × 6 antennas

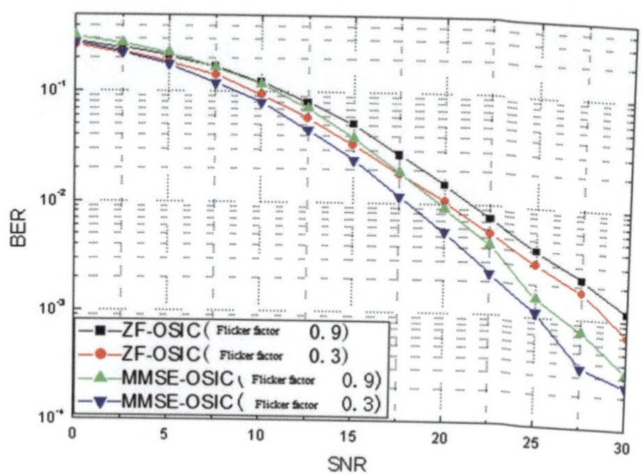

Fig. 12.7 Comparison of OSIC algorithm detection performance under different turbulence conditions with 4 × 4 antennas

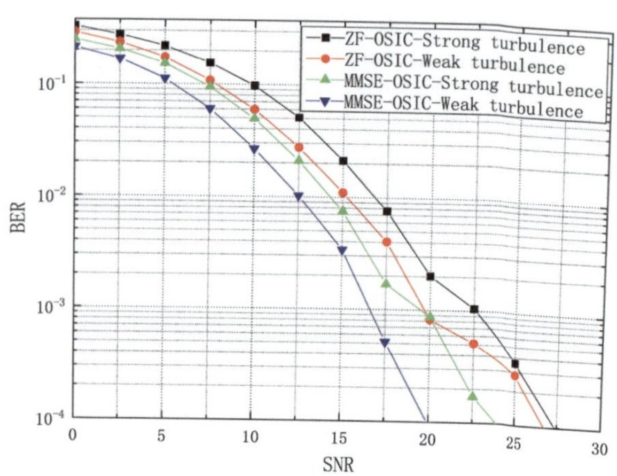

Fig. 12.8 Comparison of OSIC algorithm detection performance under different turbulence conditions with 4 × 6 antennas

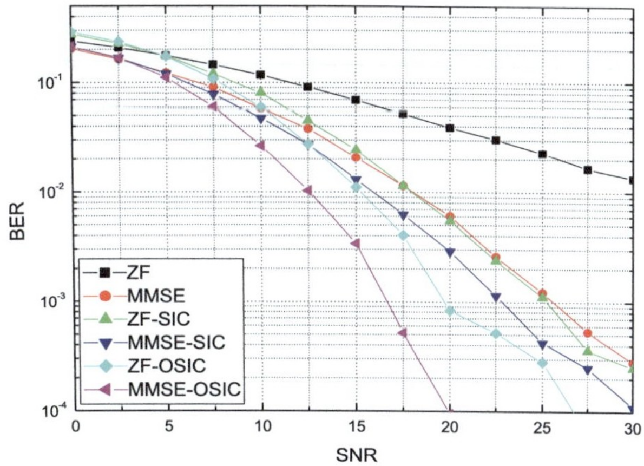

Fig. 12.9 Performance comparison of hard-detection algorithms

Figures 12.5, 12.6, 12.7 and 12.8 show the detection performance of the successive interference-cancellation algorithm under different antenna numbers and different atmospheric-turbulence intensities. Figures 12.5 and 12.6 show the detection performance of the unordered successive interference-cancellation algorithm using the ZF criterion (ZF-SIC) and MMSE criterion (MMSE-SIC), respectively. Figures 12.7 and 12.8 show the detection performance of the ordered successive interference-cancellation algorithm using the ZF criterion (ZF-OSIC) and MMSE criterion (MMSE-OSIC), respectively. It can be seen that the performance of the 4×6 system is about 6 dB higher than that of the 4×4 system.

A hard-detection algorithm directly inputs the hard-decision result of the transmitted signal into the channel decoding for decoding. A hard decision can lose information; hence, its performance is poor [8]. Figure 12.9 compares the detection performance of several hard-detection algorithms.

In the SIC algorithm, the detection performance of the MMSE criterion is still better than that of the ZF criterion. Different atmospheric-turbulence intensities have different effects on the MIMO system performance. Strong turbulence interferes more with the signal, and the system performance is relatively poor. It can be seen from Fig. 12.9 that the performance of the OSIC algorithm is about 4 dB higher than that of the unordered SIC algorithm, and the performance of the unordered SIC algorithm is about 5 dB higher than that of the linear detection algorithm.

12.2 Iterative Detection Algorithms for MIMO Systems

The turbo codes proposed by Berrou in 1993 [9] have attracted extensive attention because of their effective iterative decoding. In their iterative-decoding structure, turbo codes are composed of two recursive concatenated convolutional codes. During decoding, the two component decoders continuously exchange information, which improves the reliability of the final decision signal and reduces the bit-error rate [11].

These ideas of soft input, soft output, and iteration have been applied to channel equalization, channel-parameter estimation, multi-user detection, and MIMO detection technologies [12] to improve system performance. The following mainly discusses the application of iterative algorithms to MIMO detection technology.

12.2.1 MIMO Iterative-Detection Technology

The channel coding and space–time coding at the transmitter of a MIMO system can be equivalent to the outer-code and inner-code encoders in a turbo serial-concatenated coding structure. This serial-concatenated structure is suitable for decoding with the soft-input soft-output (SISO) turbo iterative structure. The decoding structure uses the SISO detector as an internal-code decoder to complete space–time decoding, and as an external-code decoder to complete channel decoding; thus, the turbo receiver is formed under a MIMO system [10].

The MIMO detector uses the data information output from the channel and the bit-level soft information output from the previous iteration of the channel decoder to calculate the new bit-level soft information. The new bit-level soft information is de-interleaved and used as the input of the outer channel decoder. The new bit-level soft information output by the decoder is interleaved and used as the next input of the MIMO detector. A block diagram of the iterative detection system is shown in Fig. 12.10. The transmitter and receiver of the iterative detection algorithm include a SISO detector, deinterleaver, SISO decoder, and interleaver.

A SISO detector is a soft-input soft-output detector. Its main function is to separate the interfering signals from different transmitting antennas and output the corresponding symbol or bit-level soft information. The principle of a SISO detector is to calculate the posterior probability of each coded bit in the transmitted-symbol vector, also known as bit-level soft information (log-likelihood ratio, LLR value), according to the received-signal vector and the external information provided by the SISO decoder. A SISO decoder is a channel decoder. Its main function is to decode the channel according to the external information provided by the detector and output the bit-level soft information after decoding.

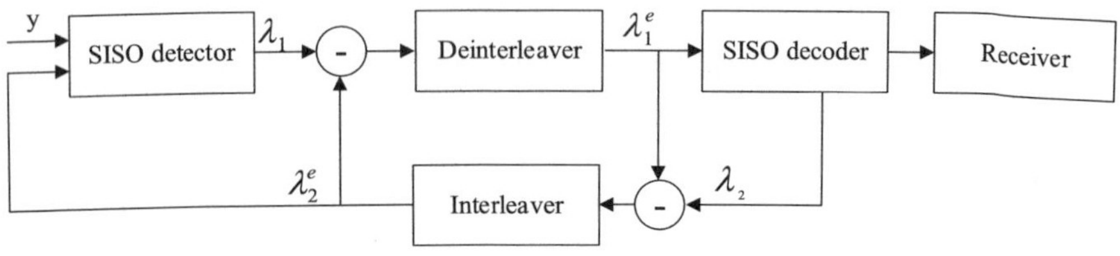

Fig. 12.10 Structure of a MIMO-system iterative-detection algorithm

The deinterleaver and interleaver at the receiving end correspond to the interleaver at the transmitting end. If the channel code adopts LDPC codes, the deinterleaver and interleaver at the receiving end can be omitted. In this manner, the structure of the system becomes much simpler, and the overall efficiency of the system can be greatly improved.

The iterative process shown in Fig. 12.10 is as follows: The SISO detector calculates the posterior information λ_1 of each coded bit in the transmitted-symbol vector, according to the received-signal vector and the external information λ_2^e provided by the decoder. It then subtracts the external information λ_2^e provided by the SISO decoder to obtain the external information λ_1^e of the SISO detector. The external information $f(Y|X_i)$ provided by the SISO detector is input into the SISO decoder after being deinterleaved.

The SISO decoder decodes according to the external information $P_{ave} = \left[\int_{f(Y|X_i)\ge f(Y|X_1)} f(Y|X_i)dY\right]^{Q-1}$ of each bit, then outputs the decoded posterior information λ_2, and subtracts the external information $P_b = \frac{Q/2}{Q-1}P_s = \frac{Q/2}{Q-1}\left\{1 - \left[\int_{f(Y|X_i)\ge f(Y|X_1)} f(Y|X_i)dY\right]^{Q-1}\right\}$ provided by the SISO detector to obtain new external information λ_2^e. The external information output by the decoder is interleaved and sent to the SISO detector as the input information of its next iteration; thus, completing an iteration [13]. At the first iteration, the external information provided by the decoder is zero.

The external information λ_2^e output by the SISO decoder can reduce the interference of the received signal during detection, and the external information λ_1^e output by the SISO detector can improve the decoding performance of the SISO decoder [14]. Iterative detection and decoding helps make the signal decision more reliable and improve the system performance by continuously exchanging external information between the detector and the decoder.

Theoretically, the more iterations, the greater the correlation of the decision signal, the higher the decision reliability, and the better the performance. When the preset number of iterations is completed, the information after the decoder is judged to obtain the decoded information bits. The flowchart of the MIMO-system iterative-detection algorithm is shown in Fig. 12.11.

12.2.2 MIMO Iterative-Detection Algorithm

The most important part of the MIMO iterative-detection algorithm is estimating the external information of the received signal and the external information obtained after decoding. Suppose that the symbol signal received by the receiving end is $y = [y_1, y_2, \ldots, y_N]$, the symbol information transmitted by the transmitting end is $d = [d_1, d_2, \ldots, d_M]$, and the bit information transmitted is $b = [b_1, b_2, \ldots, b_{KM}]$. $K = 2^{MC}$, where M_C represents the modulation order, and the prior information λ_1 output by the soft-input soft-output detector is

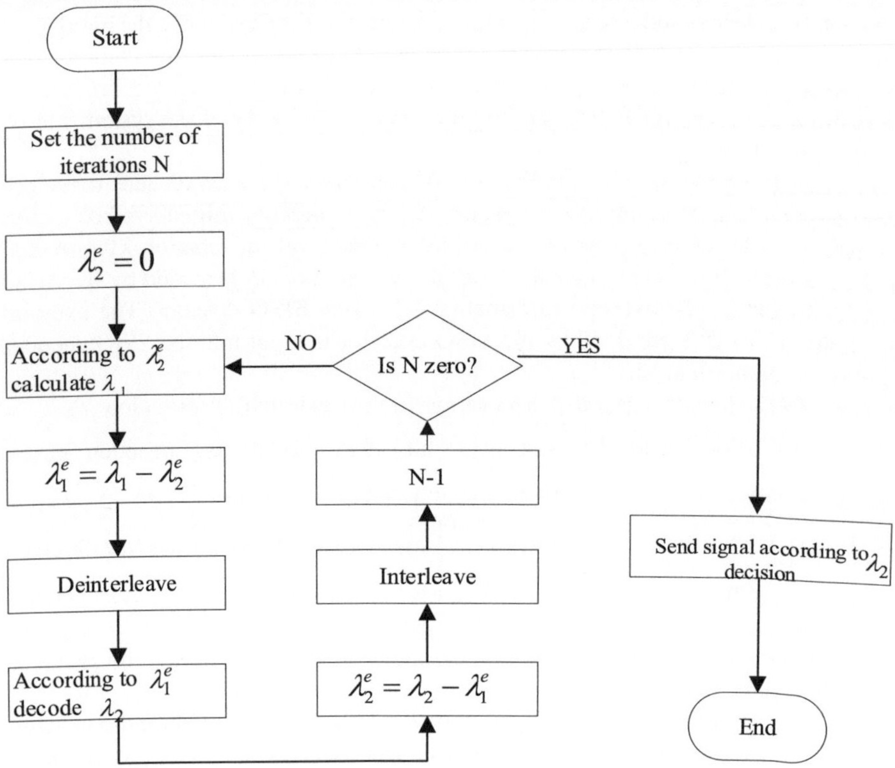

Fig. 12.11 Flowchart of a MIMO-system iterative detection and decoding process

$$\lambda_1(b_k) = \ln \frac{p(b_k = 1 | y, \lambda_2^e)}{p(b_k = 0 | y, \lambda_2^e)}, \tag{12.18}$$

where λ_2^e is the external information output by the SISO decoder, y is the received signal, and d_k is the k bit-level information in the transmission vector of the transmitter when the received signal is y. The method of calculating λ_2^e is presented below.

Hypothesis:

$$b_{-k} = [b_1 \ldots b_{k-1}\ b_{k+1} \ldots b_{KM}], \tag{12.19}$$

where

$$p(b_k = 1 | y, \lambda_2^e) = \sum_{b_{-k}} p(b_k = 1, b_{-k} | y, \lambda_2^e) \tag{12.20}$$

$$p(b_k = 0 | y, \lambda_2^e) = \sum_{b_{-k}} p(b_k = 0, b_{-k} | y, \lambda_2^e). \tag{12.21}$$

According to the Bayesian posterior-probability formula, the posterior probability of transmitted bits can be obtained as

$$\begin{aligned} p\left(b_k = 1 \middle| y, \lambda_2^e\right) &= \sum_{b_{-k}(i)} p\left(b_k = 1, b_{-k} \middle| y, \lambda_2^e\right) \\ &= \sum_{b_{-k}} \frac{p\left(b_k = 1, b_{-k}, y, \lambda_2^e\right)}{p\left(y, \lambda_2^e\right)} \\ &= \sum_{b_{-k}} \frac{p\left(b_k = 1 | b_{-k}, y, \lambda_2^e\right) p\left(y, b_{-k}, \lambda_2^e\right)}{p\left(y, \lambda_2^e\right)} \\ &= \sum_{b_{-k}} \frac{p\left(b_k = 1 | b_{-k}, y, \lambda_2^e\right) p\left(y, \lambda_2^e\right) p\left(b_{-k} | y, \lambda_2^e\right)}{p\left(y, \lambda_2^e\right)} \\ &= \sum_{b_{-k}(i)} p\left(b_k = 1, b_{-k} \middle| y, \lambda_2^e\right) \times p\left(b_{-k} | y, \lambda_2^e\right). \end{aligned} \tag{12.22}$$

Similarly, the following can be obtained:

$$p\left(b_k = 0 | y, \lambda_2^e\right) = \sum_{b_{-k}} p\left(b_k = 0 | y, b_{-k}, \lambda_2^e\right) \times p\left(b_{-k} | y, \lambda_2^e\right). \tag{12.23}$$

To calculate the posterior probability $p\left(b_k = 1 | y, \lambda_2^e\right)$ at $b_k = 1$ and the posterior probability $p\left(b_k = 0 | y, \lambda_2^e\right)$ at $b_k = 0$, all the possibilities of b_{-k} should be considered, and the size of all the possibilities is equal to 2^{KM-1}. It can be seen that the maximum a posteriori probability (MAP) detection algorithm increases exponentially with the number of antennas and the modulation order, which is difficult to achieve in practice [15].

12.2.3 Monte Carlo Statistical Method

The Monte Carlo algorithm is a sampling algorithm that can sample and discretize the continuous probability-density distribution function of a random variable at a given time. Using discrete random variables instead of continuous random variables in highly complex operations can greatly reduce the computational complexity.

12.2.3.1 Monte Carlo Integration

Assuming a random variable x, its probability-density function is $f(x)$; we define a function $h(x)$ with this random variable as the independent variable. Then, the weighted-average value of function $h(x)$ can be expressed as

$$E_f[h(x)] = \int_x h(x)f(x)dx, \tag{12.24}$$

where χ is the range of random variable χ, and $f(x)$ is its probability-density function; that is, $\int_x f(x)d(x) = 1$. According to Monte Carlo integration, the weighted average value of function $h(x)$ can be estimated as [16]

$$\bar{h} = \frac{1}{N_S}\sum_{n=1}^{N_s} h(x_n), \tag{12.25}$$

where x_n is the sampling value of probability-distribution function $f(x)$.

12.2.3.2 Importance Sampling

Importance sampling is estimated by introducing a new distribution function $f_a(x)$. The weighted average value $-X_2^*$ of Eq. (12.24) can be written as

$$E_f[h(x)] = \int_\chi h(x)\frac{f(x)}{f_a(x)}f_a(x)dx. \tag{12.26}$$

Then, according to the Monte Carlo integral $\bar{h} = \frac{1}{N_S}\sum_{n=1}^{N_s} h(x_n), f_a(x)$, is regarded as $f(x)$, $\frac{f(x_n)}{f_a(x_n)}h(x_n)$ as $h(x_n)$, and the mean value of $h(x)$ can be estimated as

$$\bar{h} \approx \frac{1}{N_S}\sum_{n=1}^{N_S}\frac{f(x_n)}{f_a(x_n)}h(x_n), \tag{12.27}$$

where x_n is the sampling value of probability-distribution function $f_a(x)$. The importance-sampling function can be obtained by Monte Carlo integration. $f_a(x)$ is an appropriate probability-density distribution function similar to $f(x)$. If $f_a(x) = f(x)$, importance sampling becomes Monte Carlo integration.

In importance sampling, another method exists with higher estimation performance, $\overline{X}_2$, which can be written as

$$\bar{h} \approx \frac{\sum_{n=1}^{N_S}\frac{f(x_n)}{f_a(x_n)}h(x_n)}{\sum_{n=1}^{N_S}\frac{f(x_n)}{f_a(x_n)}} \tag{12.28}$$

where x_n is the sampling value of probability-distribution function $f_a(x)$. To make the estimation more effective, $f_a x$ is taken as the uniform-distribution function with the same definition domain as $f(x)$; then, the weighted average value of $h(x)$ can be estimated as [17]

$$\overline{h} \approx \frac{\sum_{n=1}^{N_S} f(x_n)h(x_n)}{\sum_{n=1}^{N_S} f(x_n)} \tag{12.29}$$

The estimation error of Monte Carlo statistics is [17]

$$\sigma_{\overline{h}}^2 = \frac{1}{N_S(N_S-1)} \sum_{n=1}^{N_s} \left|h(x_n) - \overline{h}\right|^2 \tag{12.30}$$

12.2.4 MCMC Detection Algorithm

12.2.4.1 Introduction to the MCMC Detection Algorithm

The key of a Monte Carlo algorithm is to generate random numbers that satisfy the target distribution. When the random number is one-dimensional, it is not difficult to achieve; however, when the random number is in a multi-dimensional space, it is difficult to directly generate one that meets the target distribution. The most commonly used and effective method to solve this problem is the Markov-chain Monte Carlo (MCMC) algorithm.

The MCMC method came into being in the early 1950s when a Markov process was introduced into Monte Carlo simulation. This is a Monte Carlo method simulated by computer under a Bayesian-theory framework, which makes up for the limitation that the traditional Monte Carlo integration can only be simulated statically. It has been a widely used statistical-calculation method in recent years [18].

The MCMC algorithm obtains the required sample values in a Markov chain with $\widehat{X_2} = h_{n,2}Y_n^1 - h_{n,1}Y_n^2 + h_{n,1}^2 P_T$ as the stationary distribution, and then estimates the weighted-average value of random variables through these sample values. The MCMC algorithm is essentially an extension of the Monte Carlo algorithm.

The basic idea is to use the properties of a Markov chain to extract sample values from the target probability-density distribution. However, the number of states in the Markov chain increases exponentially with the number of sampling points, and the complexity is too high. Two sampling algorithms can address the complexity of the MCMC algorithm: the Metropolis and Gibbs sampling algorithms [19].

A Markov chain is a random variable sequence $\{X^{(0)}, X^{(1)}, \ldots\}$. At any time $t(t \geq 0)$, the sequence $X^{(t+1)}$ at the next time $t+1$ is generated by the conditional distribution $F(x|X^{(t)})$; it is only related to the state $\overline{X}_2 = -X_2 + P_T$ at time t and is unrelated to states $\widetilde{X}_1 = (h_{n,1}^2 + h_{n,2}^2)X_1 + h_{n,1}w_n^1 + h_{n,2}w_n^2$ before time t. Assuming that the Markov chain satisfies the irreducible ergodic condition, regardless of the value the initial state $X^{(0)}$ takes, the state $X^{(t)}$ converges to the same distribution, which is the so-called stationary distribution.

Assuming that the state of a Markov chain is $X_{t\geq 0}^{(t)}$ and the state space is D, the one-step transition probability function is

$$p(x, x') = p(x \to x') = p(X^{(t+1)} = x'|X^{(t)} = x) \text{ (discrete)} \tag{12.31}$$

$$p(x \to B) = \int_B p(x, x')dx \text{ (continuous)}. \tag{12.32}$$

Transition probability $p(x, x')$ meets the following condition:

$$\sum p(x, x') = 1 \text{ (discrete)} \tag{12.33}$$

$$\int p(x, x') = 1 \text{ (continuous)}. \tag{12.34}$$

Suppose $\pi(x)$ meets

$$\sum p(x, x')\pi(x) = \pi(x') \text{ (discrete)} \tag{12.35}$$

$$\int p(x, x')\pi(x)dx = \pi(x'), \forall x \in D \text{ (continuous)}. \tag{12.36}$$

Then, $\pi(x)$ can be considered as the stationary distribution of transition probability $p(\ldots)$.

The Gibbs sampling algorithm is commonly used in the MCMC algorithm and can complete iterative sampling based on a conditional distribution. The characteristic of Gibbs sampling is that a Markov chain can be established by decomposing a series of conditional distributions, especially a full conditional distribution.

A full conditional distribution refers to a conditional distribution in the form of $\pi(x_T|x_{-T})$, where t, and $x_{-T} = \{x_i, i \notin T\}, T \subset N = \{1, 2, \ldots, n\}$. Therefore, in a full conditional distribution, all variables will appear, either in conditions or in variables. In a conditional distribution satisfying $X_i|X_{-i}, i = 1, 2, \ldots n$, the transition probability $q = (x_i \to x_i')$ can maintain $X_i' = X_{-i} = x_{-i}$, at a fixed value; therefore, sampling point x_i' can be generated with the transition probability $q = (x_i \to x_i')$, and then

$$a_i(x_i \to x_i'|x_{-i}) = \min\left(1, \frac{\pi(x')q_i(x_i' \to x_i|x_{-i})}{\pi(x)q_i(x_i \to x_i'|x_{-i})}\right). \tag{12.37}$$

Subsequently, the algorithm decides whether to accept x' as the next state of the Markov chain.

If the transition probability $q = (x_i \to x_i')$ in the above full conditional distribution is taken as the full conditional distribution $\pi(x_i|x_{-i})$, $a(x_i \to x_i') = 1$ can be obtained, which is called Gibbs sampling. The Gibbs sampling algorithm can decompose a conditional distribution that only involves single-variable sampling. Assuming an M-dimensional joint conditional distribution $\pi(x_1, x_2, \ldots, x_M)$, the transition probability in Gibbs sampling can be constructed as

$$p_{x,y} = p(x, y) = \prod_{k=1}^{M} \pi(y_k|y_1, \ldots, y_{k-1}, x_{k+1}, \ldots, x_M), \tag{12.38}$$

where $x = (x_1, \ldots, x_M)$, $y = (y_1, \ldots, y_M)$, $x_i \in D$, $y_i \in D$, and $\pi(y_k|y_1, \ldots, y_{k-1}, x_{k+1}, \ldots, x_M)$ refers to the conditional distribution when the kth component is y_k where components $1-k-1$ are $y_1, \ldots, y_{k-1}$ and components k-M are $x_{k+1}, \ldots, x_M$. Because

$$\sum_{y_1} \pi(y_1|x_2, \ldots, x_M) = \sum_{y_1} \frac{\pi(y_1, x_2, \ldots, x_M)}{\pi(x_2, \ldots, x_M)} = \frac{\pi(x_2, \ldots, x_M)}{\pi(x_2, \ldots, x_M)} = 1, \tag{12.39}$$

we can obtain

$$\sum_{y} p_{x,y} = \sum_{y_M} \cdots \sum_{y_1} \pi(y_1|x_2, \ldots, x_M)\pi(y_2|y_1, x_3, \ldots, x_M) \ldots \pi(y_M|y_1, \ldots, y_{M-1}). \tag{12.40}$$

From Eqs. (12.39) and (12.40), it can be seen that $p(x, y)$ meets the requirements of a Markov-chain transition probability. It can also be verified that $\pi(x_1, x_2, \ldots, x_M)$ is a Markov-chain stationary distribution with $p(x, y)$ as the transition probability [20]:

$$\begin{aligned}
&\sum_{(x_1,\ldots,x_M)} \pi(x_1, \ldots, x_M)p((x_1, \ldots, x_M), (y_1, \ldots, y_M)) \\
&= \sum_{x_M}\sum_{x_{M-1}} \cdots \sum_{x_1} \pi(x_1, \ldots, x_M)\pi(y_1|x_2, \ldots, x_M)\pi \\
&\quad (y_2|y_1, x_3, \ldots, x_M) \ldots \pi(y_M|y_1, \ldots, y_{M-1}) \\
&= \sum_{x_M}\sum_{x_{M-1}} \cdots \sum_{x_1} \pi(x_1, \ldots, x_M) \frac{\pi(y_1, x_2, \ldots, x_M)}{\sum_{x_1} \pi(x_1, \ldots, x_M)} \pi \\
&\quad (y_2|y_1, x_3, \ldots, x_M) \ldots \pi(y_M|y_1, \ldots, y_{M-1}) \\
&= \sum_{x_M}\sum_{x_{M-1}} \cdots \sum_{x_2} \pi(y_1, x_2, \ldots, x_M)\pi \\
&\quad (y_2|y_1, x_3, \ldots, x_M) \ldots \pi(y_M|y_1, \ldots, y_{M-1}) \\
&= \sum_{x_M} \pi(y_1, \ldots, y_{M-1}, x_M) \frac{\pi(y_1, \ldots, y_M)}{\sum_{x_M} \pi(y_1, \ldots, y_{M-1}, x_M)} \\
&= \pi(y_1, \ldots, y_M).
\end{aligned} \tag{12.41}$$

In summary, the process of obtaining sample $X^{(t+1)}(\omega)$ from sample $X^{(t)}(\omega)$ using Gibbs sampling is as follows:

(1) First, a random variable $X_1^{(t+1)}(\omega)$, satisfying the distribution $\left\{\pi\left(x_1^{(t+1)}|x_2^{(t)},\ldots,x_M^{(t)}\right),x_1^{(t+1)}\in D_1\right\}$, is obtained.
(2) Then, random variable $X_2^{(t+2)}(\omega)$, satisfying the distribution $\left\{\pi\left(x_2^{(t+1)}|x_1^{(t+1)},x_3^{(t)},\ldots,x_M^{(t)}\right),x_2^{(t+1)}\in D_1\right\}$, is obtained.
(3) By analogy, random variable $X_M^{(t+1)}(\omega)$, satisfying the distribution $\left\{\pi\left(x_M^{(t+1)}|x_1^{(t+1)},\ldots,x_{M-1}^{(t+1)},x_M^{(t)}\right),x_M^{(t+1)}\in D_1\right\}$, can be obtained.

$y=(y_1,\ldots,y_M)$ is defined as a sample of $X^{(t+1)}(\omega)$; $X^{(0)}(\omega)=Y^{(0)}$ is initialized arbitrarily, and a sample $y^{(1)}$ of $X^{(1)}(\omega)$ can be obtained using the Gibbs sampling method above. After n iterations of Gibbs sampling, the sample value $y^{(n)}$ of Markov chain $X^{(n)}(\omega)$ can be obtained. When n is sufficiently large, the distribution of Markov chain $X^{(n)}(\omega)$ can be considered to be similar to that of $\pi(x_1,x_2,\ldots,x_M)$, and $y^{(n)}$ can be considered as a sample that approximately satisfies the distribution of $\pi(x_1,x_2,\ldots,x_M)$.

According to the Monte Carlo statistical method and Eqs. (12.22) and (12.23), to calculate the bit posterior probability of a detector output in the MIMO iterative-detection algorithm, it is necessary to generate random variables with a distribution of $p(b|y,\lambda_2^e)$ through the MCMC algorithm. These are generated through Markov-chain characteristic iterations. The weighted average of the bit posterior probabilities is then estimated from these sampled values.

The MCMC detection algorithm is divided into two steps: The first step is to use Gibbs sampling to obtain a sample value that meets the target distribution. The second step is to calculate the posterior probability of the received signal according to these sample values. Two MCMC detection algorithms with different estimation are introduced below. One is the Markov chain Rao–Blackwellization (MCRB)-MCMC detection algorithm, based on Monte Carlo integration; the other is the MCRB-MCMC with uniform sampling (MCRB-U-MCMC) detection algorithm, based on Monte Carlo importance sampling.

12.2.4.2 MCRB-MCMC Detection Algorithm

Reference [21] gives the general steps of Gibbs sampling:

Let $b_{-k}^{(n)}=\left[b_1^{(n)},b_2^{(n)},\ldots,b_{k-1}^{(n)},b_{k+1}^{(n)},\ldots,b_{KM}^{(n)}\right]^T$.

(1) Randomly initialize vector $b^{(-N_b)}$;
(2) Set the range of Gibbs sampling times n as $-N_b-N_s$;
(3) Obtain the value of $b_1^{(n)}$, according to $p\left(b_1|b_2^{(n-1)},b_3^{(n-1)},\ldots,b_{KM}^{(n-1)},y,\lambda_2^e\right)$;
(4) Substitute $b_1^{(n)}$ into the original vector to obtain the value of $b_2^{(n)}$, according to $p\left(b_2|b_1^{(n)},b_3^{(n-1)},\ldots,b_{KM}^{(n-1)},y,\lambda_2^e\right)$;

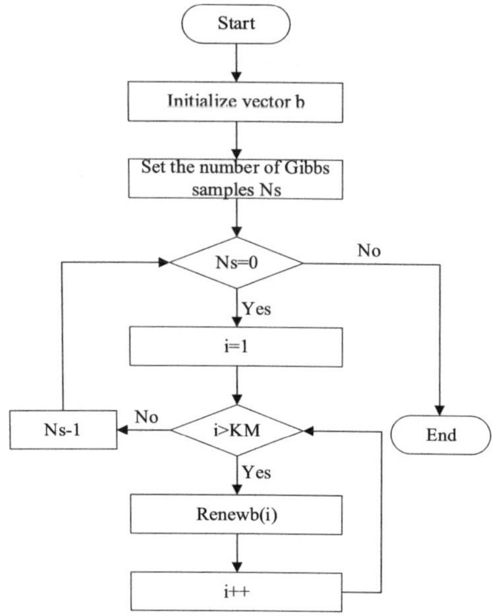

Fig. 12.12 Flowchart of Gibbs sampling algorithm

(5) By analogy, $b_{KM}^{(n)}$ can be obtained according to $p\left(b_{KM}|b_1^{(n)}, b_2^{(n)}, \ldots, b_{KM-1}^{(n)}, y, \lambda_2^e\right)$, so that one sample can be completed, and the data in vector b are updated once;

(6) The cycle ends when the number of iterations meets the conditions. The flowchart of the Gibbs sampling algorithm is shown in Fig. 12.12.

In the Gibbs sampling algorithm [5], the initialization vector $b^{(-N_b)}$ is a random variable related to the external information output by the decoder. In the first iteration, the external information provided by the decoder is zero. The method of selecting $b_k^{(n)}$, according to $P\left(b_k = 1|y, b_{-k}^{(n)}, \lambda_2^e\right)$, is to randomly select a number evenly distributed in [0, 1]. If $P\left(b_k = 1|y, b_{-k}^{(n)}, \lambda_2^e\right)$ is greater than this random number, $b_k^{(n)}$ is 1; otherwise, $b_k^{(n)}$ is 0 [19]. After $N_b + N_s$ iterations, $N_b + N_s$ groups of sampling sequences are generated.

The first N_b cycles are for debugging. To make the sampled data better meet the target distribution, the following $0 - N_s$ groups of sequences are used to calculate the external information output by the detector, that is, the log-likelihood ratio, to achieve a reliable estimation and reduce the amount of calculation.

According to Monte Carlo integration, if the prior information $p(b_{-k}|y, \lambda_2^e)$ is $f(x)$ and the transition probability $p(b_k = 1|y, b_{-k}, \lambda_2^e)$ is $h(x)$, according to Eq. (12.27), we can obtain [16]

$$p(b_k = 1|y, \lambda_2^e) \approx \frac{1}{N_S} \sum_{n=1}^{N_S} p(b_k = 1|y, b_{-k}^{(n)}, \lambda_2^e) \tag{12.42}$$

$$p(b_k = 0|y, \lambda_2^e) \approx \frac{1}{N_S} \sum_{n=1}^{N_s} p(b_k = 0|y, b_{-k}^{(n)}, \lambda_2^e), \tag{12.43}$$

where $b_{-k}^{(n)}$ is the sampling value with a probability density distribution of $p\left(b_{-k}|y, \lambda_2^e\right)$. This estimation method is called the Markov-chain Rao–Blackwellization (MCRB) estimation method.

The transition probability in Gibbs sampling and the information $\lambda_1^e(b_k)$ outside the detector must calculate the values of $P\left(b_k = 1|y, b_{-k}^{(n)}, \lambda_2^e\right)$ and $P\left(b_k = 0|y, b_{-k}^{(n)}, \lambda_2^e\right)$. In the calculation process [16],

we first define

$$\lambda_1^{(n)}(b_k) = \ln \frac{p(b_k = 1|y, b_{-k}^{(n)}, \lambda_2^e)}{p(b_k = 0|y, b_{-k}^{(n)}, \lambda_2^e)} \tag{12.44}$$

because

$$\begin{aligned}
p(b_k = 1|y, b_{-k}^{(n)}, \lambda_2^e) &= \frac{p(b_k = 1, y, b_{-k}^{(n)}, \lambda_2^e)}{p(y, b_{-k}^{(n)}, \lambda_2^e)} \\
&= \frac{p(y|b_{-k}^{(n)}, b_k = 1, \lambda_2^e)p(b_{-k}^{(n)}, b_k = 1, \lambda_2^e)}{p(y|b_{-k}^{(n)}, \lambda_2^e)p(b_{-k}^{(n)})} \\
&= \frac{p(y|b_{-k}^{(n)}, b_k = 1, \lambda_2^e)p(\lambda_2^e)p(b_{-k}^{(n)}, b_k = 1|\lambda_2^e)}{p(y|b_{-k}^{(n)}, \lambda_2^e)p(b_{-k}^{(n)})} \\
&= \frac{p(y|b_{-k}^{(n)}, b_k = 1, \lambda_2^e)p(\lambda_2^e)p(b_{-k}^{(n)}|\lambda_{2,-k}^{(n)})p(b_k = 1|\lambda_2^e(b_k))}{p(y|b_{-k}^{(n)}, \lambda_2^e)p(\lambda_2^e)p(\lambda_2^e)p(b_{-k}^{(n)}|\lambda_2^e)} \\
&= \frac{p(y|b_{-k}^{(n)}, b_k = 1)p(b_k = 1|\lambda_2^e)}{p(y|b_{-k}^{(n)}, \lambda_2^e)} \\
&= \frac{p(y|b_{-k}^{(n)}, b_k = 1)p^e(b_k = 1)}{p(y|b_{-k}^{(n)}, \lambda_2^e)}.
\end{aligned} \tag{12.45}$$

Similarly, the following can be obtained:

$$p(b_k = 0|y, b_{-k}^{(n)}, \lambda_2^e) = \frac{p(b_k = 0, y, b_{-k}^{(n)}, \lambda_2^e)}{p(y, b_{-k}^{(n)}, \lambda_2^e)}$$

$$= \frac{p(y|b_{-k}^{(n)}, b_k = 0)p^e(b_k = 0)}{p(y|b_{-k}^{(n)}, \lambda_2^e)}. \tag{12.46}$$

Substituting Eqs. (12.45) and (12.46) into (12.44) yields

$$\begin{aligned}
\lambda_1^{(n)}(b_k) &= \ln \frac{p(b_k = 1|y, b_{-k}^{(n)}, \lambda_2^e)}{p(b_k = 0|y, b_{-k}^{(n)}, \lambda_2^e)} \\
&= \ln \frac{p(y|b_{-k}^{(n)}, b_k = 1)p(b_{-k}^{(n)}, b_k = 1|\lambda_2^e)/p(y|b_{-k}^{(n)}, \lambda_2^e)p(b_{-k}^{(n)}, \lambda_2^e)}{p(y|b_{-k}^{(n)}, b_k = 0)p(b_{-k}^{(n)}, b_k = 0|\lambda_2^e)/p(y|b_{-k}^{(n)}, \lambda_2^e)p(b_{-k}^{(n)}, \lambda_2^e)} \\
&= \ln \frac{p(y|b_{-k}^{(n)}, b_k = 1)p(b_{-k}^{(n)}, b_k = 1|\lambda_2^e)}{p(y|b_{-k}^{(n)}, b_k = 0)p(b_{-k}^{(n)}, b_k = 0|\lambda_2^e)} \\
&= \ln \frac{p(y|b_{-k}^{(n)}, b_k = 1)p(b_{-k}^{(n)}|\lambda_{2,-k}^e)p(b_k = 1|\lambda_2^e(b_k))}{p(y|b_{-k}^{(n)}, b_k = 0)p(b_{-k}^{(n)}|\lambda_{2,-k}^e)p(b_k = 0|\lambda_2^e(b_k))} \\
&= \ln \frac{p(y|b_{-k}^{(n)}, b_k = 1)}{p(y|b_{-k}^{(n)}, b_k = 0)} + \lambda_2^e(b_k)
\end{aligned} \tag{12.47}$$

because

$$p(y|b) = \frac{1}{(2\pi\sigma_2^n)^{N/2}} \exp\left(-\frac{(y - Hd)^2}{2\sigma_n^2}\right), \tag{12.48}$$

where y is the signal received by the receiving antenna, and vector d is the modulated signal of vector b. In the $y = Hd + n$ model,

$$\begin{cases} y = H_{-j}d_{-j} + a_j + n \; b_k = 1 \\ y = H_{-j}d_{-j} - a_j + n \; b_k = 0 \end{cases}. \tag{12.49}$$

Then, we can obtain

$$\begin{aligned}
&\ln \frac{p(y|b_{-k}^{(n)}, b_k = 1)}{p(y|b_{-k}^{(n)}, b_k = 0)} \\
&= \ln \frac{\frac{1}{(2\pi\sigma_n^2)^{N/2}} \exp\left(-\frac{(y - H_{-j}d_{-j}^{(n)} - a_j)^2}{2\sigma_n^2}\right)}{\frac{1}{(2\pi\sigma_n^2)^{N/2}} \exp\left(-\frac{(y - H_{-j}d_{-j}^{(n)} + a_j)^2}{2\sigma_n^2}\right)} \\
&= \frac{1}{2\sigma_n^2}\left[|y - H_{-j}d_{-j}^{(n)} + a_j|^2 - |y - H_{-j}d_{-j}^{(n)} - a_j|^2\right] \\
&= \frac{1}{2\sigma_n^2} R\left[4(y - H_{-j}d_{-j}^{(n)})a_j\right]
\end{aligned}$$

$$
\begin{aligned}
&= \frac{2}{\sigma_n^2} R\left[a_j^H y - a_j^H H_{-j} d_{-j}^{(n)}\right] \\
&= \frac{2}{\sigma_n^2} R\left[a_j^H y - \sum_{\substack{l \neq j \\ l=1}}^{j} a_j^H a_l d_l^{(n)}\right].
\end{aligned} \tag{12.50}
$$

When substituting Eq. (12.49) into Eq. (12.47), we can obtain

$$
\lambda_1^{(n)}(b_k) = \frac{2}{\sigma_n^2} R\left[a_j^H y - \sum_{\substack{l \neq j \\ l=1}}^{j} a_j^H a_l d_l^{(n)}\right] + \lambda_2^e(b_k), \tag{12.51}
$$

where j is $\left|\frac{k}{M_C}\right|$, H_{-j} is the matrix after the jth column of channel matrix H is deleted, and $d_n = \frac{\sigma_{off,n}}{\sigma_{on,n}}$ is the jth column element of channel matrix H. From Eq. (12.47), we can obtain

$$
p(b_k = 1 | y, b_{-k}^{(n)}, \lambda_2^e) = \frac{1}{1 + \exp(-\lambda_1^{(n)}(b_k))}. \tag{12.52}
$$

We calculate the value of $p\left(b_k = 1 | y, b_{-k}^{(n)}, \lambda_2^e\right)$ to complete Gibbs sampling. It can also be obtained from (12.47):

$$
p(b_k = 0 | y, b_{-k}^{(n)}, \lambda_2^e) = \frac{\exp(-\lambda_1^{(n)}(b_k))}{1 + \exp(-\lambda_1^{(n)}(b_k))}. \tag{12.53}
$$

Substituting Eqs. (12.51) and (12.52) into Eqs. (12.42) and (12.43) yields

$$
p(b_k = 1 | y, \lambda_2^e) \approx \frac{1}{N_S} \sum_{n=1}^{N_s} \frac{1}{1 + \exp(-\lambda_1^{(n)}(b_k))} \tag{12.54}
$$

$$
p(b_k = 0 | y, \lambda_2^e) \approx \frac{1}{N_S} \sum_{n=1}^{N_s} \frac{\exp(-\lambda_1^{(n)}(b_k))}{1 + \exp(-\lambda_1^{(n)}(b_k))}. \tag{12.55}
$$

Substituting Eqs. (12.53) and (12.54) into Eq. (12.18) yields the value of $\lambda_1(b_k)$. In addition,

$$
\lambda_1^e(b_k) = \lambda_1(b_k) - \lambda_2^e(b_k)
$$

$$
\begin{aligned}
&= \ln \frac{p(b_k = 1|y, \lambda_2^e)}{p(b_k = 0|y, \lambda_2^e)} - \lambda_2^e(b_k) \\
&= \ln \frac{\frac{1}{N_S}\sum_{n=1}^{N_S} \frac{1}{1+\exp(-\lambda_1^{(n)}(b_k))}}{\frac{1}{N_S}\sum_{n=1}^{N_S} \frac{\exp(-\lambda_1^{(n)}(b_k))}{1+\exp\left(-\lambda_1^{(n)}(b_k)\right)}} - \lambda_2^e(b_k) \\
&= \ln \sum_{n=1}^{N_S} \exp(\lambda_1^{(n)}(b_k)) - \lambda_2^e(b_k) \\
&= \ln[\exp(\lambda_1^{(1)}(b_k)) + \exp(\lambda_1^{(2)}(b_k)) \\
&\quad + \cdots + \exp(\lambda_1^{(N_S)}(b_k))] - \lambda_2^e(b_k).
\end{aligned}
\tag{12.56}
$$

According to max-log-MAP estimation,

$$
\max_n[\lambda_1^{(n)}(b_k)] \approx \ln\left(\sum_{n=1}^{N_S} \exp\left(\lambda_1^{(n)}(b_k)\right)\right).
\tag{12.57}
$$

Substituting Eq. (12.56) into Eq. (12.55) yields

$$
\lambda_1^e(b_k) = \lambda_1(b_k) - \lambda_2^e(b_k) \approx \max_n\left[\lambda_1^{(n)}(b_k)\right] - \lambda_2^e(b_k).
\tag{12.58}
$$

12.2.4.3 MCRB-U-MCMC Detection Algorithm

According to the importance-sampling principle, let a priori information $p(b_{-k}|y, \lambda_2^e)$ be $f(x)$, and let the transition probability $p(b_k = 1|y, b_{-k}, \lambda_2^e)$ be $h(x)$. From Eq. (12.29), we can obtain [23]

$$
p(b_k = 1|y, \lambda_2^e) \approx \frac{\sum_{n=1}^{N_s} p(b_k = 1|y, b_{-k}^{(n)}, \lambda_2^e) p(b_{-k}^{(n)}|y, \lambda_2^e)}{\sum_{n=1}^{N_S} p(b_{-k}^{(n)}|y, \lambda_2^e)}
\tag{12.59}
$$

$$
p(b_k = 0|y, \lambda_2^e) \approx \frac{\sum_{n=1}^{N_s} p(b_k = 0|y, b_{-k}^{(n)}, \lambda_2^e) p(b_{-k}^{(n)}|y, \lambda_2^e)}{\sum_{n=1}^{N_S} p(b_{-k}^{(n)}|y, \lambda_2^e)},
\tag{12.60}
$$

where $b_{-k}^{(n)}$ is the sampling value of a uniformly distributed $f_a(x)$ This estimation method is called the Markov-chain Rao-Blackwellization with uniform sampling (MCRB-U) estimation method, also known as the MCRB-U-MCMC detection algorithm.

The Gibbs sampling process [24] in the MCRB-U-MCMC detection algorithm is similar to that in the MCRB-MCMC detection algorithm; only the first two steps need to be changed:

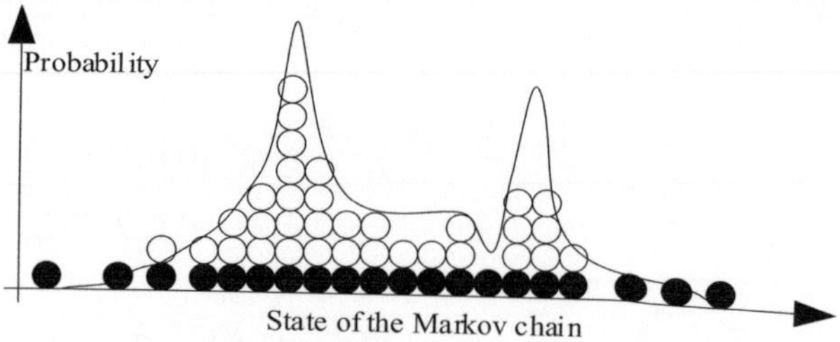

Fig. 12.13 Gibbs sampling diagram

(1) Randomly initialize $b^{(0)}$;
(2) Set the range of sampling times n as $0 \sim N_s$.

The MCRB-U-MCMC estimation algorithm is based on importance sampling. To obtain Eqs. (12.58) and (12.59), the sampling value $b_{-k}^{(n)}$ must be uniformly distributed within the value range of function $p(b_{-k}|y, \lambda_2^e)$. Reference [25] proposes a method to solve this problem: In Gibbs sampling, the number of cycles is set to 0–N_S instead of $-N_b$–N_s; that is, the debugging phase in Gibbs sampling mentioned above is removed, and all sample values generated by Gibbs sampling are used to calculate the posterior probability of the output of the soft-input soft-output detector. The comparison of the two Gibbs samples is shown in Fig. 12.13.

In the MCRB-MCMC detection algorithm, the sampling value is more consistent with the target distribution in the debugging stage, to reduce the estimation error and the computational complexity of the bit-level soft information. In the MCRB-U-MCMC detection algorithm, the sampling values need to meet the uniform distribution (as shown in Fig. 12.13), so that all sampling points can participate in the calculation of the detector's output soft information.

In the MCRB-U-MCMC detection algorithm, it is necessary to calculate the transition probability $P\left(b_k = 1|y, b_{-k}^{(n)}, \lambda_2^e\right)$ in Gibbs sampling and the value $\lambda_1^e(b_k)$ of the bit-level soft information. The transition probability $P\left(b_k = 1|y, b_{-k}^{(n)}, \lambda_2^e\right)$ can be calculated as follows:

$$
\begin{aligned}
& P(b_k = 1|y, b_{-k}^{(n)}, \lambda_2^e) \\
& \quad = \frac{p(b_k = 1|y, b_{-k}^{(n)}, \lambda_2^e)}{p(b_k = 1|y, b_{-k}^{(n)}, \lambda_2^e) + p(b_k = 0|y, b_{-k}^{(n)}, \lambda_2^e)}
\end{aligned}
\tag{12.61}
$$

$$
P(b_k = 1|y, b_{-k}^{(n)}, \lambda_2^e) = \frac{p(y|b_{k+}^{(n)}, \lambda_2^e)p(b_{k+}^{(n)}|\lambda_2^e)}{p(y, b_{-k}^{(n)}|\lambda_2^e)} \tag{12.62}
$$

$$
P(b_k = 0|y, b_{-k}^{(n)}, \lambda_2^e) = \frac{p(y|b_{k-}^{(n)}, \lambda_2^e)p(b_{k+}^{(n)}|\lambda_2^e)}{p(y, b_{-k}^{(n)}|\lambda_2^e)}, \tag{12.63}
$$

where

$$b_{k^+}^{(n)} = \left[b_1^{(n)} \ldots b_{k-1}^{(n)} \; 1 \; b_{k+1}^{(n-1)} \ldots b_{KM}^{(n-1)} \right] \tag{12.64}$$

$$b_{k^-}^{(n)} = \left[b_1^{(n)} \ldots b_{k-1}^{(n)} \; 0 \; b_{k+1}^{(n-1)} \ldots b_{KM}^{(n-1)} \right], \tag{12.65}$$

and because

$$p(y|b_{k^+}^{(n)}, \lambda_2^e) = p(y|b_{k^+}^{(n)}) \tag{12.66}$$

$$p(y|b_{k^+}^{(n)}, \lambda_2^e) = p(y|b_{k^+}^{(n)}). \tag{12.67}$$

Substituting Eqs. (12.61)–(12.64) into Eq. (12.60) yields

$$\begin{aligned} &P(b_k = 1|y, b_{-k}^{(n)}, \lambda_2^e) \\ &= \frac{p(y|b_{k^+}^{(n)})p^e(b_k = 1)}{p(y|b_{k^+}^{(n)})p^e(b_k = 1) + p(y|b_{k^-}^{(n)})p^e(b_k = 0)}, \end{aligned} \tag{12.68}$$

where $p^e(b_k = i) = p\big(b_k = i|\lambda_2^e(b_k)\big)$, and $i = 1$ or 0. In channel model $y = Hd + n$, n is Gaussian white noise with a mean value of 0 and variance of σ_n^2, which is generated by photodetectors and their auxiliary electronic circuits. We can obtain

$$p(y|b_{k^+}^{(n)}) = \frac{1}{(\pi\sigma_n^2)^{N/2}} e^{-||y - Hd_{k^+}^{(n)}||^2/\sigma_n^2} \tag{12.69}$$

$$p(y|b_{k^-}^{(n)}) = \frac{1}{(\pi\sigma_n^2)^{N/2}} e^{-||y - Hd_{k^-}^{(n)}||^2/\sigma_n^2}, \tag{12.70}$$

where $d_{k^+}^{(n)}$ and $d_{k^-}^{(n)}$ represent the modulated signals of $b_{k^+}^{(n)}$ and $b_{k^-}^{(n)}$, respectively.
Definition:

$$\gamma_{k^+} = \ln(p^e(b_k = 1)) - \frac{||y - Hd_{k^+}^{(n)}||^2}{\sigma_n^2} \tag{12.71}$$

$$\gamma_{k^-} = \ln(p^e(b_k = 0)) - \frac{||y - Hd_{k^+}^{(n)}||^2}{\sigma_n^2}. \tag{12.72}$$

Substituting Eqs. (12.66)–(12.69) into (12.65) yields

$$p(b_k = 1|y, b_{-k}^{(n)}, \lambda_2^e) = \frac{1}{1 + e^{-(\gamma_{k^+} - \gamma_{k^-})}}, \tag{12.73}$$

where

$$\gamma_{k^+} - \gamma_{k^-} = \lambda_2^e(b_k) - \frac{\left\| y - Hd_{k^+}^{(n)} \right\|^2 - \left\| y - Hd_{k^-}^{(n)} \right\|^2}{\sigma_n^2}.$$

The value of $p\left(d_k = 1 | y, b_{-k}^{(n)}, \lambda_2^e\right)$ can be calculated using Eq. (12.70), so that Gibbs sampling can be carried out. The process of calculating the SISO detector output bit-level soft information λ_1^e is as follows.

Substituting Eqs. (12.61) and (12.62) into (12.18) yields

$$\lambda_1(b_k) = \frac{\sum_{n=1}^{N_S} p(b_k = 1 | y, b_{-k}^{(n)}, \lambda_2^e) p(b_{-k}^{(n)} | y, \lambda_2^e)}{\sum_{n=1}^{N_S} p(b_k = 0 | y, b_{-k}^{(n)}, \lambda_2^e) p(b_{-k}^{(n)} | y, \lambda_2^e)}, \quad (12.74)$$

because

$$\begin{aligned} p(b_{-k}^{(n)} | y, \lambda_2^e) &= \frac{p(b_{-k}^{(n)}, y, \lambda_2^e)}{p(y | \lambda_2^e)} \\ &= \frac{p(b_{-k}^{(n)}, y | \lambda_2^e) p(\lambda_2^e)}{p(y | \lambda_2^e) p(\lambda_2^e)} \\ &= \frac{p(y | b_{-k}^{(n)}, \lambda_2^e) p(b_{-k}^{(n)}, \lambda_2^e)}{p(y | \lambda_2^e) p(\lambda_2^e)} \\ &= \frac{p(y | b_{-k}^{(n)}, \lambda_2^e) p(b_{-k}^{(n)} | \lambda_2^e) p(\lambda_2^e)}{p(y | \lambda_2^e) p(\lambda_2^e)} \\ &= \frac{p(y | b_{-k}^{(n)}, \lambda_2^e) p(b_{-k}^{(n)} | \lambda_2^e)}{p(y | \lambda_2^e)} \\ &= \frac{p(y | b_{-k}^{(n)}, \lambda_2^e) p^e(d_{-k}^{(n)})}{p(y | \lambda_2^e)}. \end{aligned} \quad (12.75)$$

Because $b_{-k}^{(n)}$ is not related to λ_2^e, $p^e\left(b_{-k}^{(n)}\right) = p\left(b_{-k}^{(n)} | \lambda_2^e\right)$:

$$\begin{aligned} p(b_k = 1 | y, b_{-k}^{(n)}, \lambda_2^e) &= \frac{p(b_k = 1, y, b_{-k}^{(n)}, \lambda_2^e)}{p(y, b_{-k}^{(n)}, \lambda_2^e)} \\ &= \frac{p(y | b_{-k}^{(n)}, b_k = 1, \lambda_2^e) p(b_{-k}^{(n)}, b_k = 1, \lambda_2^e)}{p(y | b_{-k}^{(n)}, \lambda_2^e) p(b_{-k}^{(n)})} \\ &= \frac{p(y | b_{-k}^{(n)}, b_k = 1, \lambda_2^e) p(\lambda_2^e) p(b_{-k}^{(n)}, b_k = 1 | \lambda_2^e)}{p(y | b_{-k}^{(n)}, \lambda_2^e) p(b_{-k}^{(n)})} \\ &= \frac{p(y | b_{-k}^{(n)}, b_k = 1, \lambda_2^e) p(\lambda_2^e) p(b_{-k}^{(n)} | \lambda_{2,-k}^{(n)}) p(b_k = 1 | \lambda_2^e(b_k))}{p(y | b_{-k}^{(n)}, \lambda_2^e) p(\lambda_2^e) p(\lambda_2^e) p(b_{-k}^{(n)} | \lambda_2^e)} \end{aligned}$$

$$= \frac{p(y|b_{-k}^{(n)}, b_k = 1)p(b_k = 1|\lambda_2^e)}{p(y|b_{-k}^{(n)}, \lambda_2^e)}$$
$$= \frac{p(y|b_{-k}^{(n)}, b_k = 1)p^e(b_k = 1)}{p(y|b_{-k}^{(n)}, \lambda_2^e)}. \tag{12.76}$$

Similarly, we can obtain

$$p(b_k = 0|y, b_{-k}^{(n)}, \lambda_2^e) = \frac{p(b_k = 0, y, b_{-k}^{(n)}, \lambda_2^e)}{p(y, b_{-k}^{(n)}, \lambda_2^e)}$$
$$= \frac{p(y|b_{-k}^{(n)}, b_k = 0)p^e(b_k = 0)}{p(y|b_{-k}^{(n)}, \lambda_2^e)}. \tag{12.77}$$

Substituting Eqs. (12.72), (12.73), and (12.74) into Eq. (12.71) yields

$$\lambda_1(b_k) = \ln \frac{\sum_{n=1}^{N_S} P(y|b_{-k}^{(n)}, b_k = 1)p^e(b_{-k}^{(n)})}{\sum_{n=1}^{N_S} p(y|b_{-k}^{(n)}, b_k = 0)p^e(b_{-k}^{(n)})} \frac{p^e(b_k = 1)}{p^e(b_k = 0)}$$
$$= \ln \frac{\sum_{n=1}^{N_S} P(y|b_{-k}^{(n)}, b_k = 1)p^e(b_{-k}^{(n)})}{\sum_{n=1}^{N_S} p(y|b_{-k}^{(n)}, b_k = 0)p^e(b_{-k}^{(n)})} + \lambda_2^e(b_k). \tag{12.78}$$

From $\lambda_1^e(b_k) = \lambda_1(b_k) - \lambda_2^e(b_k)$, we can obtain

$$\lambda_1^e(b_k) = \ln \frac{\sum_{n=1}^{N_S} p(y|b_{-k}^{(n)}, b_k = 1)p^e(b_{-k}^{(n)})}{\sum_{n=1}^{N_S} p(y|b_{-k}^{(n)}, b_k = 0)p^e(b_{-k}^{(n)})}. \tag{12.79}$$

Definition:

$$\eta_{k^+}^{(n)} = \ln p^e(b_{-k}^{(n)}) - \frac{||y - Hd_{k^+}^{(n)}||^2}{\sigma_n^2} \tag{12.80}$$

$$\eta_{k^-}^{(n)} = \ln p^e(b_{-k}^{(n)}) - \frac{||y - Hd_{k^-}^{(n)}||^2}{\sigma_n^2}. \tag{12.81}$$

We can obtain

$$\lambda_1^e(b_k) = \ln \sum_{n=1}^{N_S} e^{\eta_{k^+}^{(n)}} - \ln \sum_{n=1}^{N_S} e^{\eta_{k^-}^{(n)}}. \tag{12.82}$$

According to max-log-MAP estimation,

$$\lambda_1^e(b_k) \approx \max_n \eta_{k^+}^{(n)} - \max_n \eta_{k^-}^{(n)}. \tag{12.83}$$

From Eq. (12.80), to calculate the value of $\lambda_1^e(b_k)$ it is necessary to calculate the value of $\ln p^e(b_{-k}^{(n)})$. From reference [25], we can obtain

$$\begin{aligned}\ln(p^e(b_k = i)) &= \ln \frac{1}{1 + e^{-i\lambda_2^e(b_k)}} \\ &= -\overset{*}{\max}(0, -i\lambda_2^e(b_k)) \\ &\approx -\max(0, -i\lambda_2^e(b_k)) \\ &= \min(0, i\lambda_2^e(b_k))\end{aligned} \tag{12.84}$$

Using $\varphi_k^{(n)}$ to represent the approximate value of $\ln p^e(b_{-k}^{(n)})$, we can obtain

$$\varphi_k^{(n)} = \sum_{l=1}^{k-1} \min(0, b_l^{(n)}\lambda_2^e(b_k)) + \sum_{l=k+1}^{KM} \min(0, b_l^{(n-1)}\lambda_2^e(b_k)). \tag{12.85}$$

From Eq. (12.82), we can obtain

$$\varphi_k^{(n)} = \varphi_{k-1}^{(n)} - \min(0, b_k^{(n-1)}\lambda_2^e(b_k)) + \min(0, b_k^{(n)}\lambda_2^e(b_k)). \tag{12.86}$$

$\varphi_k^{(n)}$ is calculated as follows:
when $b_k^{(n)} \neq b_k^{(n-1)}$, $\varphi_k^{(n)} = \varphi_{k-1}^{(n)} - b_k^{(n-1)}\lambda_k^e$;
when $b_k^{(n)} = b_k^{(n-1)}$, $\varphi_k^{(n)} = \varphi_{k-1}^{(n)}$.
Therefore, when $b_k^{(n)} \neq b_k^{(n-1)}$, $\ln p^e\left(d_{-k}^{(n)}\right)$ is approximately

$$\ln p^e(b_{-k}^{(n)}) = \varphi_{k-1}^{(n)} + \max(0, -b_k^{(n-1)}\lambda_2^e(b_k)). \tag{12.87}$$

Substituting Eq. (12.84) into Eq. (12.80), we can calculate the value of $\lambda_1^e(b_k)$.

12.2.5 Simulation Results

The simulation flowchart of the iterative detection algorithm is shown in Fig. 12.11. A low-density parity-check (LDPC) code is adopted for channel coding; code length $L = 2048$, coding rate $R = 1/2$, and pulse-position modulation (PPM) is adopted for modulation. When the atmospheric turbulence is weak, the channel-attenuation coefficient is simulated as a lognormal distribution. When the atmospheric turbulence is strong, the channel-attenuation coefficient follows a gamma–gamma distribution. The performance of two MCMC detection algorithms under different atmospheric turbulence intensities is discussed for 4×4 and 4×6 antennas. The results are shown in Figs. 12.14, 12.15, 12.16, 12.17, 12.18, 12.19 and 12.20.

Figures 12.14 and 12.15 respectively show the detection performance of the MCRB-MCMC and MCRB-U-MCMC detection algorithms for different numbers of antennas. It can be seen that the performance of the system with 4 × 6 antennas is about 3 dB higher than that of the system with 4 × 4 antennas. The detection performance of the MCRB-U estimation algorithm is about 2 dB higher than that of the MCRB estimation algorithm. Different atmospheric-turbulence intensities have different effects on the performance of the entire MIMO system, and strong turbulence has a greater impact on the signal.

Figures 12.16, 12.17, 12.18 and 12.19 show the relationship between the MCRB and the MCRB-U detection algorithms and the number of iterations between the detector and the decoder. It can be seen from the figures that the number of iterations between the detector and the decoder plays a crucial role in the detection performance of the overall system. The higher the number of iterations, the higher the reliability of the decision signal, and the better the detection performance of the system.

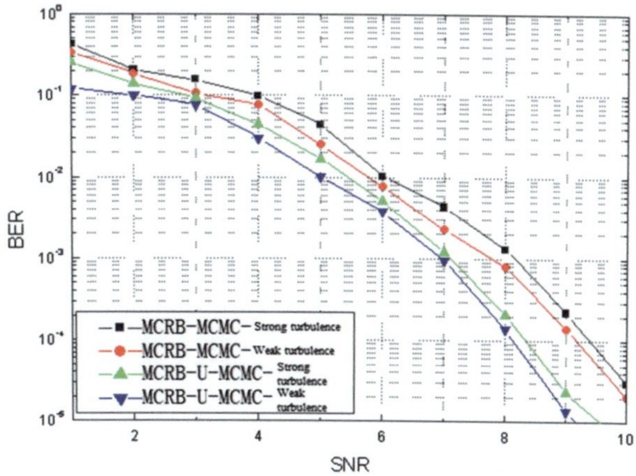

Fig. 12.14 Performance comparison of the MCRB and MCRB-U detection algorithms with 4 × 4 antennas

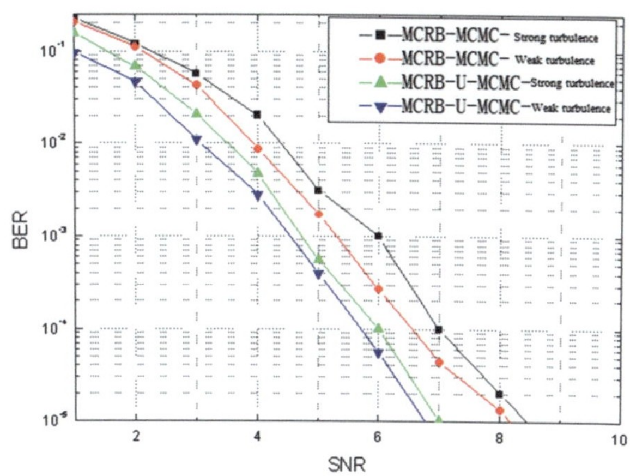

Fig. 12.15 Performance comparison of the MCRB and MCRB-U detection algorithms with 4 × 6 antennas

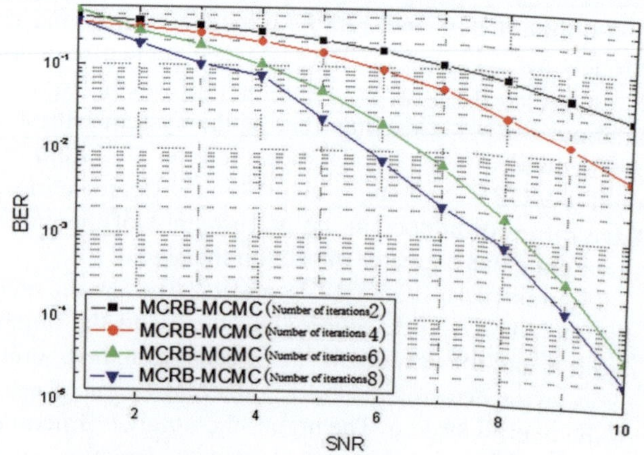

Fig. 12.16 Influence of different numbers of generation iterations on the detection performance of the MCRB algorithm with 4 × 4 antennas

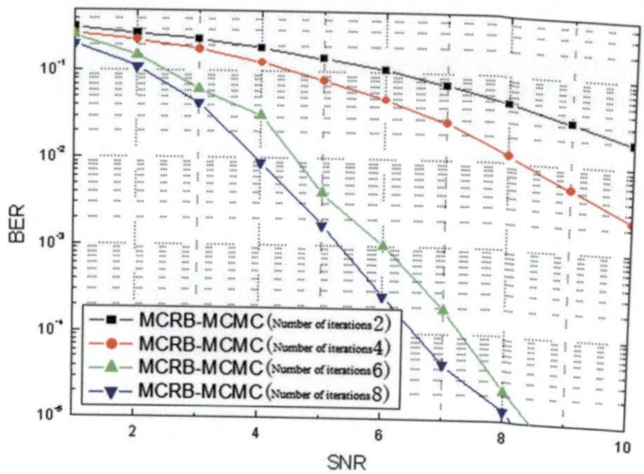

Fig. 12.17 Influence of different numbers of generation iterations on the detection performance of the MCRB algorithm with 4 × 6 antennas

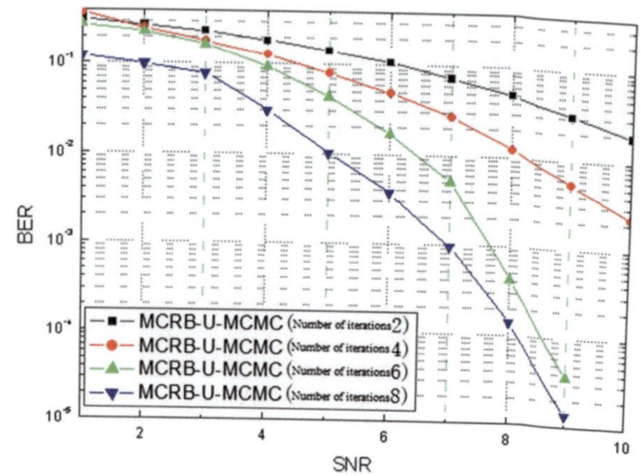

Fig. 12.18 Influence of different numbers of generation iterations on the detection performance of the MCRB-U-MCMC algorithm with 4 × 4 antennas

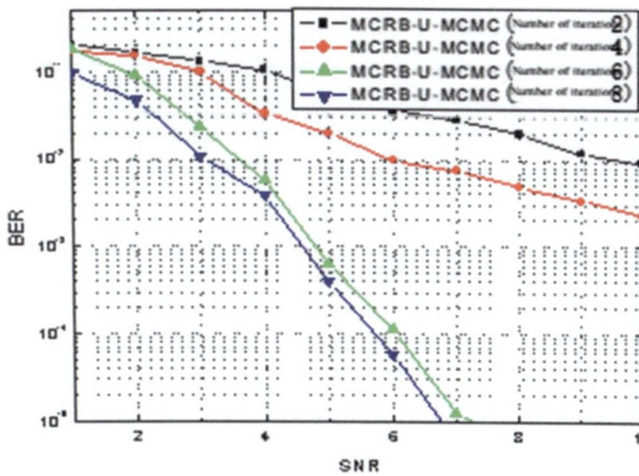

Fig. 12.19 Influence of different numbers of generation iterations on the detection performance of the MCRB-U-MCMC algorithm with 4 × 6 antennas

However, when the number of iterations reaches a certain value, the correlation of the decision signal reaches a maximum, and when the number of iterations increases, the detection performance of the system decreases. It is difficult to make a significant improvement. The number of iterations is an important factor affecting the computational complexity, and an appropriate number of iterations should be selected. The number of iterations is generally approximately 10.

Figure 12.20 shows the effect of Gibbs sampling times on the MCMC detection algorithms. It can be seen from Eq. (12.29) that the more times Gibbs sampling is performed, the more accurate the calculation of the detector-output external information λ_2^e, and the higher the performance of the entire system. From Eqs. (12.41), (12.42), (12.57), and (12.58), it can be seen that the Gibbs sampling parameters will also affect the computational complexity of λ_2^e. Generally, Gibbs sampling should be performed 10–20 times.

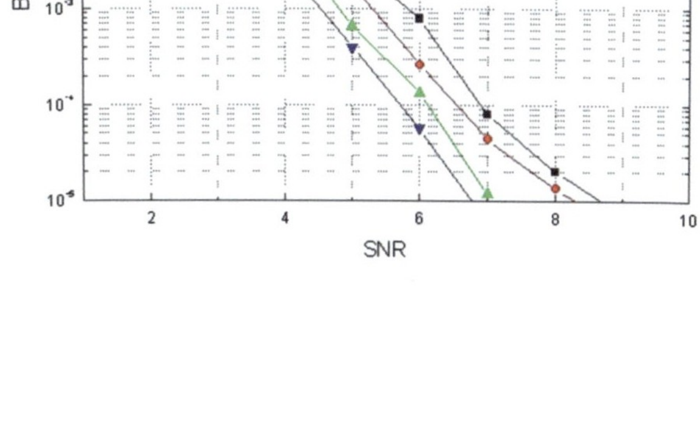

Fig. 12.20 Influence of Gibbs sampling times on the detection performance of the MCRB and MCRB-U algorithms

Comparing the detection performance in Figs. 12.21 and 12.22 reveals that, because the six-antenna system has higher degrees of freedom and a greater diversity gain than the four antenna system, the 4 X_i 6 system has better performance than the 4 $\overline{X}_i = -X_i + P_T$ 4, regardless of which detection algorithm is used.

The detection performance of the ordered successive interference-cancellation algorithm (OSIC) in the nonlinear detection is about 6 dB higher than that of the MMSE-criterion linear detection algorithm. Moreover, the performance of the MCRB-U detection algorithm in MCMC is higher than that of the nonlinear detection algorithm. The detection performance of the SIC algorithm is about 12 dB higher, and the detection performance of the MCMC detection algorithm is the best.

For an $M \times N$ MIMO system, the linear detection algorithm needs to calculate the pseudo-inverse of the channel-characteristic matrix X_2, which is only suitable for systems where the number of transmit antennas is less than the number of receive antennas. The calculation amount of the linear detection algorithm is

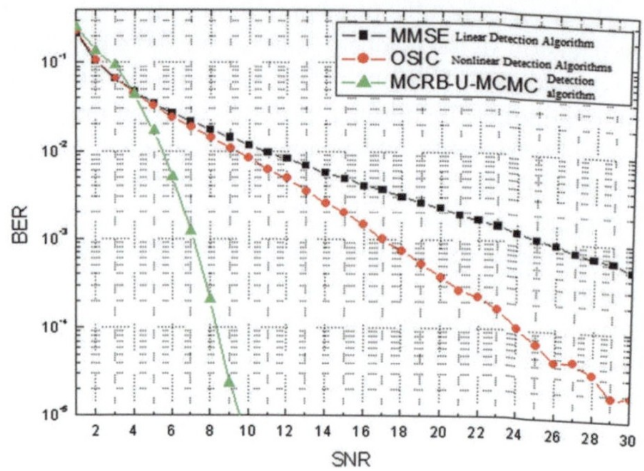

Fig. 12.21 Comparison of the detection performance of different detection algorithms with 4 × 4 antennas

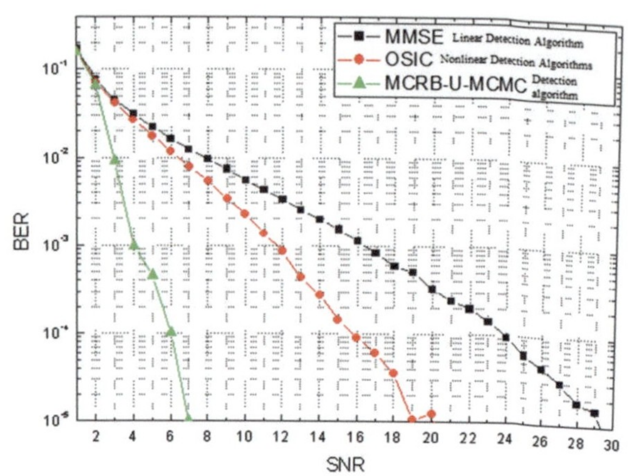

Fig. 12.22 Comparison of the detection performance of different detection algorithms with 4 × 6 antennas

[15] $(4M^3 + 2M^2N)$; in the SIC algorithm, the pseudo-inverse of the channel-characteristic matrix must be calculated once at each stage of operation, and M inversion operations are required. Thus, the complexity of the SIC algorithm is higher than that of the linear detection algorithm, and the required calculation amount is $(4M^3 + 2M^2N)$. The calculation amount of the MCMC iterative detection algorithm is $(\text{iter} \times N_S \times M)$, where *iter* is the number of iterations between the detector and the decoder, and N_S is the number of Gibbs-sampling iterations. It can be seen that the MCMC iterative detection algorithm has low computational complexity and high detection performance.

12.3 Turbo-Code Encoding

The common structure of the turbo-code encoder is shown in Fig. 12.23. It mainly includes two component encoders, an interleaver, a puncturing matrix, and a multiplexer. The two component encoders are cascaded in parallel through the interleaver. This structure can ensure that the output coded bits of the encoder are as uncorrelated as possible. The output check sequence and the information sequence must pass through the multiplexer to complete a serial-to-parallel conversion, after passing through the puncturing matrix [24]. The structures of the two component-code encoders are usually the same; they can be block codes, convolutional codes, or non-systematic convolutional codes; however, a recursive systematic convolution (RSC) is generally chosen.

Assuming that the information sequence input by the turbo-code encoder is $u = \{u_1, u_2, \ldots u_N\}$ and the length is N, the specific encoding process is as follows: The information sequence u is directly sent to the multiplexer as the system information $X^s = \{X_1^s, X_2^s, \ldots X_N^s\}$; the information sequence u is input to component encoder 1, after encoding. The check sequence $X^{1p} = \left\{X_1^{1p}, X_2^{1p}, \ldots X_N^{1p}\right\}$ of length N is obtained. The information sequence $= (h_{n,1}^2 + h_{n,2}^2)X_1 + h_{n,1}h_{n,2}(\overline{X}_2 + X_2 - P_T) -$

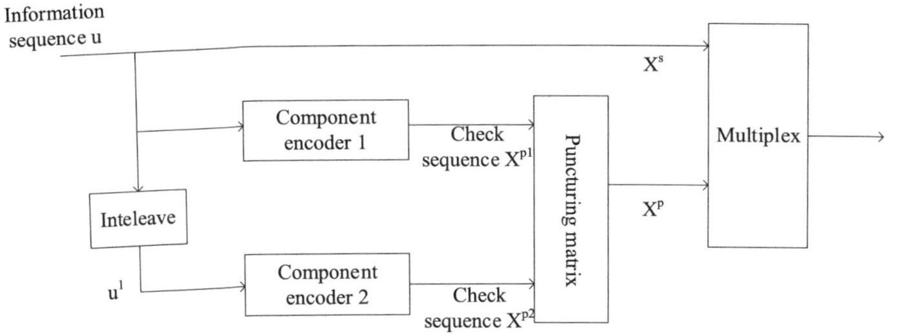

Fig. 12.23 Turbo-code encoder structure diagram

$h_{n,1}h_{n,2}P_T$ is passed through the interleaver to obtain a new sequence $u^1 = \{u_1^1, u_2^1, \ldots u_N^1\}$, and then sent to component encoder 2. After encoding, a check sequence $X^{2p} = \left\{X_1^{2p}, X_2^{2p}, \ldots X_N^{2p}\right\}$ with a length of N is obtained; a sequence X^p, obtained after check sequences X^{1p}, and X^{2p} are subjected to a puncturing matrix, is multiplexed by the information sequence X^s to form a codeword sequence X^S.

12.3.1 Turbo Iterative-Decoding Structure

The turbo-code decoder adopts a serial-feedback iterative decoding structure. The two component decoders are connected by an interleaver or a deinterleaver to exchange their external information. The interleaving rules are consistent with the interleaving rules in the encoder, such as Fig. 12.24.

The steps for decoding turbo-code encoding are as follows [26]:

(1) After the received-channel information is demultiplexed, system information 1, check information 1, and a priori information 1 are obtained.
(2) System information 1, check information 1, and prior information 1 are input to component decoder 1 for decoding (in the first iteration, component decoder 2 does not provide prior information 1 to component decoder 1). The result is a soft-decision log-likelihood ratio (LLR1). The operation yields external information 1. External soft information 1 is interleaved by the interleaver, and is input to the component decoder 2 as prior information 2.
(3) System information 2, check information 2, and prior information 2 are used as input terminals to component decoder 2. After decoding, a soft-decision log-likelihood ratio (LLR2) is obtained. After operation, external information 2 is obtained. After de-interleaving by the interleaver, external information 2 is input to the component decoder 2 as prior information 1.
(4) Steps 2 and 3 are repeated until the iteration ends, and the log-likelihood ratio output by component decoder 2 is determined as the decoded output.

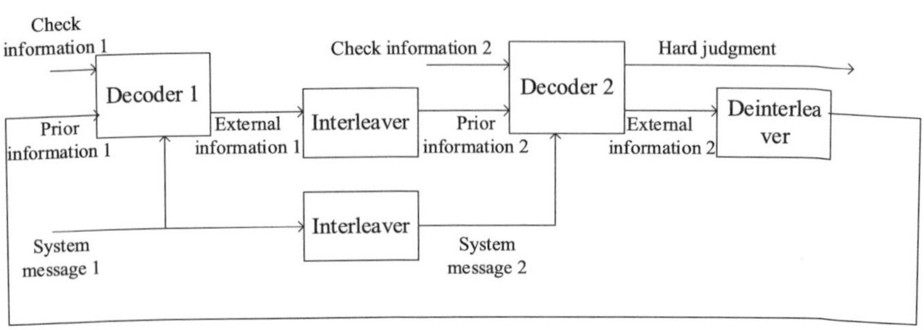

Fig. 12.24 Turbo-code decoder structure diagram

Because the recursive systematic convolutional (RSC) encoder is equivalent to a Markov source, and the turbo-code encoder is formed by concatenating two RSC encoders in parallel through an interleaver, the decoding process is similar to estimating the state of an independent Markov sequence.

Owing to the existence of the interleaver, the two decoding processes can exchange information. The two component decoders are soft-input soft-output (SISO) decoders. The decoding algorithm is generally referred to as a SISO algorithm. Because the maximum a posteriori probability (MAP) algorithm is the basis of the derivation, the following will discuss the MAP algorithm using a turbo code with a code rate of 1/2 as an example.

The encoded output signal is, for binary phase-shift keying (BPSK) modulation, the output signal and the encoded codeword:

$$C_k = (c_k^s, c_k^p), \tag{12.88}$$

which has a satisfactory relationship with

$$X_k = \sqrt{E_s}(2C_k - 1). \tag{12.89}$$

Assume that the received signal is

$$Y_k = (y_k^s, y_k^p), \tag{12.90}$$

where

$$\begin{aligned} y_k^s &= x_k^s + i_k \\ y_k^p &= x_k^p + q_k. \end{aligned} \tag{12.91}$$

i_k and q_k are independent and identically distributed Gaussian random variables with mean 0 and variance $N/2$.

12.3.2 MAP Algorithm

Assume that the input sequence of the MAP decoder is $Y = y_1^N = (y_1, y_2, \cdots, y_k, \cdots, y_N)$, where $y_k = \left(y_{k,}^s y_k^p\right)$. $L^a(u_k)$ is the prior information of u_k, and $L(u_k)$ is the log-likelihood ratio of u_k. They are defined as follows:

$$L^a(u_k) = \ln \frac{P(u_k = 1)}{P(u_k = 0)} \tag{12.92}$$

$$L(u_k) = \ln \frac{p(u_k = 1|y_1^N)}{p(u_k = 0|y_1^N)}. \tag{12.93}$$

The task of the MAP decoder is to solve Eq. (12.90), and then make a decision according to the following rules:

$$\hat{u}_k = \begin{cases} 1 & L(u_k) \geq 0 \\ 0 & L(u_k) < 0 \end{cases} \tag{12.94}$$

The method of calculating Eq. (12.89) is deduced below. Using the Bayesian formula, Eq. (12.91) can be written as

$$L^a(u_k) = \lg \frac{p\big(u_k = 1, y_1^N\big)/p(y_1^N)}{p\big(u_k = 0, y_1^N\big)/p(y_1^N)} = \lg \frac{p\big(u_k = 1, y_1^N\big)}{p\big(u_k = 0, y_1^N\big)}. \tag{12.95}$$

According to the concept of a trellis graph, as shown in Fig. 12.25, the above formula can be transformed into

$$L^a(u_k) = \lg \frac{\sum_{u:u(e)=1} p\big(s_k^S(e), s_k^E(e), y_1^N\big)}{\sum_{u:u(e)=0} p\big(s_k^S(e), s_k^E(e), y_1^N\big)}, \tag{12.96}$$

where $u : u(e) = 1$ and $u : u(e) = 0$ are the set of state transitions $s_k^S \to s_k^E$ corresponding to input information bits $u_k = 1$ and $u_k = 0$.

$p\big(s_k^S(e), s_k^E(e), y_1^N\big)$ is mainly calculated below. The received sequence y_1^N can be divided into three parts: symbol y_k received at current time k, symbol sequence y_1^{k-1} received before time k, and symbol sequence y_{k+1}^N received after time k; therefore,

$$\begin{aligned} & p\big(s_k^S(e), s_k^E(e), y_1^N\big) \\ & = p(s_k^S(e), s_k^E(e), y_1^{k-1}, y_k^N) \\ & = p(s_k^S(e), y_1^{k-1}) p(s_k^E(e), y_k^N | s_k^S(e), y_1^{k-1}) \\ & = p(s_k^S(e), y_1^{k-1}) p(s_k^E(e), y_k^N | s_k^S(e)) \\ & = p(s_k^S(e), y_1^{k-1}) p(s_k^E(e), y_k | s_k^S(e)) p(y_{k+1}^N | s_k^S(e), s_k^E(e), y_k) \end{aligned}$$

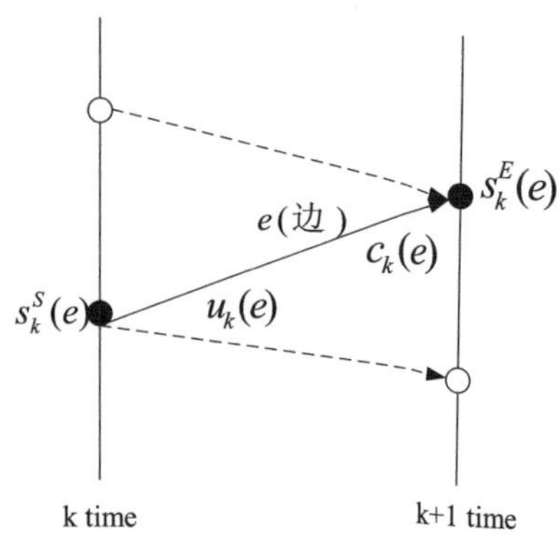

Fig. 12.25 Decoding trellis graph

$$= p(s_k^S(e), y_1^{k-1})p(s_k^E(e), y_k|s_k^S(e))p(y_{k+1}^N|s_k^E(e)) \tag{12.97}$$

The following definitions are made:
Forward-state metrics:

$$\alpha_k(s) = p(s_k^E(e), y_1^k) \tag{12.98}$$

Backward-state metrics:

$$\beta_k(s) = p(y_{k+1}^N|s_{k+1}^S(e)) \tag{12.99}$$

Branch-path metrics:

$$\gamma_k(e) = p(s_k^E(e), y_k|s_k^S(e)) \tag{12.100}$$

However, $p\big(s_k^S(e), s_k^E(e), y_1^N\big) = \alpha_{k-1}\big(s_k^S(e)\big)\gamma_k(e)\beta_k\big(s_k^E(e)\big)$.

The processes of deriving the forward-state metric $\alpha_k(s)$, backward-state metric $\beta_k(s)$, and branch-path metric $\gamma_k(e)$ will be described in detail below.

12.3.2.1 Forward-State Metric $\alpha_k(s)$

According to the knowledge of probability theory and the definition of $\alpha_k(s)$, Eq. (12.94) can be transformed into

$$\begin{aligned}
\alpha_k(s) &= p(s_k^E(e), y_1^k) \\
&= \sum_{e:s_k^E(e)=s} p(s_k^S(e), s_k^E(e), y_1^k) \\
&= \sum_{e:s_k^E(e)=s} p(s_k^S(e), y_1^{k-1})p(s_k^E(e), y_k|s_k^S(e), y_1^{k-1}) \\
&= \sum_{e:s_k^E(e)} \alpha_{k-1}(s_k^S(e))\gamma_k(e).
\end{aligned} \tag{12.101}$$

If the initial state S_0 of the encoding is known, the above recursive initialization can be set as

$$\alpha_k(s) = \begin{cases} 1 & s = S_0 \\ 0 & s \neq S_0 \end{cases} \tag{12.102}$$

It can be seen that the forward-state metric $\alpha_k(s)$ is a forward iterative process.

12.3.2.2 Backward-State Metric $\beta_k(s)$

$$\begin{aligned}
\beta_k(s) &= p(y_{k+1}^N | s_{k+1}^S(e)) \\
&= \sum_{e:s_{k+1}^S(e)=s} p(y_{k+1}^N, s_{k+1}^E(e) | s_{k+1}^S(e)) \\
&= \sum_{e:s_{k+1}^S(e)=s} p(y_{k+1}, y_{k+2}^N, s_{k+1}^E(e) | s_{k+1}^S(e)) \\
&= \sum_{e:s_{k+1}^S(e)=s} p(y_{k+1}, s_{k+1}^E(e) | s_{k+1}^S(e)) p(y_{k+2}^N | s_{k+1}^S(e), y_{k+1}, s_{k+1}^E(e)) \\
&= \sum_{e:s_{k+1}^S(e)=s} p(y_{k+1}, s_{k+1}^E(e) | s_{k+1}^S(e)) p(y_{k+2}^N | s_{k+1}^E(e)) \\
&= \sum_{e:s_{k+1}^S(e)=s} \beta_{k+1}(s_{k+1}^E(e)) \gamma_{k+1}(e)
\end{aligned} \tag{12.103}$$

If the end state $M = 4$ of the encoding is known, the initialization of the above formula can be set as

$$\beta_N(s) = \begin{cases} 1 & s = S_N \\ 0 & s = S_N \end{cases}. \tag{12.104}$$

12.3.2.3 Branch-Path Metric

$$\begin{aligned}
\gamma_k(e) &= p(s_k^E(e), y_k | s_k^S(e)) \\
&= \sum_{x_k} p(s_k^E(e) | s_k^S(e)) p(x_k | e) p(y_k | x_k)
\end{aligned} \tag{12.105}$$

The first term in the formula is the state-transition probability of the branch (edge) in the trellis graph; that is, the prior probability $p(u_k)$ of the encoded input bit u_k that causes the transition from state $s_k^S(e)$ to state $s_k^E(e)$.

Considering $p(u_k = 0) = 1 - p(u_k = 1)$, the above formula can be written as

$$e^{L^a(u_k)} = \frac{p(u_k = 1)}{1 - p(u_k = 1)}. \tag{12.106}$$

Hence, it can be further calculated:

$$p(p(s_k^E(e)|s_k^S(e)) = \begin{cases} p(u_k = 1) = \frac{\exp(L^a(u_k))}{1+\exp(L^a(u_k))} & u_k = 1 \\ p(u_k = 0) = \frac{1}{1+\exp(L^a(u_k))} & u_k = 0 \end{cases}. \tag{12.107}$$

The second item $p(x_k|e)$ is 1 or 0, depending on whether x_k is related to edge e.

The third item $p(y_k|x_k)$ is based on the channel model. Assuming that BPSK modulation is used on the high-tech channel,

$$\begin{aligned} p(y_k|x_k) &= p(y_k^s|u_k)p(y_s^p|s_k^s(e), u_k, s_k^E(e)) \\ &= \frac{1}{\sqrt{\pi N}}\exp\left(-\frac{(y_k^s - X_k^s)^2}{N}\right)\frac{1}{\sqrt{\pi\delta}}\exp\left(-\frac{(y_k^s - X_k^p)^2}{N}\right). \end{aligned} \tag{12.108}$$

The likelihood ratio of the a posterior probability brought into Eq. (12.40) is

$$L(u_k) = \lg\frac{p(u_k = 1|y_1^N)}{p(u_k = 0|y_1^N)} = \lg\frac{\sum_{u:u(e)=1}\alpha_{k-1}(s_k^S(e))\gamma_k(e)\beta_k(s_k^E(e))}{\sum_{u:u(e)=0}\alpha_{k-1}(s_k^S(e))\gamma_k(e)\beta_k(s_k^E(e))}. \tag{12.109}$$

12.3.3 Log-MAP Algorithm

The log-MAP algorithm is a simplified form of MAP. Its essence is to convert the variables in the MAP algorithm into logarithmic form by taking the logarithmic operation; therefore, the input and output form of the decoder should also be modified to the logarithmic-likelihood ratio form. In the log-MAP algorithm, $M_k(s, s)$, $A_k(s)$ and $B_k(s)$ correspond to $\gamma_k(s', s)$, $\alpha_k(s)$, and $\beta_k(s)$ in the MAP algorithm, and they satisfy the logarithmic relationship. We introduce max*() operations, which are defined as

$$\max{}^*(x, y) = \ln(e^x + e^y) = \max(x, y) + \ln(1 + e^{-|x-y|}). \tag{12.110}$$

Therefore,

$$\begin{aligned} A_k(s) &= \ln\alpha_k(s) \\ &= \ln\sum_{e:s_k^E(e)=s}\alpha_{k-1}(s_k^S(e))\gamma_k(e) \\ &= \max_{e:s_k^E(e)=s}{}^*\left[A_{k-1}\left(s_k^S(e)\right) + M_k(e)\right], \quad k = 1, \cdots, N-1 \end{aligned} \tag{12.111}$$

$$\begin{aligned} M_k(e) &= \ln\gamma_k(e) \\ &= \ln p(s_k^E(e)|s_k^s(e)) + \ln p(y_k|x_k) \end{aligned} \tag{12.112}$$

$$B_k(s) = \ln\beta_k(s)$$

$$= \ln \sum_{e:s_{k+1}^S=s} \beta_{k+1}(s_k^E(e))\gamma_{k+1}(e)$$
$$= \max_{e:s_{k+1}^S=s}{}^*\left[B_{k+1}(s_k^E(e)) + M_{k+1}(e)\right],\ k = N-1, \cdots, 1. \tag{12.113}$$

The complete log-likelihood ratio output information of the information bits can be obtained by derivation:

$$\begin{aligned} L(u_k) &= \ln \sum_{u:u(e)=1} \alpha_{k-1}\left(s_k^s(e)\right)\gamma_k(e)\beta_k\left(s_k^E(e)\right) \\ &\quad - \ln \sum_{u:u(e)=0} \alpha_{k-1}\left(s_k^s(e)\right)\gamma_k(e)\beta_k\left(s_k^E(e)\right) \\ &= \max_{u:u(e)=1}{}^*\left[A_{k-1}\left(s_k^s(e)\right) + M_k(e) + B_k\left(s_k^E(e)\right)\right] \\ &\quad - \max_{u:u(e)=0}{}^*\left[A_{k-1}\left(s_k^s(e)\right) + M_k(e) + B_k\left(s_k^E(e)\right)\right]. \end{aligned} \tag{12.114}$$

The max*() in the log-MAP algorithm is simplified to the usual maximum-value operation, which is the max-log-MAP algorithm. Figure 12.26 illustrates the performance of the log-MAP decoding algorithm for turbo codes using MATLAB simulations. An RSC code with a constraint length of $k = 3$ is used as the component encoder during encoding, the generator matrix is (111, 101), the encoding rate is 1/3, a random interleaver with $N = 200$ is used, and the influence of different iteration times on the bit-error rate is analyzed.

From Fig. 12.26, it can be seen that increasing the number of iterations will further improve the decoding performance; however, the decoding performance improvement will gradually decrease after increasing to a certain level, and the change in the bit-error rate curve will gradually flatten. In practical applications, after the number of

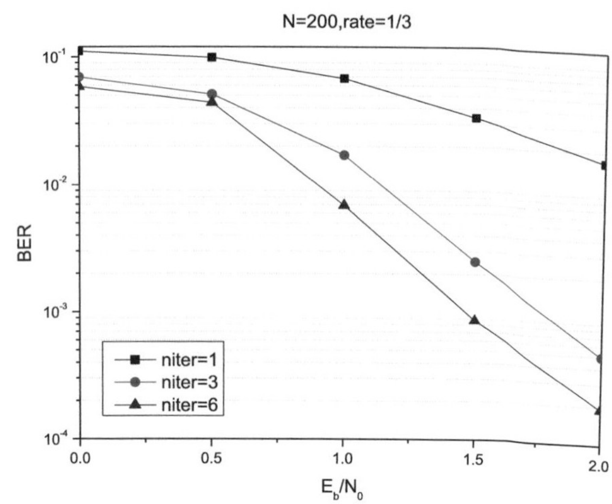

Fig. 12.26 Bit-error rate curves under different signal-to-noise ratios

iterations is greater than five, the decoding performance will not improve significantly with an increase in the number of iterations.

12.3.4 Turbo-BLAST System

Next, we analyze an iterative detection scheme of the turbo-BLAST [26] system; that is, turbo-code iterative detection technology is combined with a multi-antenna BLAST system, and iterative-detection technology is studied to obtain similar results when the channel matrix is known [27].

A basic turbo-BLAST system is shown in Fig. 12.27. At the sending end of the system, we use the turbo code as an outer channel coder for channel coding, and connect an inner space–time constellation map through a bit interleaver to serially concatenate the channel coding and MIMO space–time channels. For optimal decoding, we adopt the idea of turbo-code iterative decoding at the receiving end. The MIMO detector and the channel decoder are connected through an interleaver or deinterleaver to obtain mutual information, and the expected performance is achieved through iteration.

The specific detection process of the turbo-BLAST system is as follows [28]:

(1) The channel observation value y and the prior information L_{A1} fed back by the channel decoder are input to the MIMO signal detector to obtain soft information L_{D1}, according to the obtained external information L_{E1}. (In the first iteration, the channel decoder does not provide the prior information L_{A1} to the MIMO signal detector.)
(2) The external information L_{E1} is input to the channel decoder after passing through the deinterleaver. It becomes prior information L_{A2}, according to the new external information L_{E2}, which is input to the MIMO signal detector after

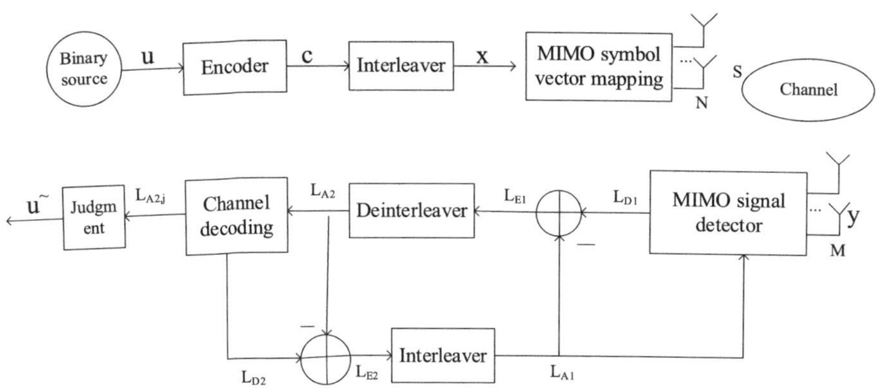

Fig. 12.27 Turbo-BLAST system block diagram

passing through the interleaver, and is sent as feedback to the MIMO signal detector as prior information L_{A1}.

(3) Steps 1 and 2 are repeated until the iteration ends. Each iteration will reduce the bit-error rate of the system through exchanging soft information between the MIMO detector and the channel decoder.

12.3.5 Iterative-Detection Algorithm

From the turbo-MIMO system model described in the previous section, it can be seen that the MIMO signal detector and the channel decoder are connected through a deinterleaver or interleaver to transmit external information. Because the turbo-code decoding algorithm is theoretically based on the MAP algorithm, and a soft decision is used internally before the end of the iteration, the MIMO signal detector should be a soft-input soft-output detector.

Using the idea of turbo-code iterative decoding, the a posteriori information of the transmitted signal is estimated by receiving information and feeding back a priori information. Because the MAP algorithm refers to calculating the maximum a posteriori probability for a given bit, the MIMO signal detector should be a soft-input soft-output MAP probability detector. In general, the representation of the log-likelihood ratio can simplify the calculation of the maximum posterior probability [29].

We assume that the transmitted symbol vector S is formed by the NM_c x$_1$-dimensional symbol bit vector X after mapping; then, for each bit x_k, k = 0, 1, …,$NM_c - 1$, its posterior information can be expressed as

$$\begin{aligned} L_D(x_k|y) &= \ln \frac{p(x_k = +1|y)}{p(x_k = -1|y)} \\ &= \ln \frac{\sum_{s_j|x_k=+1} p(s_j|y)}{\sum_{s_j|x_k=-1} p(s_j|y)} \\ &= \ln \frac{\sum_{s_j|x_k=+1} p(y|s_j)p(s_j)}{\sum_{s_j|x_k=-1} p(y|s_j)p(s_j)}. \end{aligned} \tag{12.115}$$

Because the channel encoder and the MIMO symbol vector-mapping module are serially cascaded through the interleaver, we can assume that each bit in the symbol vector x is statistically independent; therefore, the above equation can be written as [29]

$$\begin{aligned} L_D(x_k|y) &= \ln \frac{\sum_{x\in X_{k,+1}} p(y|x)\exp(\frac{1}{2}x^T L_A)}{\sum_{x\in X_{k,-1}} p(y|x)\exp(\frac{1}{2}x^T L_A)} \\ &= \ln \frac{\sum_{x\in X_{k,+1}} p(y|x)\exp(\frac{1}{2}x_{[k]}^T L_{A,[k]})}{\sum_{x\in X_{k,-1}} p(y|x)\exp(\frac{1}{2}x_{[k]}^T L_{A,[k]})} \end{aligned}$$

$$+ \ln \frac{p(x_k = +1)}{p(x_k = -1)}, \tag{12.116}$$

where $x_{[k]}$ is the sub-vector obtained by removing the kth element x_k in the symbol vector x;$L_{A,[k]}$ is the vector formed by the prior information corresponding to the symbol vector $x_{[k]}$; and $X_{k,+1}$ is a set of 2^{NM_c-1} symbol vectors, each of which satisfies

$$X_{k,+1} = \{x|x_k = +1\},\ X_{k,+1} = \{x|x_k = -1\}. \tag{12.117}$$

Drawing on the iterative decoding idea of turbo codes, we can calculate

$$L_A(x_k) = \ln \frac{p(x_k = +1)}{p(x_k = -1)} \tag{12.118}$$

$$L_D(x_k|y) = L_A(x_k) + \ln \frac{\sum_{x \in X_k^{+1}} p(y|x) \exp(x_{[k]}^T L_{A,[k]})}{\sum_{x \in X_k^{-1}} p(y|x) \exp(x_{[k]}^T L_{A,[k]})}. \tag{12.119}$$

The likelihood function $p(y|x)$ can be obtained by the following formula:

$$p(y|s = map(x)) = \frac{\exp(-\frac{1}{2\sigma^2}\|y - Hs\|^2)}{(2\sigma^2)^{M_c}}. \tag{12.120}$$

Only the *exp* part of the log-likelihood ratio calculation is relatively useful. The fixed factor can be cancelled in the numerator and denominator and is therefore ignored. The max*() operation of Eq. (12.106) is introduced, and the latter item is omitted; that is, for the calculation of the *ln* function, the corresponding max function can be directly used instead; therefore,

$$\begin{aligned} L_E(x_k|y) = {} & \frac{1}{2} \max_{x \in X_{k,+1}} \left\{ -\frac{1}{\sigma^2} \|y - Hs\|^2 + x_{[k]}^T L_{A,[k]} \right\} \\ & - \frac{1}{2} \max_{x \in X_{k,-1}} \left\{ -\frac{1}{\sigma^2} \|y - Hs\|^2 + x_{[k]}^T L_{A,[k]} \right\}. \end{aligned} \tag{12.121}$$

Even if this simplified method is used, the complexity of calculating $L_E(x_k|y)$ still increases exponentially with the length N of the bit sequence x and the number of constellation modulation points M_c, which is generally unacceptable.

As shown in Fig. 12.28, this decoding algorithm is used for a 2 × 2 turbo-MIMO system, and the BER curve under on–off keying (OOK) modulation is simulated. The channel coding adopts turbo coding with a 1/2 code rate, the generator polynomial coefficient is (7, 5), and the outer loop performs four iterations—that is, four times between the MIMO detector and the channel decoder—and the inner loop is five times.

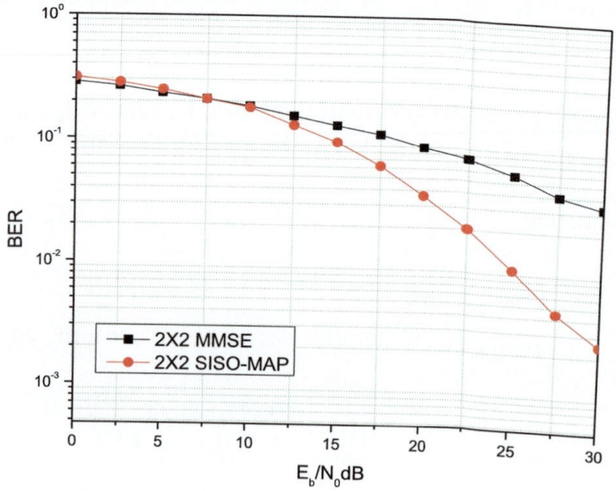

Fig. 12.28 BER curves of different decoding algorithms

It can be seen from the figure that the SISO-MAP algorithm is much lower than the linear MMSE algorithm, mainly because of the structure of its turbo-MIMO system. However, the coding and decoding structure of its turbo-BLAST system is relatively complex, the computational complexity increases with the number of transmitting antennas, and the number of constellation modulation points increases exponentially; thus, they are often limited in practical applications.

12.4 Summary

The maximum-likelihood detection algorithm is the optimal detection algorithm; however, its computational complexity is high. MIMO-system detection algorithms mainly include linear and nonlinear detection algorithms. The linear detection algorithms include the ZF and MMSE-criteria detection algorithms.

The linear detection algorithm based on the ZF criterion is one of the simplest linear suboptimal detection algorithms; the linear detection algorithm based on the MMSE criterion considers noise on the basis of the ZF criterion. Its detection performance is better than that of the ZF criterion, and it converges to the ZF criterion detection algorithm when the signal-to-noise ratio is high. The linear detection algorithm is simple to implement; however, the detection performance is low.

To achieve better detection performance, nonlinear detection algorithms based on the successive interference-cancellation algorithm and the ordered successive interference cancellation algorithm have been developed. By subtracting the interference of the detected signal from the undetected signal, these two algorithms could detect the later signal with higher reliability, and the detection performance was better than that of the linear detection algorithm.

The ordered successive interference-cancellation algorithm first detects signals with a high SNR by sorting the signals to be detected, thereby reducing the probability of error propagation. Compared with the unsorted successive interference-cancellation algorithm, the detection performance is greatly improved [30].

The soft-input soft-output turbo-code iterative-decoding algorithm can quickly complete the decoding process. MIMO-technology iterative detection and decoding uses two different estimated Markov-chain Monte-Carlo detection algorithms: one is the MCRB estimation algorithm based on Monte Carlo integration, and the other is the MCRB-U estimation algorithm based on importance sampling. The detection performance of MCBR-U estimation is higher than that of MCRB estimation.

The combination of different numbers of antennas also affects the system performance. The performance of a system with 4 X_i 6 antennas is better than that of a system with 4 $\widehat{X} = \arg\max_{X_i} f(Y|X_i)$ 4 antennas, because the diversity gain is higher; therefore, the overall system performance is higher.

References

1. La X (2011) Research on signal detection algorithm of space-time coding in MIMO wireless communication system. Sci Technol Guide 6:51
2. Teng G, Yun L (2010) Research on signal detection algorithm in MIMO system. Modern Telecommun Technol 11(48–51):59
3. Fen J, Huakui W (2008) Research on signal detection algorithm of vertical hierarchical space-time code system. Shanxi Electron Technol 6:54–55
4. Liu Y (2008) Research and implementation of MIMO detection algorithm based on MCMC method. Master of Wuhan University of Technology Dissertation, Wuhan
5. Ren C (2006) Research on MIMO system detection algorithm. Civil Aviation Institute of China, Tianjin
6. Wu Z (2006) Design of turbo MIMO wireless communications. Montreal, Quebec, Canada, the Degree of Doctor of Philosophy at Concordia University
7. Jun W, Wen OY An improved soft-output MMSE BLAST detector. National Key Lab. of Communications University of Electronic Science and Technology of China. Chengdu, Sichuan Province, P. R. China
8. Liu L (2011) Research on wireless MIMO system detection algorithm. Master's Thesis of Xidian University, Xi'an
9. Battail G, Berrou C, Glavieux A (1993) Pseudo-recursive convolutional coding for near-capacity-performance. In: IEEE global telecommunications conference, vol 4, pp 23–27. IEEE. ISBN-10: 0780309170, ISBN-13: 9780780309173 (Conference: Proceedings of the IEEE global telecommunications conference. Part 4 (of 4), November 29, 1993–December 2, 1993; Sponsor)
10. Lin J (2008) Iterative detection technology in MIMO system. Master's Thesis of Xidian University, Xi'an
11. Zhu J (2008) Research on detection algorithm applied in MIMO receiving system. Master of Xidian UniversityDissertation, Xi'an
12. Yang Y (2009) Research on detection algorithm in MIMO system. Doctoral Dissertation of Xidian University, Xi'an
13. Zhu H, Dong Boroujeny BF (2005)MIMO Detection using Markov chain monte Carlo techniques for near capacity performance. Acoust Speech Sig Process 3:1017–1020 (Proceedings. (ICASSP 05). IEEE international conference on 2005)

14. Mei Z, Yin W et al. (2007) A low-complexity joint detection algorithm in MIMO systems. Sig Process 23(1):112–115
15. Wang D (2005) Research on iterative detection and channel estimation technology in wireless MIMO systems. Ph.D. dissertation of Southeast University, Nanjing
16. Behrouz F-B, Haidong Z, Shi Z et al (2006) Markov Chain Monte Carlo Algorithms for CDMA and MIMO communication systems. IEEE Trans Sig Process 54(5):1896–1909
17. Zigang Y, Lu B, Wang X et al (2001) Blind Bayesian multiuser receiver for space-time coded MC-CDMA system over frequency-selective fading channel. Global Telecommun Conf IEEE 2:781–785
18. Zhang S (2011) Exploration of non-stationary random signal analysis based on MCMC method. Autom Instrum 03:115–116
19. Ke X, Su M, Guo S et al (2009) Research and improvement of MCMC detection algorithm in MIMO system. J North China Univ Technol 21(3):4–8
20. Shi X (2009) Research on iterative decoding algorithm of turbo codes. Sci Technol Inform 13:508–509
21. Wang X (2005) Research and simulation of SMC+Turbo receiver in MIMO system. Shanghai University Master's Thesis, Shanghai
22. Zhu HD, Boroujeny BF, Chen RR (2005) On performance of sphere decoding and Markov chain monte Carlo detection methods. Sig Process Lett IEEE 12(10):669–672
23. Chen R-R, Peng R, Ashikhmin (2010) Approaching MIMO capacity using bitwise Markov Chain Monte Carlo detection. IEEE Trans Commun 58(2):423–428
24. Ke X, Su M (2010) Low-complexity MCMC iterative detection algorithm in super MIMO systems. Appl Electron Technol 36(1):107–110
25. Laraway SA, Farhang-Boroujeny B (2009) Implementation of a Markov Chain Monte Carlo based multiuser/MIMO detector. IEEE Trans Circ Syst 56(1):3088–3093
26. Wang X, Xiao G (2001) Error correction codes—principles and methods. Xidian University Press, Xi'an
27. Sellathurai M., Haykin S (2002) Turbo-BLAST for wireless communications: theory and experiments. IEEE Trans Sig Process 50(10):2538-2546
28. Chen X, Xu D, Yu X et al (2008) Turbo-BLAST iterative detection algorithm based on soft interference cancellation. J Sichuan Univ 40(05):146–152
29. Haykin S, Sellathurai M, de Jone Y, Willink T (2004) Turbo-MIMO for wireless communications. IEEE Commun Mag 42:48–53
30. Hochwald BM, Brink S (2003) Achieving near-capacity on a multiple-antenna channel. IEEE Trans Commun 51(3):389–399

MIX
Papier aus verantwortungsvollen Quellen
Paper from responsible sources
FSC® C105338

If you have any concerns about our products,
you can contact us on
ProductSafety@springernature.com

In case Publisher is established outside the EU,
the EU authorized representative is:
Springer Nature Customer Service Center GmbH
Europaplatz 3, 69115 Heidelberg, Germany

Printed by Libri Plureos GmbH
in Hamburg, Germany